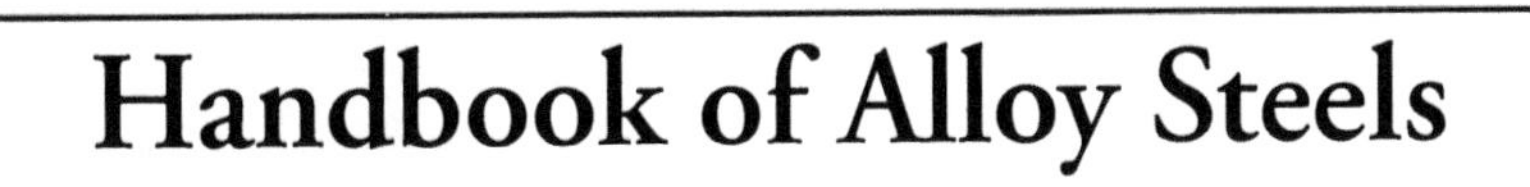

Handbook of Alloy Steels

Handbook of Alloy Steels

Editor: Preston Green

New York

Published by NY Research Press
118-35 Queens Blvd., Suite 400,
Forest Hills, NY 11375, USA
www.nyresearchpress.com

Handbook of Alloy Steels
Edited by Preston Green

International Standard Book Number: 978-1-64725-438-4 (Hardback)

Cataloging-in-Publication Data

Handbook of alloy steels / edited by Preston Green.
p. cm.
Includes bibliographical references and index.
ISBN 978-1-64725-438-4
1. Steel alloys. 2. Steel--Metallurgy. 3. Alloys. I. Green, Preston.
TA478 .H36 2023
620.17--dc23

Contents

Preface

The purpose of the book is to provide a glimpse into the dynamics and to present opinions and studies of some of the scientists engaged in the development of new ideas in the field from very different standpoints. This book will prove useful to students and researchers owing to its high content quality.

Alloy refers to a mixture of two or more chemical elements, of which at least one is a metal. Alloy steels are the steels that are made by combining steel with other alloying elements in order to change their properties such as hardness, toughness, strength and wear resistance. Some of the common alloying elements include chromium, nickel, manganese, tungsten and vanadium. Alloy steels are classified into three types including low alloy steels, high alloyed steels and micro-alloyed steels. They are used for producing heating elements for appliances such as pans and pots, toasters, corrosion-resistant containers, and silverware. Alloy steel are also used as skeleton frames in the construction of skyscrapers, bridges, stadiums, and airports. This book explores all the important aspects of alloy steels in the present day scenario. It will also provide interesting topics for research, which interested readers can take up. This book is a resource guide for experts as well as students.

At the end, I would like to appreciate all the efforts made by the authors in completing their chapters professionally. I express my deepest gratitude to all of them for contributing to this book by sharing their valuable works. A special thanks to my family and friends for their constant support in this journey.

Editor

1

Influence of Coiling Temperature on Microstructure, Precipitation Behaviors and Mechanical Properties of a Low Carbon Ti Micro-Alloyed Steel

Mingxue Sun [1,*], Yang Xu [1] and Wenbo Du [2]

1 School of Mechanical and Automotive Engineering, Qingdao University of Technology, Qingdao 266520, China; xy45269026@gmail.com
2 National Key Laboratory for Remanufacturing, Academy of Army Armored Forces, Beijing 100072, China; dwbneu@163.com
* Correspondence: sunmingxue@qut.edu.cn.

Abstract: The microstructural evolution, nanosized precipitation behaviors and mechanical properties of a Ti-bearing micro-alloyed steel at different coiling temperatures were studied using optical microstructure (OM), scanning electron micrograph (SEM), transmission electron microscopy (TEM), Vickers hardness and tensile tests. When the coiling temperature was 500 °C, the specimen showed mainly bainitic structure, whereas polygonal ferrite was visible as the coiling temperature increased to 650 °C and 700 °C. The Vickers hardness of tested steel reached the maximum, which can be attributed to the largest number of nanosized precipitates in ferrite at the coiling temperature of 650 °C. A coiling temperature of 650 °C was optimal for the formation of TiC because of the high diffusion rate of alloying elements and kinetics of precipitation. In the laboratory rolling experiment, when the coiling temperature was 630 °C, the steel with yield strength of 682 ± 2.1 MPa and tensile strength of 742 ± 4.9 MPa was produced. The fine-grain strengthening and precipitation strengthening were 262 MPa and 268 MPa, respectively.

Keywords: coiling temperature; microhardness; nanosized precipitation; precipitation strengthening

1. Introduction

Nanosized precipitation of titanium (Ti), niobium (Nb), molybdenum (Mo) and vanadium (V) carbonitrides has been known to be effective for strengthening high-strength low-alloy (HSLA) steels, which are widely used in industrial equipment, bridge, pipeline and automobile. High strength for micro-alloyed steel is one of the researching hotspots in recent years [1–5]. Ti, Nb, V and Mo elements with individually or combined added have been employed to refine microstructure and enhance precipitation strengthening, which could significantly improve the strength of steel. Previous studies showed that the precipitation could provide more than 300-MPa strength increment based on Orowan mechanism when the particle size was less than 10 nm [6–11].

With the development of micro-alloyed steels, cost reduction becomes a challenge in industrial production due to the high price of micro alloys. Hence, the addition of Ti has attracted increasing attention as a main micro-alloying element to produce high-strength steel because of its abundant resources and the low price. Some studies have attempted to investigate precipitation behaviors of Ti-bearing steels [10–15]. Funakawa and Shiozaki [1] developed a Ti–Mo-based hot-rolled steel with a yield strength over 700 MPa which was coiled at 620 °C by use of nanosized (Ti, Mo)C particles in ferrite. Kong [10] and Wang [11] studied the influence of hot-rolling process parameters on the precipitation behaviors of nanosized (Ti, Mo)C particles and found that uniform distributed particles with nanometer size could be obtained in ferrite based on rolling in recrystallization region.

Xu [12] reported that the mechanism between dislocation and precipitates for random precipitation and interphase precipitation of a Nb–Ti micro-alloyed steel which was formed at different coiling temperature, was different. Natarajan [15] studied the effect of microstructural evolution, precipitation behaviors and dislocation structure at two coiling temperature in a Nb–Ti high-strength steel. The results showed that there was insufficient time for the dissolved microalloying elements to precipitate during cooling process at high cooling rate, so precipitates mainly formed during coiling stages. Furthermore, in most of the reported literature, the volume fraction of precipitates was artificially determined by transmission electron micrograph (TEM) and the precipitation strengthening was calculated according to theoretical model [16,17]. Therefore, detailed investigation on precipitation behaviors at different coiling temperatures should be conducted in Ti micro-alloyed steels.

In this study, the effect of coiling temperature on the microstructure and precipitation behavior of a Ti-bearing micro-alloyed steel was investigated. The optimum precipitation temperature was confirmed by Vickers hardness and TEM. Meanwhile, strengthening mechanism was discussed at hot-rolled condition. The findings from the present study are expected to provide some technical supports for the development of high-strength hot-rolled steels.

2. Experimental Procedure

The chemical composition of the tested steel was C 0.05, Si 0.39, Mn 1.5, Ti 0.2, P 0.005, S 0.002, N 0.005 and Fe balance (mass%). The steel was prepared by vacuum melting and forged into billets with a size of 80 mm × 80 mm × 120 mm, which were then used for hot-rolling experiments. The billets were rolled into 12 mm in thickness by use of a Φ450 mm mill of the state key laboratory of rolling and automation (Shenyang, China). The plates were soaked at 1200 °C for 24 h followed by water quenching in order to homogenize the micro-alloying elements. Specimens for thermal simulation experiments were cut from the plates along the rolling direction and then machined into a dimension of Φ8 mm × 15 mm.

Thermal simulation experiments were conducted on a Gleeble 3800 thermal simulation machine (DSI, Poestenkill, NY, USA). As shown in Figure 1, the specimens were heated to 1200 °C with a heating rate of 10 °C/s and held for 300 s to homogenize the micro-alloyed elements and dissolve carbonitrides. The specimens were then cooled with a cooling rate of 10 °C/s to1050 and 850 °C to be deformed by 0.3 with a strain rate of 5/s. After deformation, specimens were cooled to 700, 650, 600, 550 and 500 °C with a cooling rate of 10 °C/s and then cooled to room temperature at a cooling rate of 0.1 °C/s to simulate the coiling process.

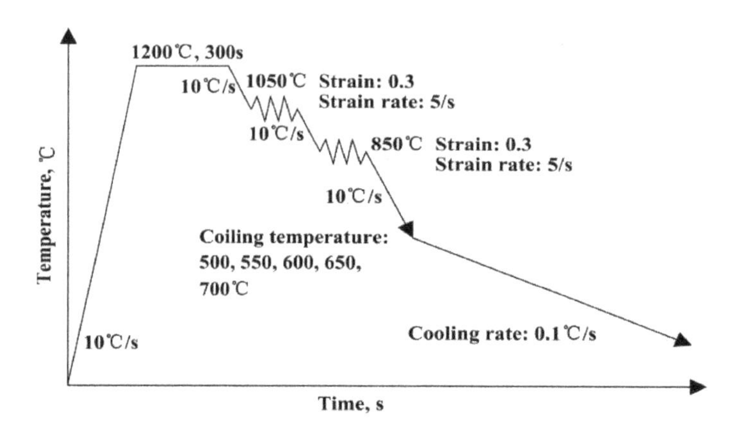

Figure 1. Schedule of thermal simulation experiments.

Hot-rolling experiments were carried out on a Φ450 mm mill with a variety of cooling devices. The billets for hot rolling were isothermally treated at 1200 °C for two hours in a K010 box-shaped furnace, then rolled to 4 mm through nine-pass rolling. Rolling process was carried out in two stages (roughing rolling and finish rolling). The corresponding rolling reduction was 80% and

75%, respectively. The reduction schedule was: 80–53–36–24–16–12–9–7–5–4 (mm). The billet was rough-rolled through four-pass rolling with average reduction of 33% and deformation temperature in the range 1100–1050 °C. The intermediate billet was rolled to the desired thin strips through five-pass rolling with average reduction of 24% and finish rolling temperature in the range 870–850 °C. Finally, the product was laminar cooled with a cooling rate of 20 °C/s to 630 °C and then put into the asbestos (which was used in fire protection and insulation, because of its fiber strength and heat resistance) to cool to room temperature.

Samples for optical metallography of the thermal simulation experiments were cut by wire cutting along the radial direction, while samples for hot-rolling experiments were cut along the perpendicular to the rolling direction. Microstructures of the thermal simulation experiments were observed by use of a LEICAQ550 metalloscope (LEICA Microsystems, Wetzlar, Germany) after etched with 4% nital. A QUANTA 600 SEM (FEI, Hillsboro, United States) was used to observe the microstructure of the hot-rolled specimen, which was polished and etched using 4% nital solution. Vickers hardness of different processes was detected with a load of 50 g and a load time of 10 s. Ten measurements were made on each sample.

Specimens for tensile experiments were cut from hot-rolled slabs along the rolling direction and then machined to the dimension according to the GB/T2975 standard, as shown in Figure 2. Tensile strength, yield strength and elongation were obtained on a CMT-5105 tensile testing machine (MTS, Shenzhen, China) at ambient temperature, and the tensile rate was 5 mm/min. Three tensile experiments were conducted per condition.

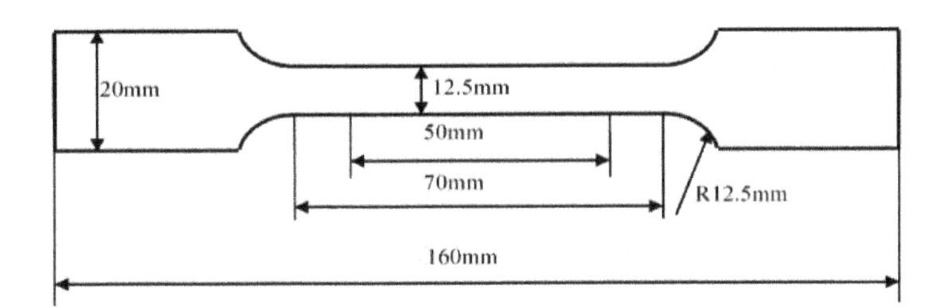

Figure 2. Dimension of specimens for tensile experiments.

Carbon replica and thin foil for TEM observation were prepared to analyze the microstructure of precipitates. Specimens for carbon replicas were cut from the thermal simulation samples along the radial direction. Qualitative analysis of precipitates was performed using an energy-dispersive spectrometer (EDS). The specimen of the hot-rolled steel for TEM was cut into a thickness of 0.5 mm along perpendicular to the rolling direction, mechanically thinned to 0.05 mm by abrasion SiC papers and then twin-jet electro polished to perforation using a mixture of 20% perchloric acid and 80% ethanol at −20 °C, using a potential of 35 V. The characterization of precipitates was carried out on a TecnaiG2 F20 field-emission-gun TEM (FEI, Hillsboro, OR, USA).

3. Results and Discussion

3.1. Optical Metallography

The optical microstructures of the tested steel at different coiling temperatures during thermal simulation experiment are shown in Figure 3. The microstructure was composed of granular bainite (GB) and acicular ferrite (AF) when coiled at 500, 550 and 600 °C. The microstructure was mainly polygonal ferrite (PF) when the coiling temperatures were 650 and 700 °C. Bainite and ferrite become finer with the decreasing of coiling temperature. Using a linear interception method, the average ferrite grain sizes at 650 and 700 °C were measured to be 12.2 ± 2.2 μm and 18.4 ± 3.7 μm, respectively. It is important to note that coiling temperature was a significant factor to the microstructural evolution. As the coiling temperature decreased, diffusibility of carbon was reduced, resulting in the inhibition of ferrite formation. Further, low coiling temperature which caused larger undercooling, contributes to ferrite grain refinement.

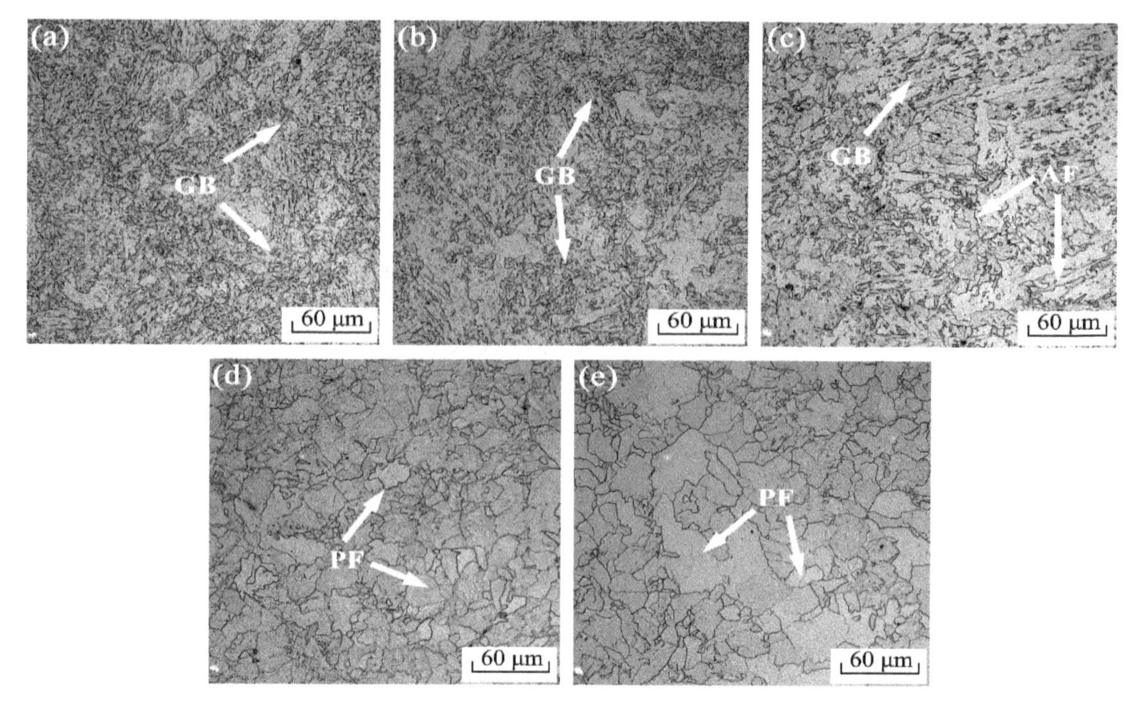

Figure 3. Optical microstructures of tested steel at different coiling temperatures. (**a**) 500 °C; (**b**) 550 °C; (**c**) 600 °C; (**d**) 650 °C; (**e**) 700 °C.

3.2. Vickers Hardness

Figure 4 shows the variation of Vickers hardness at different coiling temperatures. The Vickers hardness increased with the increasing of coiling temperature from 500 °C to 650 °C and rose from HV 250 ± 5 to HV 304 ± 11. As the coiling temperature increased from 650 °C to 700 °C, Vickers hardness decreased. In generally, Globular bainite should have a higher hardness than polygonal ferrite. Combining with the microstructural evolution, the higher hardness cannot be obtained when the microstructure was granular bainite at low coiling temperature. The significant difference in Vickers hardness could be mainly attributed to the precipitation strengthening effect due to the addition of Ti. Precipitation behaviors will be discussed to reveal the hardness difference at different coiling temperatures in the following sections.

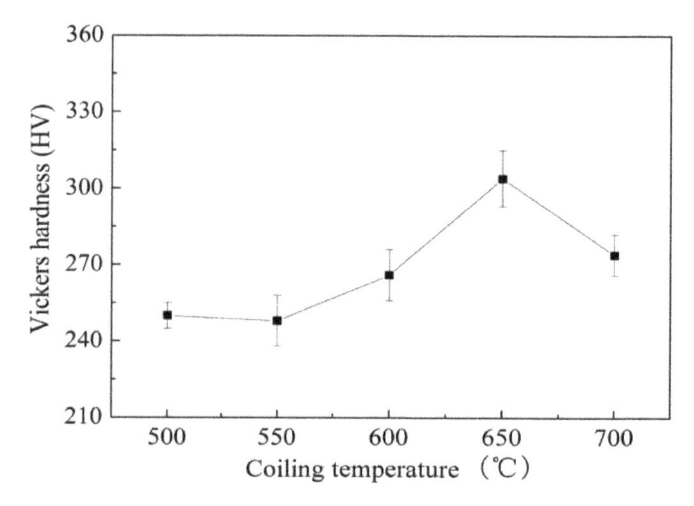

Figure 4. Vickers hardness of the tested steel at different coiling temperatures.

3.3. Precipitation Behavior

In the studied steel, Ti was added for grain refinement and precipitation strengthening. The characterization of precipitates at different coiling temperatures and EDS results is shown in Figure 5. When the coiling temperature was 500 °C, particles with the size larger than 100 nm were observed according to the replica technique. With the increasing of coiling temperature, besides the particles with the size larger than 100 nm, increasing particles, which did not exceed 50 nm, could be observed in ferrite. Two representative particles with the size of 200 and 30 nm were selected for energy spectrum analysis as shown in Figure 5f. The cubic precipitate in Figure 5a and the spherical precipitates in Figure 5d were confirmed by EDS analysis to be Ti(C, N) and TiC, respectively. Nitrogen preferentially precipitates in the form of large TiN particles, which are formed at high temperature. Ti and C could accumulate by attaching on TiN, thus the formation of Ti(N, C) particles occurs during deformation process. In addition to this, it is evident from Figure 5 that the tested steel contains a large number of spherical precipitates in the form of TiC. The solid solubility of TiN and TiC precipitation could be expressed in Equation (1) through (3) [4]:

$$\lg\{[Ti][N]\}_{\gamma} = 3.94 - 15190/T \tag{1}$$

$$\lg\{[Ti][C]\}_{\gamma} = 5.33 - 10475/T \tag{2}$$

$$\lg\{[Ti][C]\}_{\alpha} = 4.40 - 9575/T \tag{3}$$

where $[Ti]$, $[N]$ and $[C]$ are the solid solution amount of element Ti, N and C in austenite and ferrite, T is the solid solution temperature, γ and α represent austenite and ferrite, respectively. When the samples were soaked at temperature of 1200 °C, the solid solubility of TiN and TiC were calculated as 4.24×10^{-7} and 1.65×10^{-2}. It is indicated that N almost fully precipitated as TiN at high temperature. Based on precious studies [17], TiN particles could be neglected for the strengthening effect because of their large size. The remaining Ti of 0.183 wt% at 1200 °C mainly precipitated in the form of TiC precipitation during the subsequent deformation, cooling and coiling process, which had strong effect on strength.

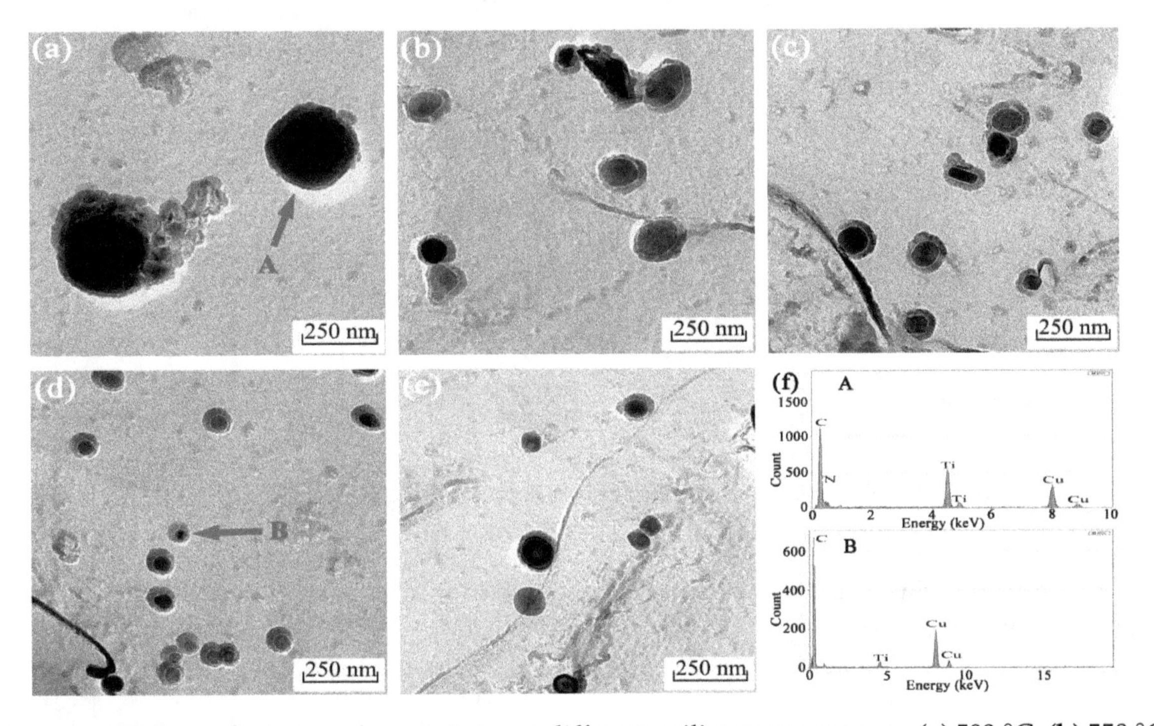

Figure 5. TEM morphologies of precipitates at different coiling temperatures. (**a**) 500 °C; (**b**) 550 °C; (**c**) 600 °C; (**d**) 650 °C; (**e**) 700 °C; (**f**) EDS patterns of different precipitated particles indicated as the red label "A" in (**a**) and "B" in (**d**).

Figure 6 presents the size distribution of precipitation for carbon replica specimens at different coiling temperatures. More than eight images were used to determine the size of precipitation using Image-Pro Plus software. It is obvious that when the coiling temperature was 500 °C, the main precipitates were large ones with the size over 100 nm. With the increasing of coiling temperature, the solute Ti became reactive for nucleation, and the volume fraction of small particles increased. When coiled at 650 °C, it is obvious that the TEM morphology is dominated by precipitates with the size less than 50 nm. As the coiling temperature increased to 700 °C, the volume fraction of large precipitates increased, which can be attributed to the coarsening of TiC precipitates.

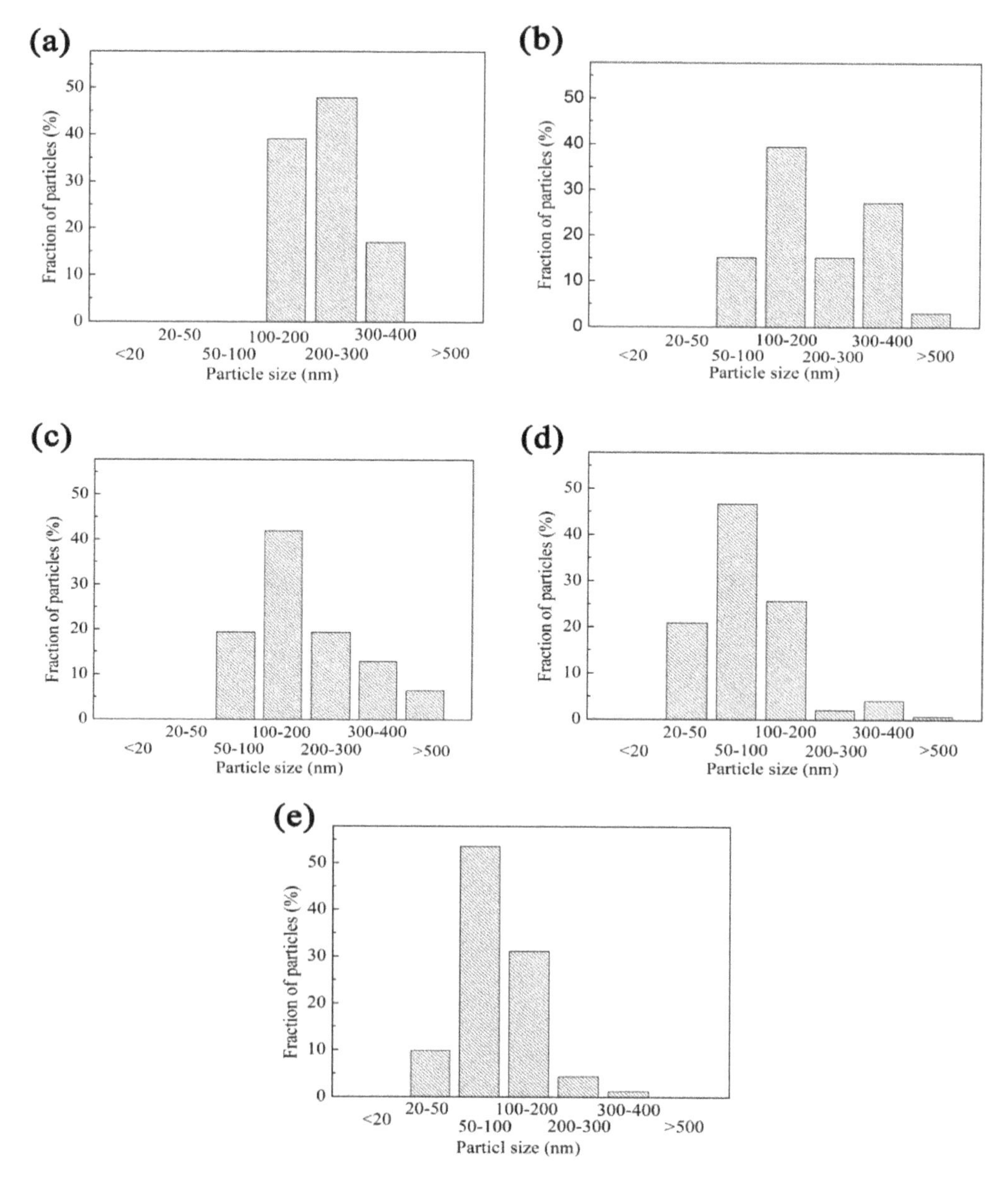

Figure 6. Frequency distribution of precipitates at different coiling temperatures. (**a**) 500 °C; (**b**) 550 °C; (**c**) 600 °C; (**d**) 650 °C; (**e**) 700 °C.

According to the classical nucleation theory, nucleation rate can be described in Equation (4) through (7) [4,18]:

$$J^* = N_V \beta^* Z \exp\left(-\frac{\Delta G^*}{kT}\right) \tag{4}$$

$$\beta^* = \frac{16\pi\sigma^2_{TiC-Matrix} D_M}{a^4_{TiC} \Delta G^2_V} \tag{5}$$

$$Z = \frac{V_a \Delta G_V^2}{8\pi \left(kT\sigma_{TiC-Matrix}^3\right)^{1/2}} \quad (6)$$

$$\Delta G^* = \frac{16\pi\sigma_{TiC-Matrix}^3}{3\Delta G_V^2} \quad (7)$$

where J^* denotes the nucleation rate of TiC, N_V is the number of potential nucleation sites per unit volume, β^* is the frequency factor, Z is the Zeldovich non-equilibrium factor, ΔG^* is the activation energy of nucleation, k is the Boltzmann coefficient, T is the temperature, $\sigma_{TiC-Matrix}$ is the interfacial energy between TiC and the matrix, D_M is the volume diffusion coefficient of component Ti in the matrix, a_{TiC} is the lattice parameter of TiC, ΔG_V is the driving force of nucleation and V_a is the atomic volume of a substitutional atom of TiC.

In the present study, the heating temperature and deformation factors were all consistent in thermal simulation experiments. It is concluded that N_V, V_a, D_M and $\sigma_{TiC-Matrix}$ were all the same. The difference in the precipitation behavior could be attributed to the kinetics of precipitation at different coiling temperatures. The nucleation and coarsening of TiC were depended on the diffusion rate of Ti and C. As the coiling temperature decreased, atomic diffusion was limited, resulting in the restriction of formation and growth of TiC. Thus, a large number of Ti and C atoms were kept in solution at low coiling temperature, while they precipitated in the form of TiC particles at high coiling temperature. When the coiling temperature increased to 700 °C, the diffusion of Ti and C was promoted, which caused a larger size of TiC particles. On the comparing of the precipitation behavior at different coiling temperatures, it is clear that the largest volume fraction of finer precipitates could be obtained at coiling temperature of 650 °C for Ti-bearing steels.

Based on previous studies, nanosized precipitates could extremely improve the strength of matrix by impeding the movement of dislocations, which can be described by the Ashby–Orowan model according to the theory of Gladman as shown in Equation (8) [4]:

$$\Delta\sigma_{ppt} = 0.3728\frac{Gb}{1-\nu} \times \frac{f^{1/2}}{d}\ln\left(\frac{1.2d}{2b}\right) \quad (8)$$

where $\Delta\sigma_{ppt}$ denotes the strength increment attributed to precipitation strengthening (MPa), G is the shear modulus and equal to 81,600 MPa, b is the Burgers vector of magnitude 0.246 nm, ν is the Poisson ratio, f is the volume fraction of precipitates, and d represents the diameter of precipitates. According to Equations (8), precipitation strengthening is proportional to the volume fraction and inversely proportional to the size of precipitates. As shown in Figures 5 and 6, with the increasing of coiling temperature, the high volume fraction of finer TiC precipitates was obtained, which could provide larger contribution to the strength. Thus, precipitation strengthening was the highest at the coiling temperature of 650 °C, while it decreased at lower and higher coiling temperature. Since the volume fraction of precipitates was not exactly inferred from the replica technique, the precipitation strengthening could not be precisely calculated by Equation (8). In order to quantitatively analysis the effect of TiC particles on strength increment, a further rolling experiment was conducted.

3.4. Rolling Experiment

The rolling process parameters and mechanical properties are shown in Table 1. The yield strength, tensile strength and elongation were 682 ± 2.1 MPa, 742 ± 4.9 MPa and 24.2 ± 1.7%, respectively. The SEM and TEM of the tested steel are presented in Figure 7. The hot-rolled steel predominately consisted of polygonal ferrite and quasi-polygonal ferrite, which is consistent with the thermal simulation experiment. The average diameter of ferrite grain was 4.4 ± 0.8 μm measured by Image-Pro Plus software. It can be concluded that the ferrite grain in hot-rolling process is significant finer than that in thermal simulation experiment, which can be mainly attributed to larger deformation and higher cooling rate. There existed many spherical and cell-like nanosized precipitates in ferrite matrix, which were identified to be TiC in the range of 5–30 nm.

Table 1. Rolling parameters and mechanical properties of the hot-rolled steel.

Heating Temperature °C	Cooling Rate °C/s	Coiling Temperature °C	Yield Strength MPa	Tensile Strength MPa	Elongation %
1200	20–30	630	682 ± 2.1	742 ± 4.9	24.2 ± 1.7

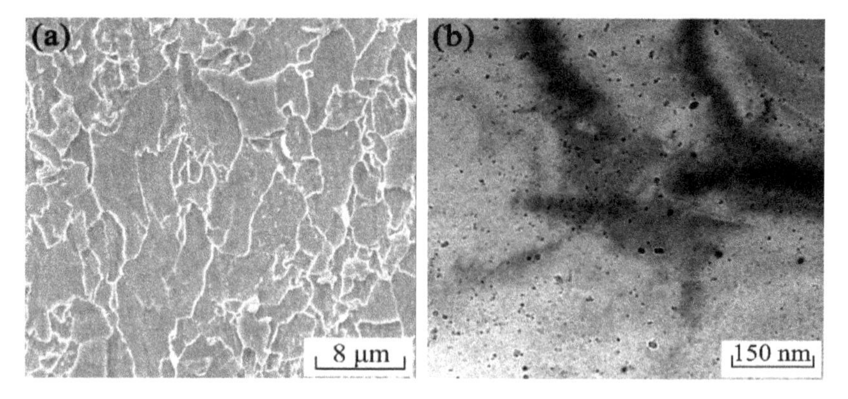

Figure 7. (**a**) SEM and (**b**) TEM of the hot-rolled steel.

The yield strength of Ti-bearing steel can be expressed by Equation (9) [19,20]:

$$\sigma_y = \Delta\sigma_0 + \Delta\sigma_{ss} + \Delta\sigma_{GB} + \sqrt{\Delta\sigma_{dis}^2 + \Delta\sigma_{ppt}^2} \quad (9)$$

where σ_y is the yield strength, $\Delta\sigma_0$ is the crystal lattice strengthening (in the present study, the value was assumed to be 54 MPa), $\Delta\sigma_{ss}$ is the solution strengthening, $\Delta\sigma_{GB}$ is the fine grain strengthening, $\Delta\sigma_{dis}$ is the dislocation strengthening. The individual contribution of each strengthening mechanism can be evaluated by Equation (10) through (12):

$$\Delta\sigma_{GB} = 17.4d^{-1/2} \quad (10)$$

$$\Delta\sigma_{SS} = 360[C] + 32[Mn] + 83[Si] + 700[P] \quad (11)$$

$$\Delta\sigma_{dis} = \alpha MGb\rho^{1/2} \quad (12)$$

where d is the average size of ferrite grain, $[M]$ (M = C, Mn, Si, P) is the mass fraction of element M in solute state, ρ is the dislocation density in per unit area (in present study, the value was assume to be 10^{13} m^{-2}), α is equal to 0.38, M is the Taylor factor and equals to 2.2 for ferrite grain. The grain refinement strengthening and precipitation strengthening values were calculated as 262 MPa and 268 MPa. Although, bainite and acicular ferrite were not obtained at high coiling temperature, the tested steel exhibited high yield strength owing to the grain refinement strengthening and precipitation strengthening because of the Ti addition.

4. Conclusions

(1) The samples were composed of granular bainite and acicular ferrite at coiling temperature of 500, 550 and 600 °C, while mainly polygonal ferrite was observed at 650 and 700 °C. Coiling temperature has a significant effect on the microstructure evolution and grain size for Ti-bearing steels.

(2) Coiling temperature had an important influence on Vickers hardness attributed to TiC particles. Vickers hardness was the highest at coiling temperature of 650 °C.

(3) The nucleation rate of TiC was depended on the diffusion rate of Ti and C, which was determined by coiling temperature. A large volume fraction of fine TiC particles could be formed in ferrite at the optimal coiling temperature of 650 °C.

(4) In the hot-rolling condition, yield strength and tensile strength were 682 ± 2.1 MPa and 742 ± 4.9 MPa, respectively. The intense precipitation strengthening effect increased the yield strength by 268 MPa attributed to the large number of nanosized TiC precipitates in ferrite.

Author Contributions: Conceptualization, M.S. and Y.X.; methodology, M.S.; software, Y.X.; validation, Y.X.; formal analysis, Y.X.; investigation, M.S.; resources, Y.X.; data curation, M.S.; writing—original draft preparation, M.S.; writing—review & editing, M.S. and W.D.; visualization, M.S.; supervision, Y.X. and W.D.; project administration, Y.X.; funding acquisition, M.S. All authors have read and agreed to the published version of the manuscript.

References

1. Yoshimasa, F.; Tsuyoshi, S. Development of high strength hot-rolled sheet steel consisting of ferrite and nanometer-sized carbides. *ISIJ Int.* **2004**, *44*, 1945–1951.
2. Kim, Y.W.; Kim, J.H.; Hong, S.G.; Lee, C.S. Effects of rolling temperature on the microstructure and mechanical properties of Ti–Mo microalloyed hot-rolled high strength steel. *Mater. Sci. Eng. A* **2014**, *605*, 244–252. [CrossRef]
3. Liu, D.S.; Cheng, B.G.; Chen, Y.Y. Strengthening and toughening of a heavy plate steel for shipbuilding with yield strength of approximately 690 MPa. *Metall. Mater. Trans. A* **2013**, *44*, 440–455. [CrossRef]
4. Yong, Q.L.; Ma, M.T.; Wu, B.R. *Microalloy Steel-the Physical and Mechanical Metallurgy*; Machinery Industry Press: Beijing, China, 2006.
5. Senuma, T. Present status of future prospect s f or precipitation research in the steel industry. *ISIJ Int.* **2002**, *4*, 1–12. [CrossRef]
6. Misra, R.D.K.; Tennetis, K.K.; Weatherly, G.C.; Tither, G. Microstructure and texture of hot-rolled Cb-Ti and V-Cb microalloyed steels with differences in formability and toughness. *Metall. Mater. Trans. A* **2001**, *34*, 2041–2051. [CrossRef]
7. Craven, A.J.; He, K.; Garvie, L.A.J.; Baker, T.N. Complex heterogeneous precipitation in titanium-niobium microalloyed Al-killed HSLA steels-(Ti, Nb)(C, N)particles. *Acta Mater.* **2000**, *48*, 3857–3868. [CrossRef]
8. Han, Y.; Shi, J.; Xu, L.; Cao, W.Q.; Dong, H. TiC precipitation induced effect on microstructure and mechanical properties in low carbon medium manganese steel. *Mater. Sci. Eng. A* **2011**, *530*, 643–651. [CrossRef]
9. Kazuhiro, S.; Yoshimasa, F.; Shinjiro, K. Hot rolled high strength steels for suspension and chassis parts "NANOHITEN" and "BHT® steel". *JFE Technical. Rep.* **2007**, *10*, 19–25.
10. Sun, C.F.; Cai, Q.W.; Wu, H.B.; Mao, H.Y.; Chen, H.Z. Effect of controlled rolling processing on nanometer-sized carbonitride of Ti-Mo ferrite matrix microalloyed steel. *Acta Metall. Sin.* **2012**, *48*, 1415–1421. [CrossRef]
11. Wang, X.N.; Di, H.S.; Du, L.X. Effects of deformation and cooling rate on nano-scale precipitation in hot-rolled ultra-high strength steel. *Acta Metall. Sin.* **2012**, *48*, 621–628. [CrossRef]
12. Xu, Y.; Zhang, W.N.; Sun, M.X.; Yi, H.L.; Liu, Z.Y. The blocking effects of interphase precipitationon dislocations' movement in Ti-bearing micro-alloyed steels. *Mater. Lett.* **2015**, *139*, 177–181. [CrossRef]
13. Xu, L.X.; Wu, H.B.; Tang, Q.B. Effects of Coiling Temperature on microstructure and precipitation behavior in Nb–Ti microalloyed steels. *ISIJ Int.* **2018**, *58*, 1086–1093. [CrossRef]
14. Xu, Y.; Sun, M.X.; Zhou, Y.L.; Liu, Z.Y.; Wang, G.D. Effect of Mo on nano-precipitation behavior and microscopic mechanical characteristics of ferrite. *Steel Res. Int.* **2015**, *86*, 1056–1062. [CrossRef]
15. Natarajan, V.V.; Challa, V.S.A.; Misra, R.D.K. The determining impact of coiling temperature on the microstructure and mechanical properties of a Titanium-Niobium ultrahigh strength microalloyed steel: Competing effects of precipitation and bainite. *Mater. Sci. Eng. A* **2016**, *665*, 1–9. [CrossRef]
16. Xu, Y.; Sun, M.X.; Yi, H.L.; Liu, Z.Y. Precipitation behavior of (Nb,Ti)C in coiling process and its effect on micro-mechanical characteristics of ferrite. *Acta Metal. Sin.* **2015**, *51*, 31–39.
17. Gan, X.L.; Yuan, Q.; Zhao, G.; Ma, H.W.; Liang, W.; Xue, Z.L.; Qiao, W.W.; Xu, G. Quantitative analysis of microstructures and strength of Nb-Ti microalloyed steel with different Ti additions. *Metall. Mater. Trans. A* **2020**, *51*, 2084–2096. [CrossRef]
18. Toshio, M.; Hitoshi, H.; Goro, M.; Tadashi, F. Effects of ferrite growth rate on interphase boundary precipitation in V microalloyed steels. *ISIJ Int.* **2012**, *52*, 616–625.

19. Jose, C.M.; Guillermo, E.; Estela, P.T. Characterization of microalloy precipitates in the austenitic range of high strength low alloy steels. *Steel Res.* **2002**, *73*, 340–345.
20. Yen, H.W.; Chen, P.Y.; Huang, C.Y.; Yang, J.R. Interphase precipitation of nanometer-sized carbides in a titanium-molybdenum-bearing low-carbon steel. *Acta Mater.* **2011**, *59*, 6264–6274. [CrossRef]

2

Microstructural Control and Properties Optimization of Microalloyed Pipeline Steel

Mohamed Soliman †

Institute of Metallurgy, Clausthal University of Technology, Robert-Koch-Straße 42, 38678 Clausthal-Zellerfeld, Germany; mohamed.soliman@tu-clausthal.de.
† Faculty of Engineering, Galala University, 43511 Galala City, Egypt.

Abstract: A series of physical simulations, with parameters resembling those of industrial rolling, were applied using a thermo-mechanical simulator on microalloyed bainitic pipeline steel to study the influence of varying the processing parameters on its microstructure evolution and mechanical properties. In this study, the austenitization temperature and roughing parameters were kept unchanged, whereas the parameters of the finishing stage were varied. The developed microstructures were studied using scanning electron microscopy (SEM) and transmission electron microscopy (TEM). It is illustrated that selecting the appropriate cooling strategy (without altering the deformation schedule) can produce an optimized microstructure that breaks through the strength–ductility trade-off. Increasing the cooling rate after the finishing stage from 10 $K{\cdot}s^{-1}$ to 20 $K{\cdot}s^{-1}$ activated the microstructure refinement by effective nucleation of acicular ferrite and formation of finer and more dispersed martensite/austenite phase. This resulted in a remarkable enhancement in the ductility without compensating the strength. Furthermore, a pronounced strength increase with a slight ductility decrease was observed when selecting the appropriate coiling temperature, which is attributed to the copious precipitation associated with locating the coiling temperature near the peak temperature of precipitation. On the other hand, it was observed that the coiling temperature is the predominant parameter affecting the strain aging potential of the studied steel. Higher strain aging potentials were perceived in the samples with lower yield strength and vice versa, so that the differences in yield strength after thermo-mechanical treatments evened out after strain aging.

Keywords: microalloyed pipeline steel; hot deformation parameters; microstructural control; thermo-mechanical simulation; precipitation kinetics; strain aging

---

1. Introduction

The World Energy Outlook (WEO, 2019) [1] forecast for world energy supply specified that natural gas will play a significant role in the world's energy supply over the next two decades and that the world is poised to enter a "Golden Age of Gas". In many advanced economies, coal is being increasingly replaced by renewables and natural gas. Natural gas will gradually overtake coal and by 2040 the gas demand will have grown by over a third [1]. The demand for development of cost-effective steel with high strength and toughness for oil and gas pipeline-construction is driven by the increased worldwide consumption of both of the fossil fuels, particularly gas-consumption. High strength pipelines facilitate operating at higher pressures, while reducing the wall thickness and lowering the construction costs. In addition to the high operating pressures, extreme external effects, like ground movement or buffering exerted forces on the pipe during construction, play an important role that urges pipeline constructors to utilize increasingly demanding high-strength and ultra-high-strength steel grades [2].

The strength, toughness and weldability of the pipelines have been extended by combining the designed alloying in conjunction with specific thermomechanical processing routes. This development is based on a design strategy that combines both new alloying concepts along with thermo-mechanical rolling schedules tailored for the designed alloy. In this strategy, carbon is not the main strengthening source; it is kept below 0.09 wt.% to improve the weldability and toughness [3]. The strength–ductility balance in the pipeline steels is attained via thermo-mechanical processing (TMP), which strengthens the steel through microstructure refinement, precipitation hardening and microstructural modification. The TMP comprises three stages, namely rough rolling, finish rolling and accelerated cooling. In the course of the rough stage, the repeated cycles of work hardening and the recrystallization process results in the refining of the austenite grains. The finish rolling begins following the roughing stage. The deformation during this stage takes place in the non-recrystallization region, which results in substantial refinement to the final microstructure. The accelerated cooling step starts subsequently to the finishing step with the aim of suppressing the polygonal ferrite formation and urging the formation of non-equilibrium, non-equiaxed ferrite microstructures [4]. The latter transformation products are known to contribute to increasing strength, while maintaining a reasonable level of toughness through both small effective grain sizes and increased dislocation densities [5]. The developed microstructure is complex, consisting of mixtures of different morphologies, and accordingly, wide combinations of mechanical properties can be achieved by controlling them.

Beside the microstructure-control, controlling the precipitation process is also essential for achieving the desired mechanical properties in pipeline steels, which are considered as one of the most successful applications of microalloying [6,7]. The amount of the used microalloying elements, for single and multi-elements, are up to 0.15 wt.%. They could be Nb, V and/or Ti alloy [8]. Nb can be used as a single microalloying element or combined with V. The Ti is added in combination with one of the two microalloying elements. The interactions between the microalloying elements are complex, but in general terms Nb precipitates more readily in austenite than does V, so it is relatively more effective as a grain refiner [9].

In the present study, thermo-mechanical simulations were carried out on samples of a low-carbon CMnMoNbTi pipeline steel following a close-to-industrial schedule. In these simulations, parameters in the finishing stage, namely the finish rolling temperature, coiling temperature and cooling rates between the finish rolling stage and the coiling temperature, were varied. The effect of varying the simulation parameters on the microstructure development and mechanical properties were studied. The dependence of the developed microstructures and the mechanical properties on the applied thermo-mechanical processing parameters was analyzed. The effect of the processing parameters on the strain-aging behavior was also investigated.

2. Materials and Methods

2.1. Material and Preparation of Specimens

Flat compression samples for the physical simulations were machined out a 52 mm thickness industrially rough rolled transfer slab with the chemical composition given in Table 1. This composition corresponds to an API X80-grade. The low C level is for enhancing weldability. The addition of Mn, Mo and Cr is to compensate for the hardenability loss due to the low carbon content. Microalloying with Nb and Ti is conducted to enhance the strength through precipitation strengthening. Additionally, an undissolved quantity of TiN precipitates remain in the solution during the austenitization stage. These particles limit the coarsening of the austenite grains by the strong pinning effect. Figure 1 shows the geometry of the flat compression samples. The samples serve as tensile testing samples after applying the TMP on them. The longitudinal axes of the specimens were taken parallel to the rolling direction and their thicknesses parallel to the transfer bar thickness. The specimens have shoulders of 42 mm for clamping the specimens after TMP during tensile testing and two Ø 6 mm holes for decreasing the heat-dissipation from the testing-zone towards the shoulders. A centerline segregation of the Mn solute element is expected to occur at the mid-thickness region of the 52 mm slab [10]. This segregation zone was avoided by machining the flat compression specimens out of the upper and lower third of the slab.

Table 1. Composition in wt.% of the studied material.

C	Mn	Si	Mo	Cr	Ti	Nb	N
0.055	1.84	0.3	0.26	0.18	0.0256	0.101	0.006

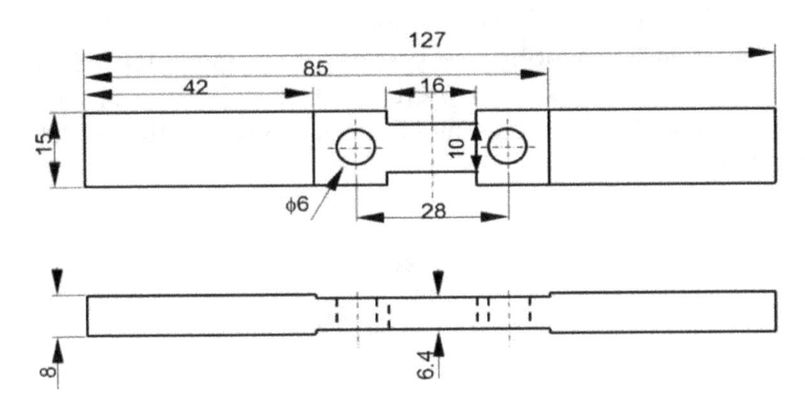

Figure 1. The flat compressing sample (dimensions in mm). The sample was used for tensile testing after applying the TMP.

2.2. Thermo-Mechanical Simulation

The thermo-mechanical simulation was carried out using the flat compression setup of the TTS 820 thermo-mechanical simulator (TA Instruments, Huellhorst, Germany). The temperature during the simulation was monitored using a thermocouple spot-welded on the specimen. The specimen was placed on two ceramic rollers and fixed from the upper side by two ceramic rods. The deformation steps were carried out via two deformation stamps, which upset the specimen in its centre. A detailed description of the flat compression setup of the TTS820 is given in [11]. The flat compression specimens were subjected to the TMP schedule shown in Figure 2. The process variables were selected based on industrial rolling parameters. The focus of this study lies in the consideration of the finishing stage. In this schedule, specimens were heated up to the austenitization temperature (T_A) of 1235 °C. The samples were subsequently cooled to 1100 °C and deformed at this temperature applying a true strain value φ of 0.3. The delay-time between the roughing and finishing stages was selected to be 5 s. The successive three deformation steps were performed to simulate the finish rolling stage. Subsequently, the specimens were cooled from the finish rolling temperature "FRT" of the last deformation step to the coiling temperature "CT" applying cooling rate "a". The accelerated cooling is considered to prevent the formation of the polygonal ferrite (PF). Finally, the cooling in the coil was simulated by applying a very slow cooling rate of 30 K/h (0.0083 K/s), starting from the coiling temperature CT. A strain rate of 15 s^{-1} was applied for all the upsetting steps. The investigated TMP-parameters are given in Figure 2. Additionally, dilatometric investigations employing the TMP of Figure 2, one time with FRT = 840 °C and another time without applying the deformation steps, were carried out using the Dil 805A/D deformation dilatometer (TA Instruments, Huellhorst, Germany). A detailed description of this dilatometer is given in [12].

2.3. Microstructure Characterization

Microstructure characterization was carried out using scanning electron microscopy (SEM) and transmission electron microscopy (TEM). CAMSCAN 44 (Cambridge scanning Company Ltd., Cambridge, UK) SEM investigations were carried out on central sections of the deformation region cut parallel to the deformation direction. The samples were prepared applying the conventional metallographic preparation procedures comprising mounting, grinding and polishing. A quantity of 0.3 μm oxide-polish silica suspension was used for the final polishing. Nital etchant of 2% concentration was used for revealing the microstructure. After etching, the samples were rinsed with ethyl alcohol and dried under a warm air drier. Stereological measurements were carried out on the scanning electron micrographs to evaluate the obtained phase fractions using the manual point count method, and to evaluate the martensite/austenite particle size by applying the linear intercept method.

Carbon extraction replicas and thin foils for TEM investigation were prepared from a slice mechanically thinned to 0.3 mm. The slices were taken from the central zone of the deformed region parallel to the direction of deformation. The thin foils were prepared by electropolishing in a solution of 10% perchloric acid and 90% ethanol. The process of single-sided extraction replication was used for preparing the extraction replicas. A JEOL 2000 EX II (JEOL Lt., Tokyo, Japan) and an FEI Tecnai 20F microscope (Philips, Amsterdam, The Netherlands) operating at 200 kV were used for the microstructural analyses.

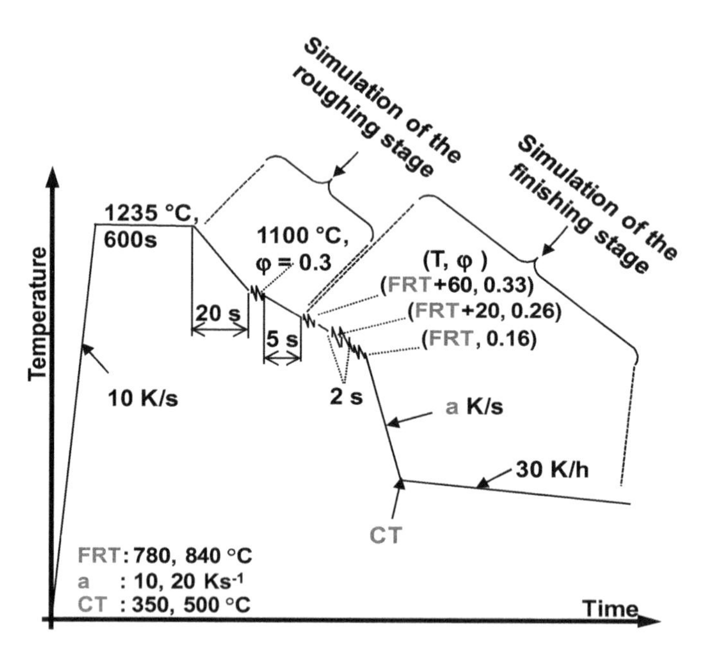

Figure 2. Schematic of the applied TMP.

2.4. Tensile Testing

After thermo-mechanical simulation processes, the samples were machined in the form of mini-tensile specimens, as previously illustrated in [11]. Three samples were tested for each processing condition. Pre-straining and loading until fracture of the tensile samples were conducted in a computerized universal testing machine UTS (Zwick-Roell, Ulm, Germany) with a 250 kN load cell. A video extensometer (Zwick-Roell, Ulm, Germany) was used for measuring the strain during testing.

2.5. Evaluating the Strain Aging Potential

The aging treatment was performed after applying a pre-strain of 2% on the TMP samples. The pre-strained samples were aged at 170 °C for 20 min in a recirculating air furnace. The strain aging potential was assessed by the value "S_2". S_2 is evaluated as $S_2 = R_{eL} - R_2$, where R_{eL} is the lower yield strength after aging and R_2 is the measured stress at the applied 2% pre-strain [12,13].

3. Results and Discussion

3.1. Classification and Characteristics of the Obtained Micro-Constituents

The microstructures of pipeline steels consist of mixtures of different morphologies, and therefore, several nomenclatures have arisen over the years for the identification of such microstructures. Figure 3 shows an example of the obtained microstructure together with key-illustrations for classifying its different micro-constituents [4,11,14]. The microstructure of Figure 3 consists mainly of a mixture of acicular ferrite (AF) and granular bainite (GB). The AF has a featureless interior area with a jagged outer boundary. Likewise, the GB is characterized by a non-smooth outer boundary but its interior area is featured with retained austenite and carbide phases, dispersed within the grain. A further observable micro-constituent in the microstructure is the bainitic ferrite (BF). The BF has a feathery form of ferritic laths. Additionally, a very limited quantity of polygonal ferrite (PF), below 3%, appears in the microstructure. The PF is characterized by its featureless interior areas and a smooth outer boundary. The AF poses an assemblage of parallel or non-parallel interwoven ferrite laths. This form of ferrite laths could not be distinguished in the scanning electron micrographs SEM. However, a deeper insight using TEM revealed an interwoven structure, as shown in Figure 4b,c. In line with this observation, Zhao et al. [15] and Yakubtsov el al. [16] revealed the interwoven structure of the AF under TEM, which was not distinguishable using SEM. Figure 4a illustrates that the disorganized microstructure of the AF gives it its larger ability to deflect cracks, thus it is the most favorable structure in pipeline steels [15]. Two types of AF are observed in this study, with crossing laths and with parallel laths, as shown in Figure 4b,c, respectively. The high dislocation density in the AF laths that is observable under TEM is an essential characteristic of intermediate transformation products and is principally caused by heavy accommodated strain occurred due to the transformation process [17]. The high dislocation density contributes considerably in achieving high strength in the final product. The finer AF laths of Figure 4c exhibit higher dislocation density than that of the coarser ones of Figure 4b. It is also observable in Figure 4b,c that acicular ferritic laths are separated by strips of martensite/austenite (M/A) islands, which are distributed at their grain boundaries and have elongated shapes. Wang et al. [18] reported that the M/A islands in acicular ferritic pipeline steels distributed at the grain boundary of AF are formed during the continuous cooling and that they become thinner and shorter by increasing the cooling rate after hot deformation. The occurrence of the AF in the studied structure is attributed to both the chemistry of the steel and the applied processing route. It was previously reported that the low-carbon Mn-Mo-Nb-Ti microalloyed pipeline steel promotes the AF transformation [4,19]. Furthermore, hot deformation also strongly promotes the formation of AF through enlarging the temperature zone of its formation in the continuous cooling transformation (CCT) diagram [4].

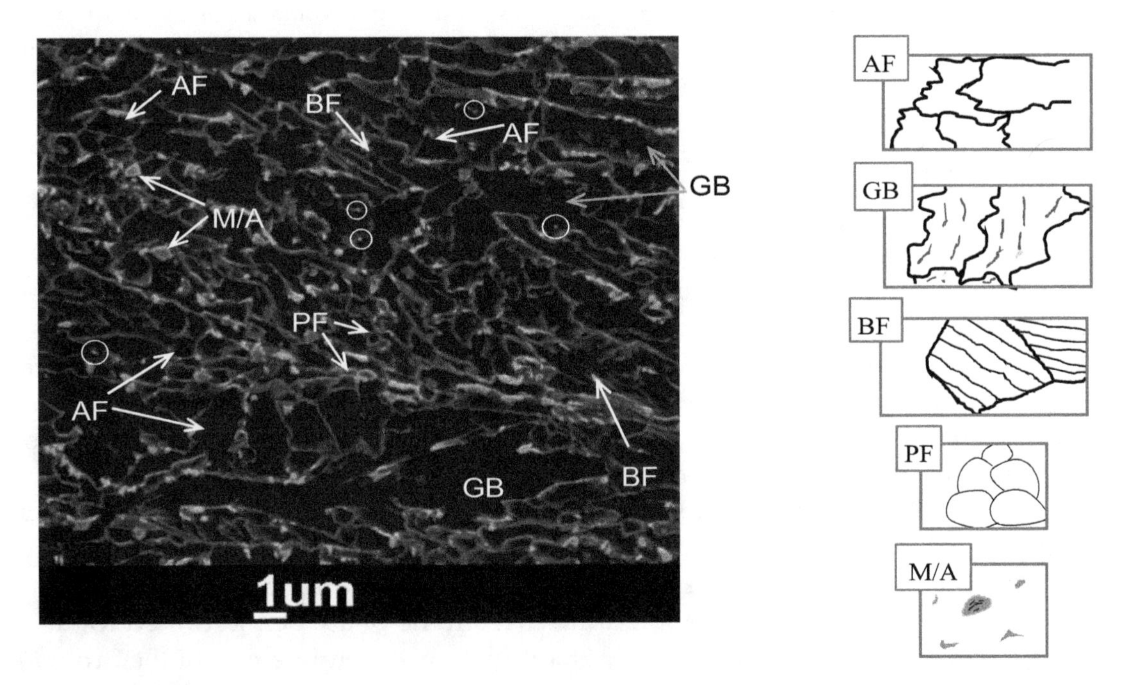

Figure 3. Scanning electron micrograph of steel processed with the parameters: FRT = 780 °C, a = 20 $K{\cdot}s^{-1}$, CT = 350 °C together with schematic illustrations of the different obtained micro-constituents.

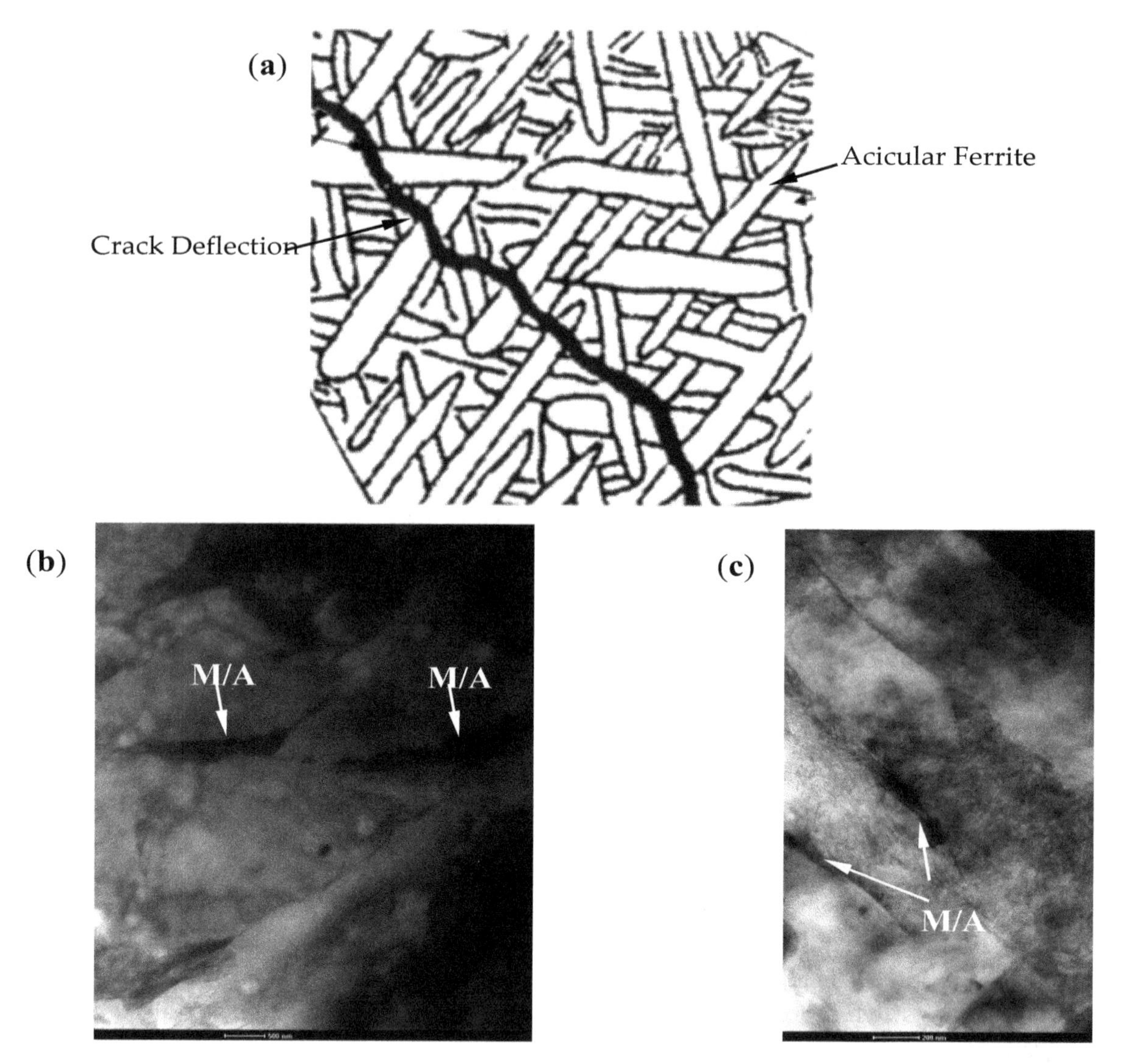

Figure 4. (**a**) Schematic feature of cleavage crack deflection at the lath boundaries of acicular ferrite [15]. (**b**,**c**) TEM micrographs of acicular ferrite developed under the processing parameters: FRT = 780 °C, a = 20 $K{\cdot}s^{-1}$, CT = 350 °C. The AF in (**b**) is of crossing interwoven ferrite laths and in (**c**) is of parallel ones.

Figure 5 shows TEM micrographs of GB and PF. Under TEM, GB appears as equiaxed ferrite grains but with much higher dislocation density than that of the PF and with islands of M/A within it [20,21]. The PF exhibits regions with slight dislocation densities at its neighborhood-boundary of the M/A island (designated with arrows in Figure 5b). This dislocations were created during the martensite formation, which is accompanied by a change in the shape of the transformed region [22]. Because the martensite formation occurs in a temperature range where the shape change cannot be accommodated elastically, the plastic deformation that is driven by the shape change causes the accumulation of dislocations in both the produced phase and in the surrounding ferrite phase [23]. Thus, plastic deformation is induced in the PF to accommodate the geometrical changes that are associated with martensite and bainite transformation.

The dispersed small M/A phase appears as shiny elevated phase in Figure 3 and as black islands in Figures 4 and 5. This M/A phase results from carbon partitioning at the interface between the parent phase, i.e., the austenite, and the phases being formed during continuous cooling of steel. During the phase transformation, the carbon diffuses from the phase being formed to the austenite sites, which results in an increase in the austenite stability. Consequently, a part of the carbon-enriched austenite remains in the transformed microstructure as retained austenite and the other part transforms to martensite during the continuous cooling. This results in coexistence of retained austenite with martensite in a single micro-constituent, which is generally considered as an M/A micro-constituent. The M/A attains two morphologies, namely the blocky one, which is observable in Figures 3 and 5 and the slender one, which is observable as thin layers between acicular ferrite laths, as distinguishable in Figure 4b,c.

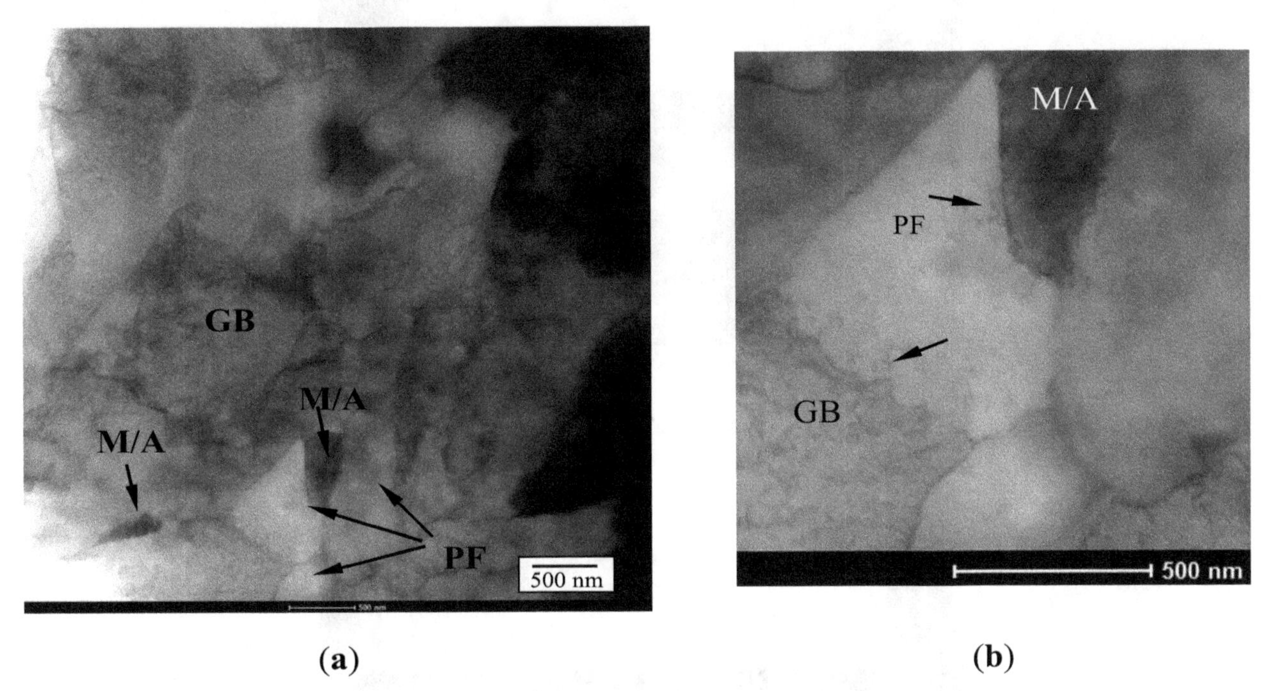

Figure 5. (**a**) TEM micrographs showing GB and PF. GB has much higher dislocation density than PF. (**b**) shows an enlarged view of PF; the arrows show the slight dislocations developed in PF at the interface with M/A and GB. The processing parameters are as that of Figure 4.

3.2. Classification and Characteristics of the Obtained Precipitates

Beside the abovementioned micro-constituents, precipitates of size ranging from ~50 nm up to ~250 nm are observable as well (some of them are white circled in Figure 3). High-resolution scanning electron investigations combined with energy-dispersive X-ray microanalysis (EDX) performed on JEOL JSM7000F (JEOL Lt., Tokyo, Japan) revealed three types of precipitates. Figure 6 illustrates scanning electron micrographs of these different types together with their EDX-spectrum. The first precipitate-type, which has rectangular non-equilateral form, like that shown in Figure 6a, was identified using EDX analysis as (Ti, Nb) (C, N). The microalloying with Ti and Nb is believed to play a particular role. The TiN has a high solubility temperature, therefore an undissolved quantity of its particles remains in the microstructure during the austenitization stage, which limits the coarsening of the austenite grains by the pinning effect [11]. Conversely, NbC completely dissolves during austenitization. The subsequent deformation and cooling processes result in the precipitation of NbC on the pre-existing TiN, i.e., NbC forms a shell around TiN [24]. The precipitates identified as (Ti, Nb) (C, N) are believed to be formed according to this scenario. Ma et al. [24] showed, using electron energy loss spectrometry (EELS), that these precipitates have a core that is Ti rich, or in some cases Ti and Nb rich, and a shell that is Nb rich. The tendency of the precipitation of NbC on the pre-existing TiN is proved by means of thermodynamic calculations using the software MatCalc® version 6.00 (MatCalc Engineering, Vienna, Austria) as will be shown in Section 3.5. It is basically shown using these calculations that, in case of the pre-existing TiN precipitates, the NbC fraction precipitating at dislocations is almost two to three times that of NbC fraction nucleating on the TiN surface. The precipitates nucleating at the dislocations will be designated by NbC-n and the ones nucleating at the surface of the TiN will be designated by NbC-s.

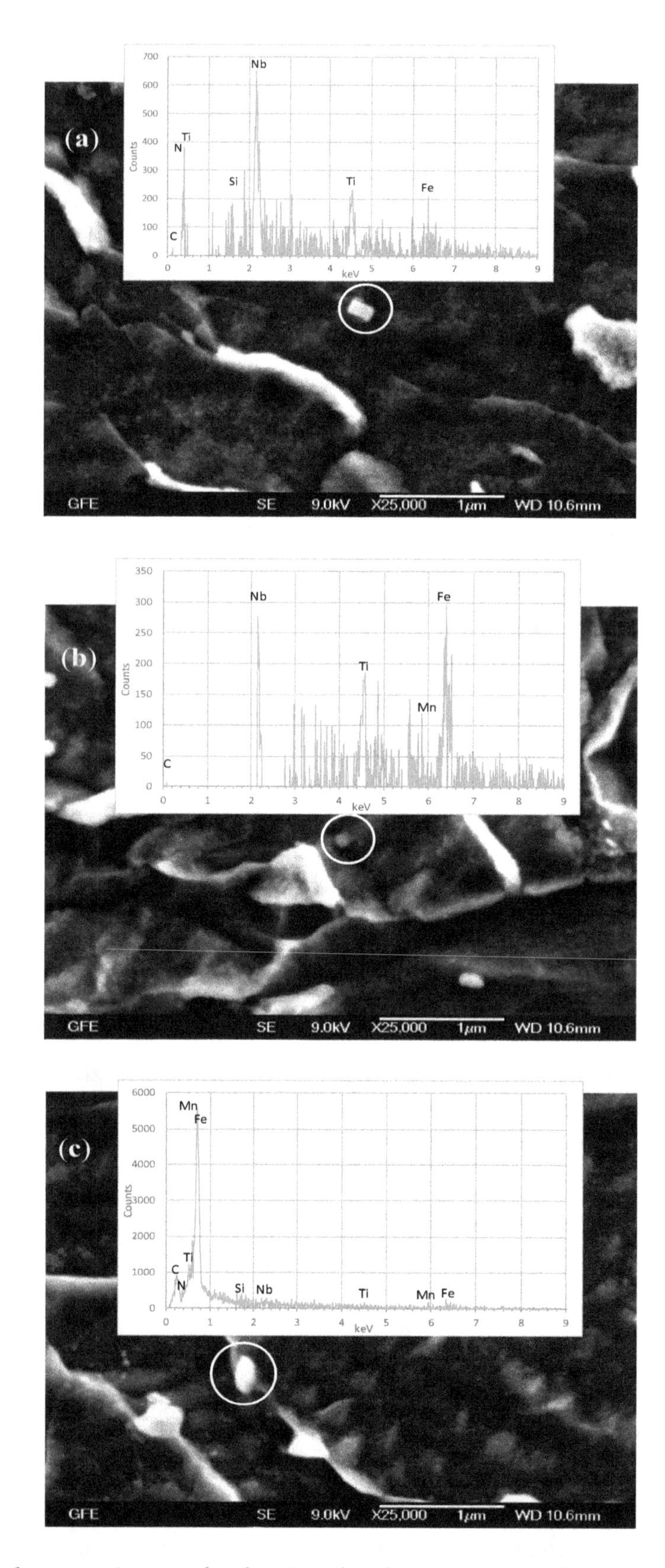

Figure 6. Scanning electron micrographs showing the observed types of precipitates together with the EDX-spectrum of the circled precipitates; (**a**) is (Ti, Nb)(C, N), (**b**) is NbC and (**c**) is cementite.

Figure 6b reveals an example of the NbC-n type of precipitates. These precipitates are smaller in size than the (Ti,Nb) (C,N) and are characterized by an equilateral rectangular shape. The EDX analysis of these precipitates did not reveal N; however, it revealed Ti. As shown in the predicted composition of the NbC precipitates of Table 2, Ti remains as an accompanying element in this precipitate-type. The absence of the N and the observation of Ti in lower intensity in the EDX together with their dissimilar shape to the (Ti,Nb)(C,N) precipitates is the reason for judging these precipitates as NbC-n and not TiN shelled with NbC-s.

Table 2. The predicted precipitates compositions (in mass fraction) using MatCalc® and considering the processing parameters: FRT = 780 °C, a = 20 $K{\cdot}s^{-1}$ and CT = 350 °C.

Element	Ti	Nb	C	Mn	Fe	N
TiN	0.499	-	0.029	-	-	0.470
NbC-n	0.115	0.390	0.464	-	-	0.031
Cementite	-	0.061	0.250	0.233	0.378	-
NbC-s	0.115	0.392	0.466	-	-	0.026

Figure 6c shows the cementite precipitates, which are characterized by their elliptical shape. The EDX analysis of these particles revealed their enrichment with Mn. This is also confirmed from their predicted composition given in Table 2. It was previously reported that the cementite often does not consist only of pure iron carbide; it contains, in most cases, varying amounts of manganese carbide [25]. The reason for the high Mn level in cementite is alloying with Mn above the level required to form MnS; the excess Mn combines with some of the carbon to produce carbide of manganese, which is formed in conjunction with iron carbide. It is worth mentioning here that the predicted composition given in Table 2 was found to be dependent on the coiling temperature and insignificantly affected by the other investigated processing parameters. The thermodynamic calculations predicted that the mass-fraction of Mn in cementite increased to 0.31 and its iron content decreased to 0.25 when increasing the CT to 500 °C.

The existing precipitates resulting from addition of microalloying elements enhance the steel-strength through:

1. Limiting the coarsening of austenite grains during austenitization by the pinning effect.
2. Refining the austenite grain via retarding the recrystallization of austenite during hot-rolling.
3. Providing nanostructure phases, which are the precipitates themselves.

An extension of this investigation, in which an extensive TEM study is to be carried out, is the main goal of a further investigation. This study will incorporate both stereological measurements on the precipitates to determine the volume fraction and the distribution of each precipitate-type, and studying of the elements' distribution within the precipitates.

3.3. Microstructure Evolution

Figure 7 shows representative scanning electron photomicrographs comparing the combined effect of cooling rate and coiling temperature on the developed microstructures. The main micro-constituents of the microstructure are the AF and GB. The BF is observed only in the samples with CT = 350 °C and a = 20 $K{\cdot}s^{-1}$ (Figure 7d). The BF grows in the form of packets consisting of clusters of thin lenticular plates or laths. The transformation of BF is a displacive one, and therefore it cannot cross the austenite grain boundaries [26]. The formation of PF is avoided/minimized, thanks to the combined effect of the alloy design and the applied thermo-mechanical schedule. Zhao et al. [4] found that the additions of Mn, Nb, Ti and/or Mo significantly suppress PF formation and promote the formation of AF. The formation of AF is also strongly promoted by hot deformation [4]. The formation of AF starts at a temperature slightly above the upper bainite region and below that of PF. GB, on the other hand, is formed at lower temperatures than that of AF [5,17,27]. The last micro-constituent to form before the martensite transformation is the BF [4].

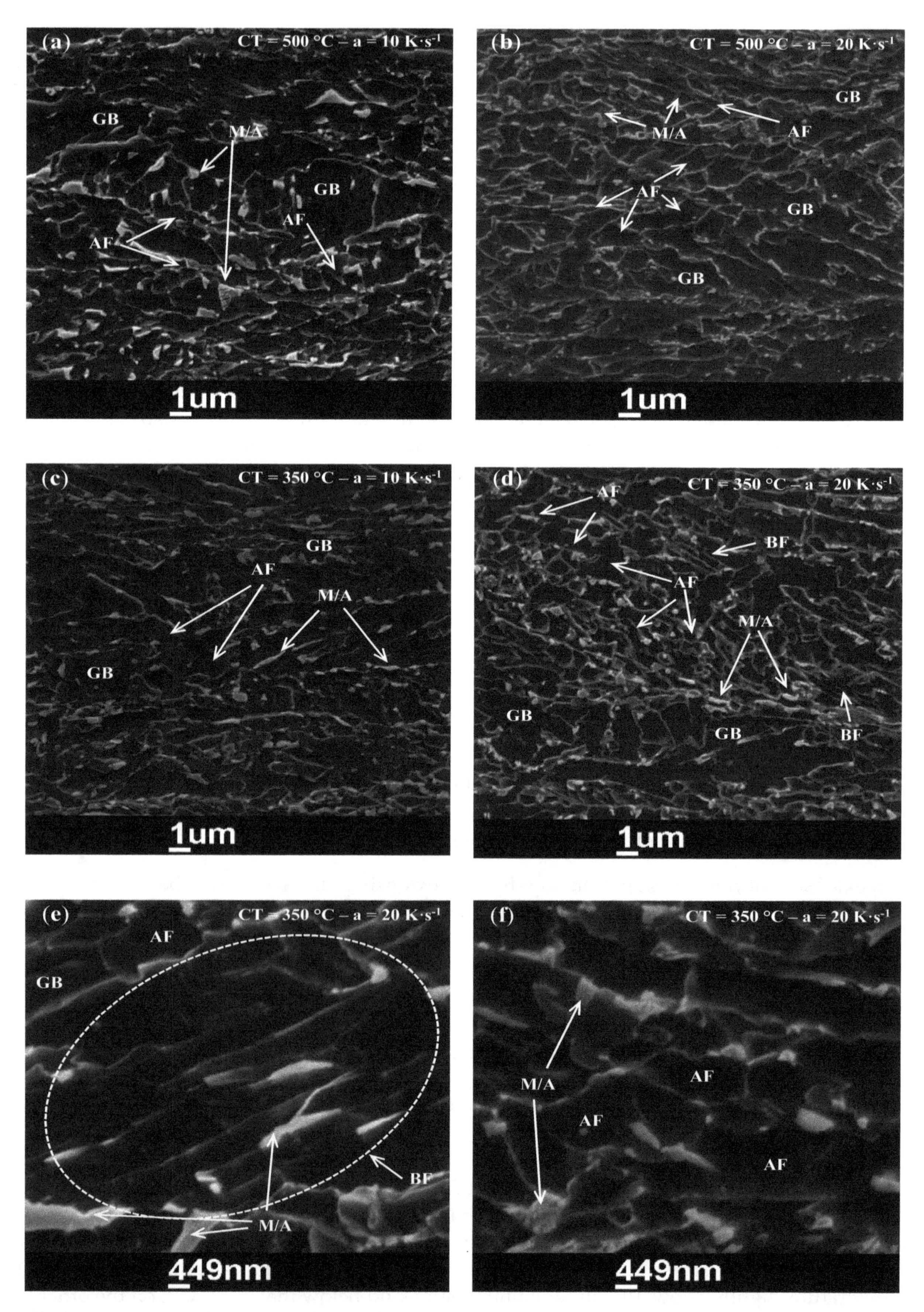

Figure 7. Scanning electron micrographs of samples processed under the prescribed conditions and having FRT = 780 °C. (**a**,**b**) for CT = 500 °C, (**c**,**d**) for CT = 350 °C and (**e**,**f**) higher magnification images of structure with same processing conditions as that of (**d**).

The metallographic investigations on the samples with FRT of 780 °C showed for both of the investigated coiling temperatures that higher AF fraction and finer and more dispersed martensite/austenite (M/A) phase are observed for the samples cooled applying a = 20 $K{\cdot}s^{-1}$. The volume fraction of the AF increased from ~0.27 to ~0.48 when increasing the cooling rate "a" from 10 $K{\cdot}s^{-1}$ to 20 $K{\cdot}s^{-1}$. Increasing the cooling rate has not significantly affected the M/A fraction, which is observed to be about 0.05. However, its particle size decreased from ~0.65 μm to ~0.30 μm with the increasing of the cooling rate. The CT has an insignificant effect on the obtained AF fraction. The latter observation is attributed to the fact that the AF is formed at a temperature range above the CT. Therefore, the AF fraction can only be dependent on the "FRT" and the cooling rate "a". Similarly, Hwang et al. [28] showed for their studied pipeline steels that they underwent little microstructural change with varying the CT from 400 °C to 600 °C.

On the other hand, increasing the cooling rate stimulates the intra-granular nucleation and subsequently the acicular ferrite transformation is promoted [29,30]. Furthermore, Anijdan and Yau [31] reported that a higher cooling rate is an important parameter that determines the formation of non-equiaxed ferrite-based micro-constituents, a desired microstructure for high strength pipeline steels. In the context of the observed significant grain refinement of M/A after applying higher cooling rates, Yi el al. [32] attributed this observation to the faster movement of the ferrite/austenite interface. Thus, the high cooling rate leaves isolated small austenite particles at temperatures above the martensite start temperature (Ms). Below Ms, the austenite particles form the M/A phase. Although the lower cooling rate produces larger retained austenite particles, their enrichment with carbon would be higher due to the availability of longer time for the carbon to diffuse from the ferrite and bainite to the retained austenite [33,34]. Similar refinement of the M/A phase is also observed by Anijdan and Yau [31].

The higher fraction of AF and the refinement of M/A with increasing "a" from 10 $K{\cdot}s^{-1}$ to 20 $K{\cdot}s^{-1}$ is also observable for the FRT 840 °C, as shown in Figure 8. Increasing FRT from 780 °C to 840 °C resulted in a marginal decrease in the developed AF fraction to 0.24 for a = 10 $K{\cdot}s^{-1}$ and to 0.46 for a = 20 $K{\cdot}s^{-1}$. Indeed, lowering the FRT activates the interior dislocation in the austenite phase, which promotes the formation nucleation sites for AF. The M/A particle size is slightly increased by the rise of the FRT; they recorded ~0.74 μm and ~0.38 μm for "a" values of 10 $K{\cdot}s^{-1}$ and 20 $K{\cdot}s^{-1}$, respectively.

3.4. Transformation Kinetics

Figure 9a,b illustrates the transformation kinetics in terms of the length change during cooling after the last deformation step in the TTS 820 simulator. The observed high noise in the dilatometric curves are due to employing a relatively loose fixing method of the samples in the simulator. However, this noise does not affect the identification of the transformation point. The PF, AF, GB and BF takes place during cooling from FRT to CT. The rapid increase in the length change at CT indicates that the transformation continues at the beginning of coiling simulation. This transformation can be considered as isothermal transformation owing to the applied very slow cooling rate of 0.0083 K/s. It is noted in Figure 9 that the sample with a = 20 $K{\cdot}s^{-1}$ – CT = 350 °C precedes to higher degree of transformation at the beginning of coiling simulation than the one subjected to a = 10 $K{\cdot}s^{-1}$ – CT = 350 °C. It is suggested that this significantly higher dilatation of the former case is due to the formation of BF, since it is the last phase to form during cooling before the martensite transformation [4]. This dilatometric observation is in agreement with the metallographic observation that the BF is observable for the case a = 20 $K{\cdot}s^{-1}$ – CT = 350 °C (Figure 7d) and it is absent in the microstructures of the other TMP conditions.

The martensite transformation takes place during continuous cooling in the coil, as revealed by the deviation from linearity in the dilatometric curves below the CT, as shown in Figure 9a,b. It is obvious from the dilatometric curves that increasing the cooling rate "a" from 10 $K{\cdot}s^{-1}$ to 20 $K{\cdot}s^{-1}$ shifts the martensite start temperature (Ms) to a lower value, which is an indication of increased austenite stability. The parameters affecting the austenite stability are its grain size and solute enrichment. The dependence of the Ms on the austenite grain size is experimentally confirmed. The smaller the particle size from which the martensite is formed, the lower the Ms [35,36]. Additionally, Yang and Bhadeshia developed a model that estimates the Ms variation as a function of the austenite grain size [37]. It is evident from Figures 7 and 8 that the M/A particles of the samples cooled with a = 20 $K{\cdot}s^{-1}$ have smaller size compared to those cooled with a = 10 $K{\cdot}s^{-1}$. Accordingly, this decrease in particle size can promote the decrease in Ms. According to the empirical equation of Lee and Park [38], decreasing the M/A particle size (formed from untransformed austenite grains with similar grain size) from 0.68 to 0.3 μm results in a decreasing of the Ms by about 12 °C. That the decrease of Ms with increasing "a" recorded about 40 °C indicates an increase in the solute elements in the small particles, which plays an additional role in decreasing Ms. Additionally, the expected better carbon distribution in smaller grain sizes, because of the reduced diffusion distance, can also shift Ms to a lower value. Martensite transformation in the austenite starts at the regions where the carbon content is the lowest. Thus, the more homogeneous distribution of carbon in austenite phase, the lower would be its Ms. The apparent absence of the martensite start in Figure 9a for samples processed with CT = 350 °C may be attributed to the occurrence of Ms near the start of the coiling simulation. The deviation from linearity due to martensite formation is thus concealed within the transformation just at the beginning of coiling simulation at 350 °C.

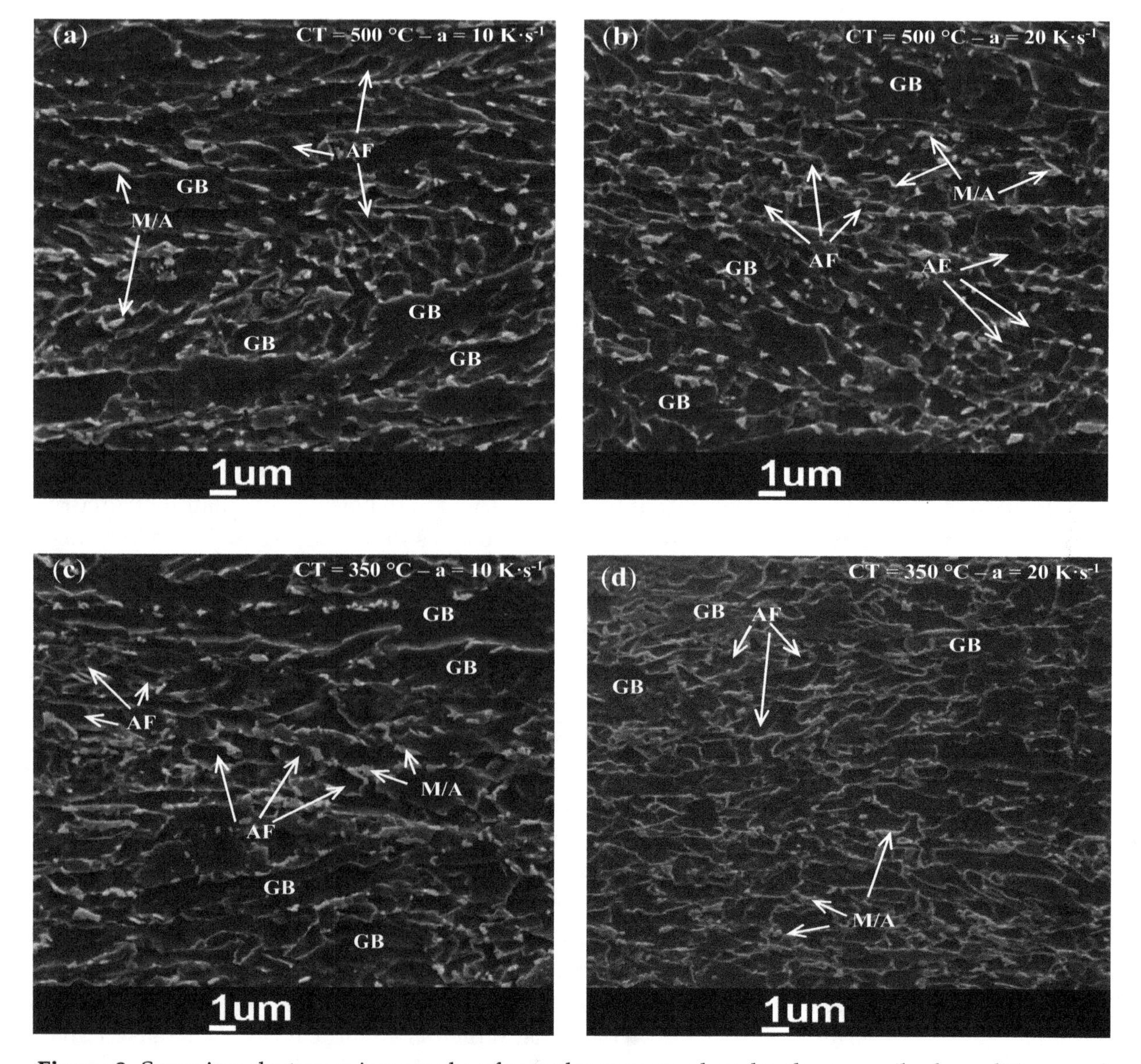

Figure 8. Scanning electron micrographs of samples processed under the prescribed conditions and having FRT = 840 °C. (**a**,**b**) for CT = 500 °C and (**c**,**d**) for CT = 350 °C.

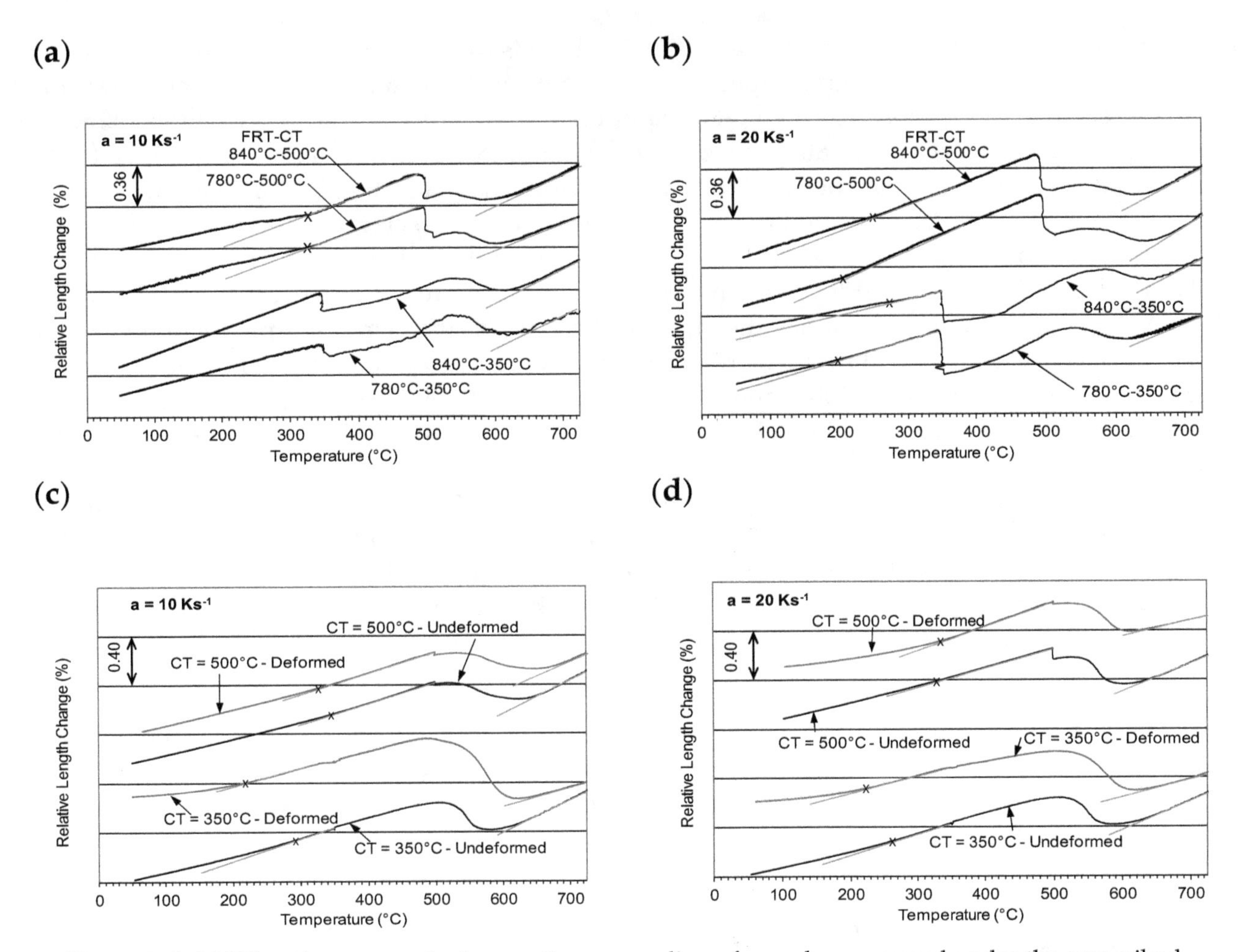

Figure 9. (**a**,**b**) Dilatation curves during continuous cooling of samples processed under the prescribed conditions. (**c**,**d**) Dilatation curves observed using Dil 805A/D for undeformed samples and deformed samples according to the TMP with FRT = 840 °C.

For comparison, Figure 9c,d show the transformation kinetics in undeformed samples and deformed ones with FRT = 840 °C observed using Dil 805A/D deformation dilatometer. These figures show that the beginning of austenite decomposition during cooling starts ~15 degrees below that of the deformed steel. Deformation of the steel moves the phase curves of the CCT diagram to the top left corner [4]. By contrast, the Ms of the deformed steels showed lower values compared to the undeformed ones. The latter effect is more pronounced when CT = 350 °C. This can be attributed to the smaller grains resulting from the structure refinement due to deformation. The austenite decomposition temperatures of the cylindrical compressed samples are comparable to those of the flat compressed ones. A temperature of ~670 °C was recorded when cooling with 10 $K \cdot s^{-1}$, and ~ 655 °C when cooling with 20 $K \cdot s^{-1}$. On the other hand, the transformation at the beginning of coiling for the flat compression samples is significantly higher than that of the cylindrical compressed ones.

3.5. Precipitation Process

In order to gain insight into the precipitation process, thermodynamic simulations of the TMP were performed be means of the software MatCalc® version 6.00 using the database mc_fe_V2.058. In these simulations, four types of precipitates, corresponding to those observed in the microstructures (see Section 3.2) were considered. These are: TiN, NbC-n and NbC-s and cementite. The simulations integrated the perceived change in the precipitation domain that occurred due to the phase transformation from austenite to ferrite at about 650 °C (Figure 9). The simulation results are presented in Figures 10 and 11. It is shown in these figures that the TiN precipitates pre-exist just at the beginning of the simulation at T_A of 1235 °C, which is due to the occurrence of its dissolution temperature above this reheating temperature. The NbC-s starts to form with the precipitation of TiN, whereas the NbC-n starts to precipitate later on at about 1070 °C, i.e., between the simulated roughing and finishing stages. In all cases, the predicted number density of NbC-s developed to attain a value that is equal to that one of the TiN. This behaviour indicates that, in the final microstructure, all the TiN precipitates are predicted to be covered with NbC-s precipitates. A slight increase in the number density of TiN, NbC-n and NbC-s are observable with the beginning of ferrite formation as well as the start of coiling at 500 °C. The cementite precipitation starts after the ferrite formation (below 650 °C). The coiling temperature has the strongest impact on the cementite precipitation. Coiling at 500 °C is predicted to result in the increasing of the cementite volume fraction by three orders of magnitude, as shown in Figure 10b,d and Figure 11b,d, respectively. These figures show also that the cementite number densities exhibit an increase by one order of magnitude at the coiling temperature. Additionally, a significant increase in the volume fraction of NbC-n and NbC-s when coiling at 500 °C is also predicted. However, there is no significant increase in the predicted precipitates when coiling at 350°C, as shown in Figure 10a,c and Figure 11a,c. This effect of CT on the precipitation process can be related to the precipitation–time–temperature (PTT) of the studied alloy. Coiling nearby to the peak temperature of precipitation activated the precipitation kinetics and therefore resulted in increasing the number density and volume fraction of the formed precipitates.

On the other hand, the attained cementite volume fraction is predicted to double when decreasing the cooling rate "a" from 20 $K{\cdot}s^{-1}$ to 10 $K{\cdot}s^{-1}$. The thermodynamic calculations showed an insignificant effect of decreasing FRT from 840 °C to 780 °C on the predicted precipitation process, as observed when comparing Figures 10 and 11.

An extensive TEM study to experimentally determine the volume fraction and number density of each type of the developed precipitates, including the very tiny ones, is the aim of a future study. The results are then to be compared with the predicted ones.

3.6. *Tensile Properties*

Figure 12 shows representative stress–strain curves for the TMP conditions. Indeed, the tensile results showed a very high level of consistency with divergence of less than ±2%, due to avoiding the segregation zone of the transfer bar during the machining of samples, and to the high reproducibility of the processing conditions using the TTS820 simulator. It is obvious in Figure 12 that the most dominating factor affecting the strength (Rm and Rp values) of the produced TMP-specimens is the coiling temperature. The specimens with CT = 500 °C recorded higher strength compared to those with CT = 350 °C. Nevertheless, this pronounced effect of CT on the tensile behaviour is not observable in the microstructures of Figures 7 and 8. This indicates that the strength-increase of this steel is influenced by the precipitation kinetics rather than the micro-constituents. It is presented Figures 10 and 11 and explored in the discussion of the previous section that increasing the coiling temperature from 350 °C to 500 °C is predicted to activate the kinetics of precipitation, particularly the precipitation of cementite, which showed an increase in its fraction and number density by two to three orders of magnitude. The TiN, NbC-n and NbC-s are also predicted to increase by order of 1.5 to two times by increasing CT. A similar increase in strength is reported by Xiao el al. due to increasing the CT of their studied microalloyed AF pipeline steel from 500 °C to 600 °C [39]. TEM investigations of Xiao et al. showed two typical carbonitrides precipitates with different size distributions. They indicated that the amount of the smaller precipitates is increased by the rising of the temperature towards the peak temperature for precipitation [39]. Therefore, selecting the proper CT is vital for precipitation strengthening, which increases both the strength and toughness.

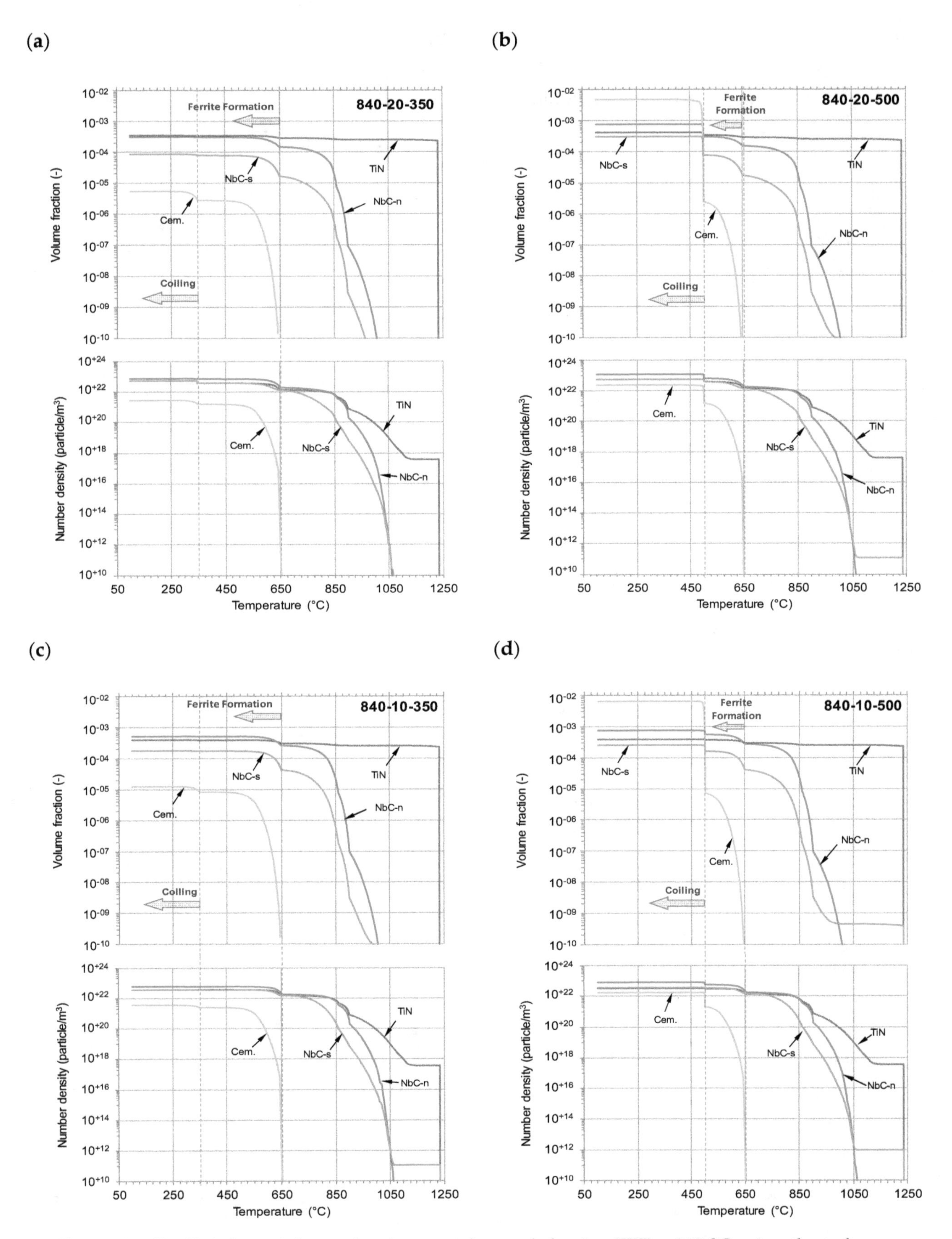

Figure 10. Predicted precipitates development for steels having FRT = 840 °C using the software MatCalc® and applying the parameters (**a**) 20 $K \cdot s^{-1}$-350 °C, (**b**) 20 $K \cdot s^{-1}$-500 °C, (**c**) 10 $K \cdot s^{-1}$-350 °C and (**d**) 10 $K \cdot s^{-1}$-500 °C.

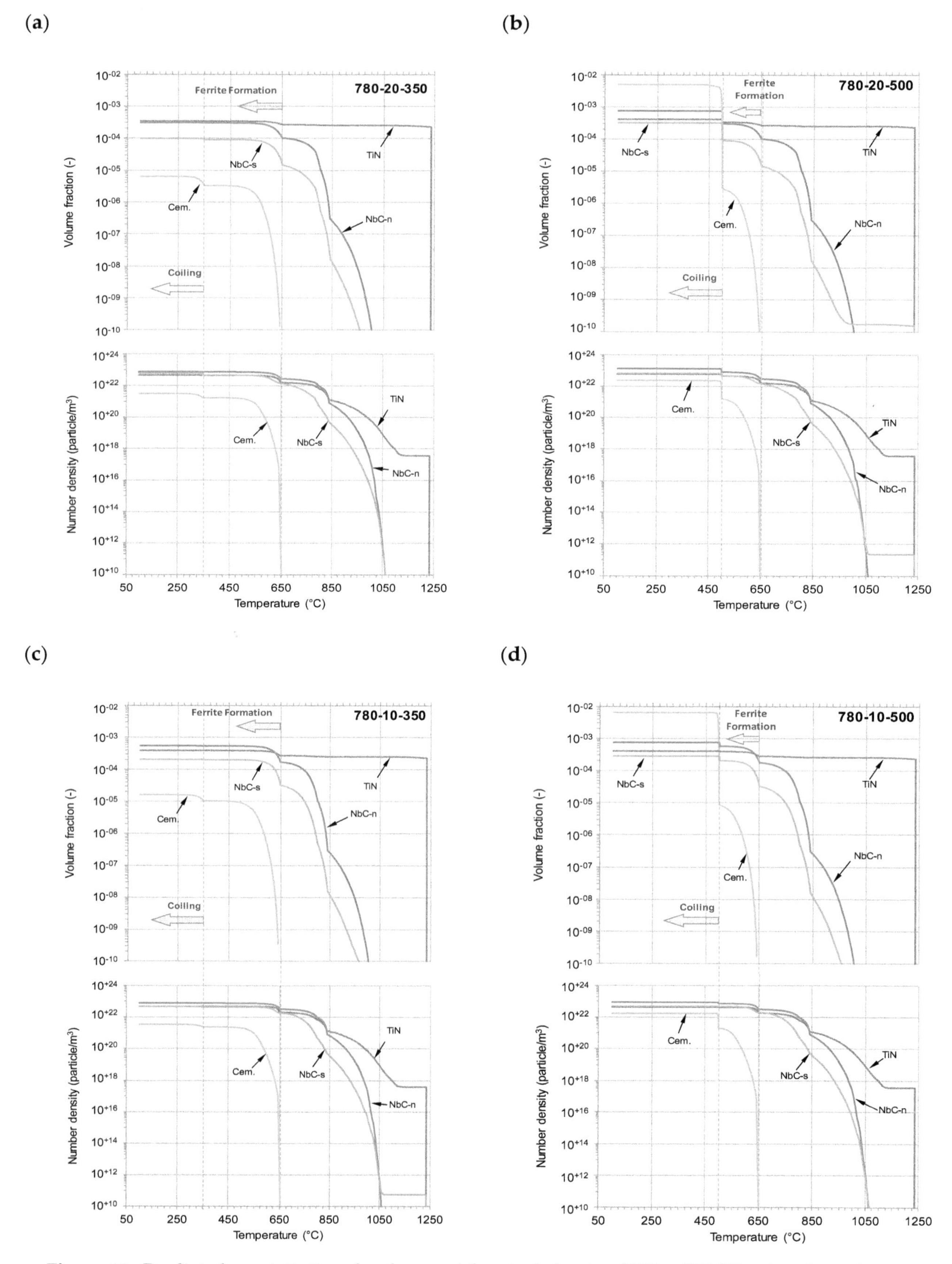

Figure 11. Predicted precipitation development for steels having FRT = 780 °C using the software MatCalc® and applying the parameters (**a**) 20 K·s^{-1}-350 °C, (**b**) 20 K·s^{-1}-500 °C, (**c**) 10 K·s^{-1}-350 °C and (**d**) 10 K·s^{-1}-500 °C.

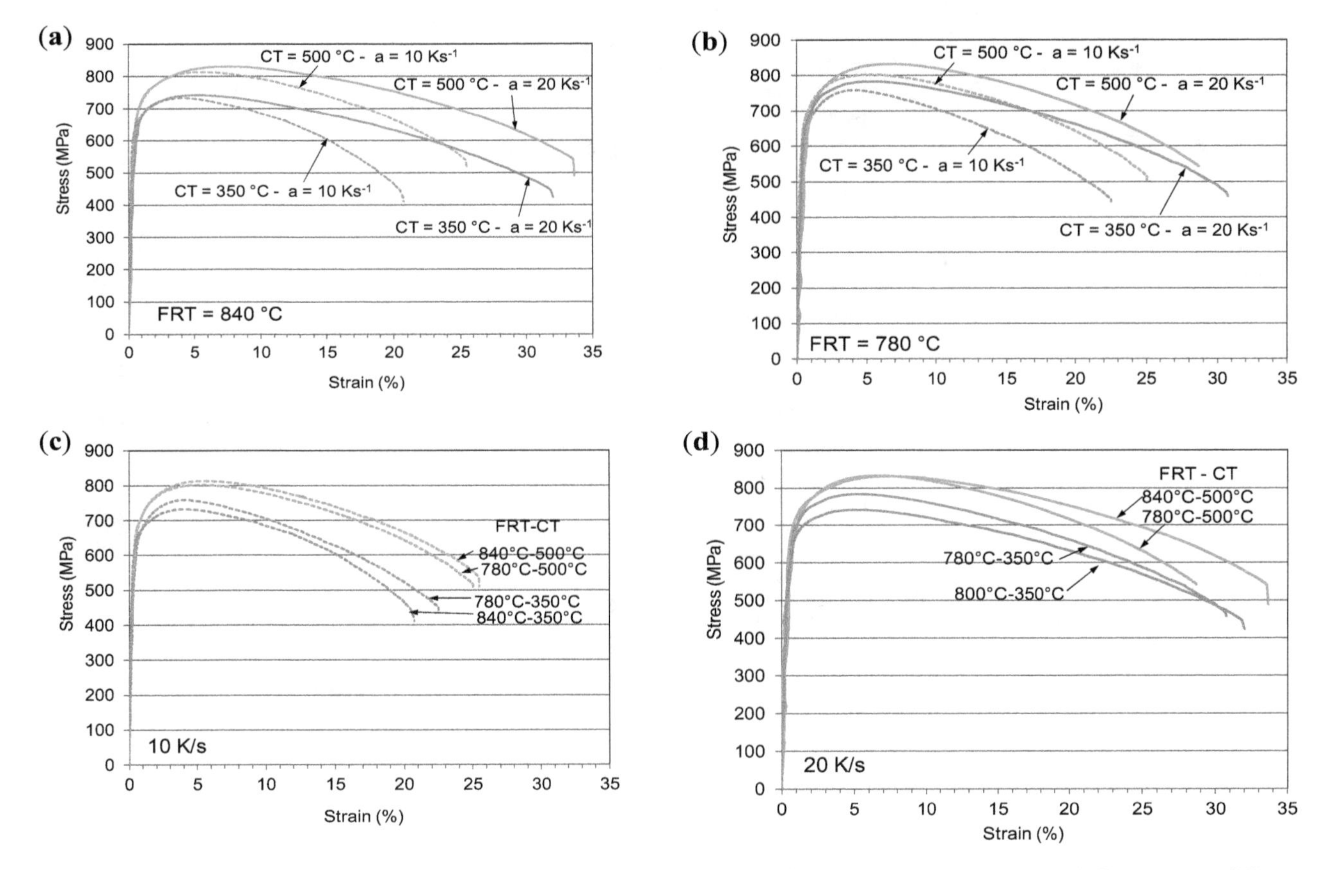

Figure 12. Stress–strain curves of samples with the prescribed TMP parameters; (**a**) FRT = 840 °C, (**b**) FRT 780 °C, (**c**) a = 10 K·s^{-1} and (**d**) a = 20 K·s^{-1}.

On the other hand, the predominant factor dictating the fracture strain is the cooling rate "a". Increasing the cooling rate from 10 K·s^{-1} to 20 K·s^{-1} results in a pronounced increase in the ductility (in terms of total elongation) associated with a slight increase in R_m, as shown in Figure 12a,b. This behaviour can be correlated with the greater AF fraction with increasing the cooling rate, as observed in Figures 7 and 8. The ability of AF to deflect the crack during its propagation (see Figure 4a) results in an enhancing of the tensile ductility with the increasing of the cooling rate. Furthermore, the M/A films at the AF lath-boundaries (Figure 4b,c) present a hindrance for crack propagation [40]. The rise of the AF fraction results in increasing these M/A films. Additionally, the finer and well distributed M/A particles of the fast cooled steel would contribute to the ductility increase.

Referring to Figure 12c,d, it is obvious that decreasing FRT has a significant effect in enhancing the Rm of the specimens with CT = 350 °C and a slight/insignificant effect for CT = 500 °C. Decreasing the FRT results in activating the formation of the substructure and probably increasing the dislocations' density, and therefore the nucleation sites of the AF are stimulated, as observed when comparing Figure 7 with Figure 8.

3.7. *Effect of the TMP Parameters on Strain Aging*

Figure 13 shows the stress–strain curves of the samples after aging. Table 3 lists the different S_2-values assessed according to Section 2.5. It was interesting to observe that the yield strength after strain aging recorded about 800 MPa, irrespective of the applied TMP parameters. The samples with lower yield strength showed higher strain aging potential, and vice versa, so that the differences in their original yield strengths (after TMP) are compensated. It becomes clear that the strain aging potential is considerably affected by the TMP parameters. The predominant parameter affecting the S_2-value is the CT, as shown in Table 3. Applying CT of 350 °C results in substantially higher S_2 value.

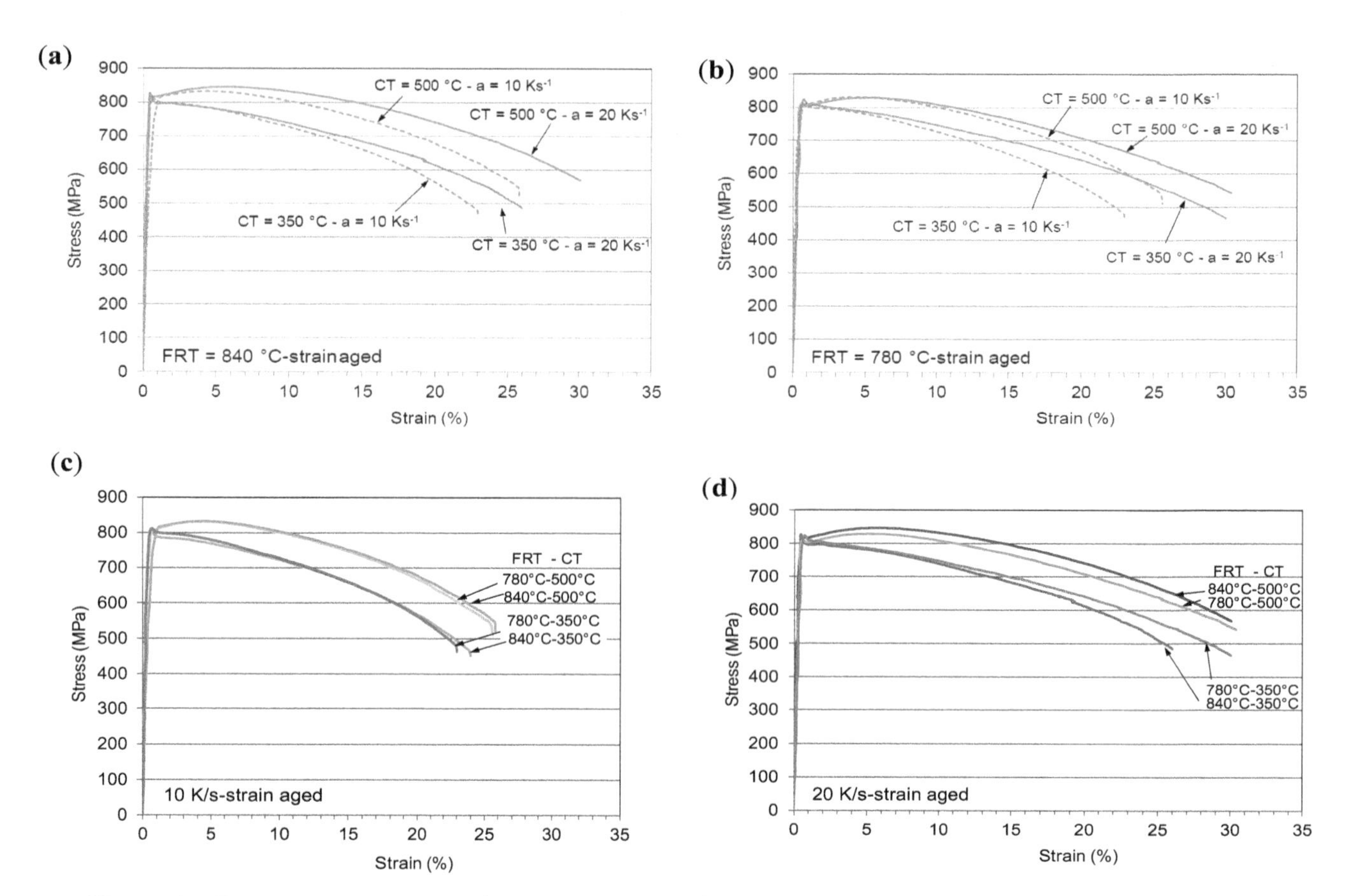

Figure 13. Stress–strain curves, after strain aging, of samples with the prescribed TMP parameters; (**a**) FRT = 840 °C, (**b**) FRT 780 °C, (**c**) a = 10 K·s^{-1} and (**d**) a = 20 K·s^{-1}.

Table 3. Strain aging potential (MPa).

a-Value	10 K·s^{-1}		20 K·s^{-1}	
FRT \ CT	500 °C	350 °C	500 °C	350 °C
840 °C	32	76	48	79
780 °C	43	72	30	71

During aging, the free C and N atoms available in the structure diffuse to the dislocations. The stabilizing of the dislocations due to the locking-effect with these interstitial atoms results in the increasing of the force necessary to cause their movement, and consequently the yield strength increases [41]. With the increasing of the available free C and N in steel, aging response should also increase because of the availability of more atoms pinning the mobile dislocations [42]. Taking into account the previous discussion regarding the formation of additional finer precipitates when CT = 500 °C (coiling near the peak temperature for precipitation, see Figure 10b,d and Figure 11b,d), one can expect that this leads to the availability of lower amounts of C and N atoms and consequently lower strain aging potential. Figure 14 shows the thermo-dynamic simulation of the precipitation process during aging after 2% pre-straining of the steels with FRT = 840 °C. The steels with FRT = 780 °C have similar precipitation behavior as those given in Figure 14 because the FRT has an insignificant effect on the predicted attained precipitates after the TMP, as previously illustrated under Section 3.5. After strain ageing, the available cementite precipitates locking the dislocations of the steels with CT = 350 °C, and those with CT = 500 °C, even out, as shown in Figure 14, and therefore the yield strengths equalize. This point needs further investigation.

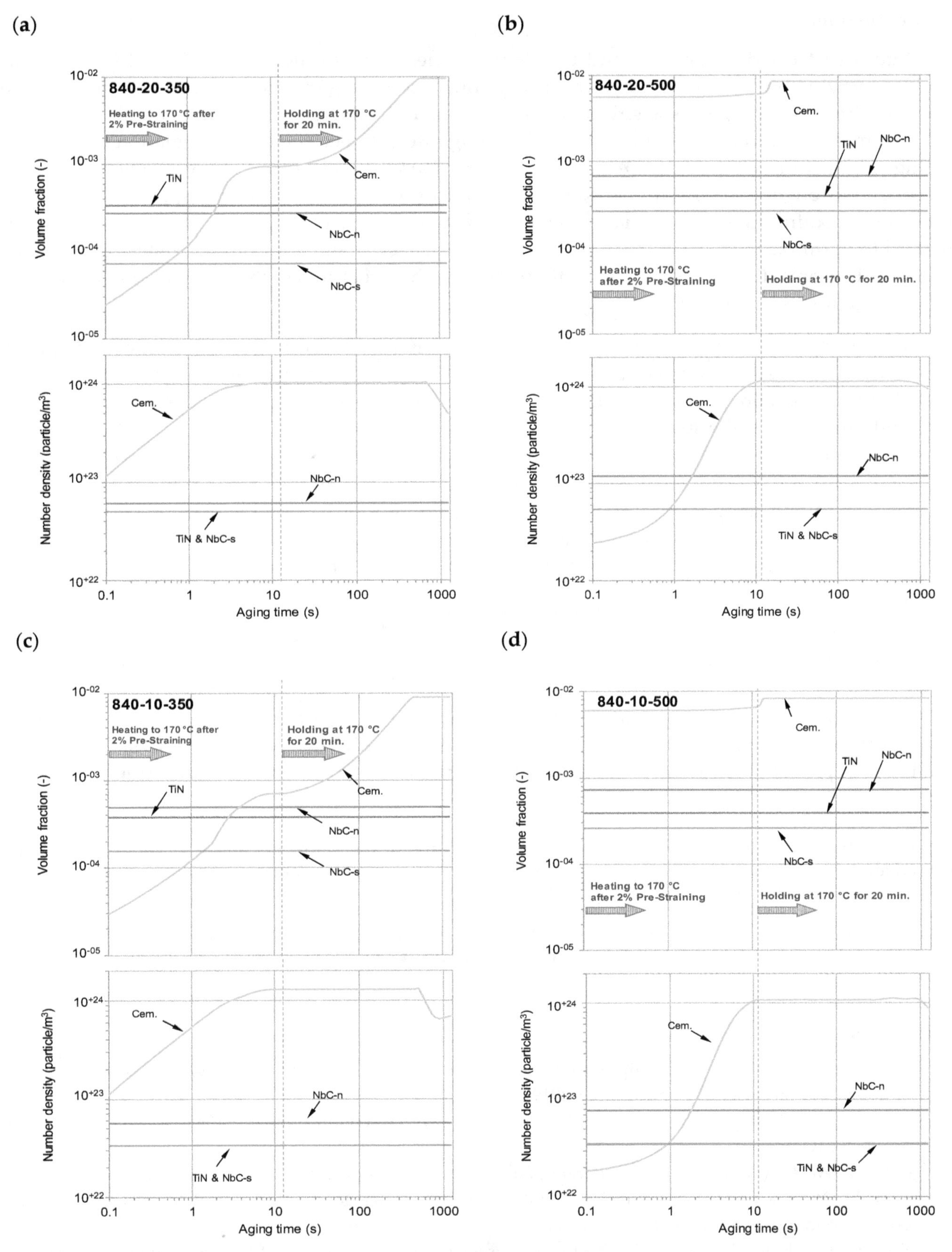

Figure 14. Precipitation kinetics during aging at 170 °C after 2% pre-straining, predicted using the software MatCalc®. The steel is thermo-mechanically processed prior to strain aging by applying FRT = 840 °C and the "a-CT" parameters of (**a**) 20 K·s^{-1}-350 °C, (**b**) 20 K·s^{-1}-500 °C, (**c**) 10 K·s^{-1}-350 °C and (**d**) 10 K·s^{-1}-500 °C.

4. Conclusions

Microstructure development and tensile properties were studied for a microalloyed pipeline steel by carrying out a series of physical simulations resembling the roughing and finishing stages of the rolling process. In these simulations, parameters in the finishing stage, namely the finish rolling temperature, coiling temperature and cooling rate between the last deformation step and the coiling temperature were varied. Austenitization temperature and roughing parameters were kept unchanged throughout all the applied simulation processes. It is shown that the appropriate selection of both the cooling rate after the last deformation step and the coiling temperature is an effective method to optimize the mechanical properties and is proven to break through the strength–ductility trade-off. This breakthrough is realized without refining the structure via changing the hot deformation parameters. It was achieved by:

1. Activating the formation of both the acicular ferrite and finer and more dispersed martensite/austenite phase by increasing the cooling rate after the last deformation step. It is shown in this study that stimulating this structure results in an appreciable increase in the ductility without reducing the strength.
2. Promoting the formation of finer and more dispersed cementite and carbonitride precipitates, which are of particular importance for enhancing the precipitation strengthening. Selecting the proper coiling temperature plays the central role by which the precipitation process can be promoted effectively. The appropriate selection of the coiling temperature in this study resulted in enhancing the strength without compensating the ductility.

Accordingly, the optimized tensile properties in this study were observed in the samples having a cooling rate after the last deformation step of 20 $K{\cdot}s^{-1}$ and a coiling temperature of 500 °C. The finish rolling temperature has a lesser effect on the tensile properties compared to that of the two above-mentioned factors. Furthermore, the coiling temperature can greatly influence the strain aging potential of the pipeline steel by affecting the available amount of free carbon and nitrogen atoms. An important observation of this study is that the samples with lower yield strength showed higher strain aging potential, and vice versa, so that their initial difference in R_P-values due to applying different TMPs evened out after strain aging.

Author Contributions: The author confirms sole responsibility for the following: study conception and design, data collection, analysis and interpretation of results, and manuscript preparation. All authors have read and agreed to the published version of the manuscript.

Acknowledgments: The author acknowledges Salzgitter Mannesmann Forschung GmbH for the material supply. The author expresses his gratitude to Heinz Palkowski for providing the facilities to perform the experimental work and for the useful discussions.

References

1. World Energy Outlook. Flagship Report. Available online: https://www.iea.org/reports/world-energy-outlook-2019 (accessed on 4 November 2019).
2. Das, A.K. The Present and the Future of Line Pipe Steels for Petroleum Industry. *Mater. Manuf. Process.* **2010**, *25*, 14–19. [CrossRef]
3. Jeffus, L. *Welding Principles and Applications*, 7th ed.; Delmar: Clifton Park, NY, USA, 2012.
4. Zhao, M.; Yang, K.; Xiao, F.-R.; Shan, Y.-Y. Continuous Cooling Transformation of Undeformed and Deformed Low Carbon Pipeline Steels. *Mater. Sci. Eng. A* **2003**, *355*, 126–136. [CrossRef]
5. Tamura, I.; Sekine, H.; Tanaka, T.; Ouchi, C. *Thermomechanical Processing of High Strength Low-Alloy Steels*; Butterworth & Co. Ltd.: London, UK, 1988.
6. Jiang, M.; Chen, L.-N.; He, J.; Chen, G.-Y.; Li, C.; Lu, X.-G. Effect of Controlled Rolling/Controlled Cooling Parameters on Microstructure and Mechanical Properties of The Novel Pipeline Steel. *Adv. Manuf.* **2014**, *2*, 265–274. [CrossRef]

7. Xu, J.; Misra, R.; Guo, B.; Jia, Z.; Zheng, L. Understanding Variability in Mechanical Properties of Hot Rolled Microalloyed Pipeline Steels: Process–Structure–Property Relationship. *Mater. Sci. Eng. A* **2013**, *574*, 94–103. [CrossRef]
8. Villalobos, J.C.; Del Pozo, A.; Campillo, B.; Mayén, J.; Serna, S. Microalloyed Steels through History until 2018: Review of Chemical Composition, Processing and Hydrogen Service. *Metals* **2018**, *8*, 351. [CrossRef]
9. Bhadeshia, H.; Honeycombe, R. *Steels: Microstructure and Properties*, 4th ed.; Butterworth-Heinemann: Oxford, UK, 2017.
10. Guo, F.; Wang, X.; Liu, W.; Shang, C.; Misra, R.D.K.; Wang, H.; Zhao, T.; Peng, C. The Influence of Centerline Segregation on the Mechanical Performance and Microstructure of X70 Pipeline Steel. *Steel Res. Int.* **2018**, *89*, 1800407. [CrossRef]
11. Soliman, M.; Palkowski, H. Influence of Hot Working Parameters on Microstructure Evolution, Tensile Behavior and Strain Aging Potential of Bainitic Pipeline Steel. *Mater. Des.* **2015**, *88*, 759–773. [CrossRef]
12. Soliman, M.; Palkowski, H. Tensile Properties and Bake Hardening Response of Dual Phase Steels with Varied Martensite Volume Fraction. *Mater. Sci. Eng. A* **2020**, *777*, 139044. [CrossRef]
13. Standard Testing Method DIN-EN 10325. Available online: https://www.mystandards.biz/standard/dinen-10325-1.10.2006.html (accessed on 4 November 2020).
14. Tafteh, R. Austenite Decomposition in a X80 Line Pipe Steel. Master's Thesis, The University of British Columbia, Vancouver, BC, Canada, May 2011.
15. Zhao, M.; Yang, K.; Shan, Y. The Effects of Thermo-Mechanical Control Process on Microstructures and Mechanical Properties of A Commercial Pipeline Steel. *Mater. Sci. Eng. A* **2002**, *335*, 14–20. [CrossRef]
16. Yakubtsov, I.; Boyd, J. Bainite Transformation During Continuous Cooling of Low Carbon Microalloyed. *Steel. Mater. Sci. Technol.* **2001**, *17*, 296–301. [CrossRef]
17. Lee, J.-L.; Hon, M.-H.; Cheng, G.-H. The Intermediate Transformation of Mn-Mo-Nb Steel During Continuous Cooling. *J. Mater. Sci.* **1987**, *22*, 2767–2777. [CrossRef]
18. Wang, C.; Wu, X.; Liu, J.; Xu, N. Transmission Electron Microscopy of Martensite/Austenite Islands in Pipeline Steel X70. *Mater. Sci. Eng. A* **2006**, *438–440*, 267–271. [CrossRef]
19. Zuo, X.; Zhou, Z. Study of Pipeline Steels with Acicular Ferrite Microstructure and Ferrite-bainite Dual-phase Microstructure. *Mater. Res.* **2015**, *18*, 36–41. [CrossRef]
20. Rodrigues, P.; Pereloma, E.V.; Santos, D. Mechanical Properities of an HSLA Bainitic Steel Subjected to Controlled Rolling with Accelerated Cooling. *Mater. Sci. Eng. A* **2000**, *283*, 136–143. [CrossRef]
21. Bramfitt, B.L.; Speer, J.G. A Perspective on the Morphology of Bainite. *Met. Mater. Trans. A* **1990**, *21*, 817–829. [CrossRef]
22. Young, C.H.; Bhadeshia, H.K.D.H. Strength of Mixtures of Bainite and Martensite. *Mater. Sci. Technol.* **1994**, *10*, 209–214. [CrossRef]
23. Lei, T.C.; Shen, H.P.; Zhang, J. Phase-Hardening and Phase Softening Phenomena in Dual-Phase Steels. In Proceedings of the International Conference on Martensitic Transformations, The Japan Institute of Metals, Nara, Japan, 26–30 August 1986; pp. 465–470.
24. Ma, X.; Langelier, B.; Gault, B.; Subramanian, S. Effect of Nb Addition to Ti-Bearing Super Martensitic Stainless Steel on Control of Austenite Grain Size and Strengthening. *Met. Mater. Trans. A* **2017**, *48*, 2460–2471. [CrossRef]
25. Sauveur, A. *The Metallography and Meat Treatment of Iron and Steel: Forgotten Books*; Ripol Classic Publishing House: London, UK, 2013; pp. 148–149.
26. Zhao, H.; Wynne, B.; Palmiere, E. Effect of Austenite Grain Size on the Bainitic Ferrite Morphology and Grain Refinement of A Pipeline Steel after Continuous Cooling. *Mater. Charact.* **2017**, *123*, 128–136. [CrossRef]
27. Xu, Y.-B.; Yu, Y.-M.; Xiao, B.-L.; Liu, Z.-Y.; Wang, G.-D. Microstructural Evolution in an Ultralow-C and High-Nb Bearing Steel during Continuous Cooling. *J. Mater. Sci.* **2009**, *44*, 3928–3935. [CrossRef]
28. Hwang, B.; Kim, Y.G.; Lee, S.; Kim, Y.M.; Kim, N.J.; Yoo, J.Y. Effective Grain Size and Charpy Impact Properties of High-Toughness X70 Pipeline Steels. *Met. Mater. Trans. A* **2005**, *36*, 2107–2114. [CrossRef]
29. Manohar, P.A.; Chandra, T. Continuous Cooling Microalloyed Steels Transformation for Behaviour of High Strength Linepipe Applications. *ISIJ Int.* **1998**, *38*, 766–774. [CrossRef]
30. Edmonds, D.V.; Cochrane, R.C. Structure-Property Relationships in Bainitic Steels. *Met. Mater. Trans. A* **1990**, *21*, 1527–1540. [CrossRef]
31. Anijdan, S.H.M.; Yue, S. The Effect of Cooling Rate, and Cool Deformation Through Strain-Induced

Transformation, on Microstructural Evolution and Mechanical Properties of Microalloyed Steels. *Met. Mater. Trans. A* **2011**, *43*, 1140–1162. [CrossRef]

32. Yi, J.J.; Yu, K.J.; Kim, I.S.; Kim, S.J. Role of Retained Austenite on The Deformation of an Fe-0.07 C-1.8 Mn-1.4 Si Dual-Phase Steel. *Met. Mater. Trans. A* **1983**, *14*, 1497–1504. [CrossRef]
33. Thompson, S.W.; Colvin, D.J.; Krauss, G. Austenite Decomposition During Continuous Cooling of an HSLA-80 Plate Steel. *Met. Mater. Trans. A* **1996**, *27*, 1557–1571. [CrossRef]
34. Goel, N.C.; Chakravarty, J.P.; Tangri, K. The Influence of Starting Microstructure on The Retention and Mechanical Stability of Austenite in an Intercritically Annealed-Low Alloy Dual-Phase Steel. *Met. Mater. Trans. A* **1987**, *18*, 5–9. [CrossRef]
35. Huang, J.; Xu, Z. Effect of Dynamically Recrystallized Austenite on The Martensite Start Temperature of Martensitic Transformation. *Mater. Sci. Eng. A* **2006**, *438–440*, 254–257. [CrossRef]
36. Lee, S.J.; Lee, Y.K. Effect of Austenite Grain Size on Martensitic Transformation of a Low Alloy Steel. *Mater. Sci. Forum* **2005**, *475–479*, 3169–3172. [CrossRef]
37. Yang, H.-S.; Bhadeshia, H. Austenite Grain Size and The Martensite-Start Temperature. *Scr. Mater.* **2009**, *60*, 493–495. [CrossRef]
38. Lee, S.-J.; Park, K.-S. Prediction of Martensite Start Temperature in Alloy Steels with Different Grain Sizes. *Met. Mater. Trans. A* **2013**, *44*, 3423–3427. [CrossRef]
39. Xiao, F.-R.; Liao, B.; Shan, Y.-Y.; Qiao, G.-Y.; Zhong, Y.; Zhang, C.; Yang, K. Challenge of Mechanical Properties of An Acicular Ferrite Pipeline Steel. *Mater. Sci. Eng. A* **2006**, *431*, 41–52. [CrossRef]
40. Shanmugam, S.; Misra, R.D.K.; Hartmann, J.; Jansto, S. Microstructure of High Strength Niobium-Containing Pipeline Steel. *Mater. Sci. Eng. A* **2006**, *441*, 215–229. [CrossRef]
41. Cottrell, A.H.; Bilby, B.A. Dislocation Theory of Yielding and Strain Ageing of Iron. *Proc. Phys. Soc. Sect. A* **1949**, *62*, 49–62. [CrossRef]
42. van Snick, A.; Lips, K.; Vandeputte, S.; de Cooman, B.C.; Dilewijns, J. Effect of Carbon Content, Dislocation Density and Carbon Mobility on Bake Hardening. In *Modern LC and ULC Sheet Steels for Cold Forming: Processing and Properties*; Institute of Ferrous Metallurgy: Aachen, Germany, 1998; pp. 413–424.

3

Effects of Sn and Sb on the Hot Ductility of Nb+Ti Microalloyed Steels

Chunyu He, Jianguang Wang *, Yulai Chen, Wei Yu and Di Tang

National Engineering Research Center for Advanced Rolling Technology, University of Science and Technology Beijing, Beijing 100083, China; chunyuhe@163.com (C.H.); yulaic@ustb.edu.cn (Y.C.); yuwei@nercar.ustb.edu.cn (W.Y.); tangdi@nercar.ustb.edu.cn (D.T.)

* Correspondence: g20189146@xs.ustb.edu.cn.

Abstract: Referencing the composition of a typical Nb+Ti microalloyed steel (Q345B), two kinds of steels, one microalloyed with Sn and Sb, and the other one only microalloyed with Sb were designed to study the effects of Sn and Sb on the hot ductility of Nb+Ti microalloyed steels. The Gleeble-3500 tester was adopted to determine the high-temperature mechanical properties of the two test steels. Fracture morphologies, microstructures and interior precipitation status were analyzed by SEM, CLSM (Confocal laser scanning microscope) and EDS, respectively. Results revealed that within the range of 950–650 °C, there existed the ductility trough for the two steels, which were mainly attributed to the precipitation of TiN and Nb (C, N). Additionally, precipitation of Sn and Sb were not observed in this research and the hot ductility was not affected by the addition of Sn and Sb, as compared with the Nb+Ti microalloyed steel. Therefore, addition of a small amount of Sn and Sb (≤0.05 wt.%) to the Nb+Ti microalloyed steel is favorable due to the improvement on corrosion resistance.

Keywords: microalloyed steel; hot ductility; precipitation; phase transition; microstructure

1. Introduction

Hot ductility metal materials can have a great impact on continuous casting, rolling, and other processes. It is directly related to alloy composition, heating and cooling conditions, sampling direction of the tensile plate, strain rate, and inclusion [1–6]. Researchers have done extensive research on the continuous casting of steel. However, some industrial problems still exist, such as surface defects prone to lateral and angular cracks [5–7]. Additionally, the addition of microalloys Nb, Ti, V, and other elements is regarded as the main reason leading to these problems [8–10].

According to extensive previous reports, hot ductility curves of microalloyed steels obtained by the high-temperature tensile tests have obvious ductility troughs, and brittle regions occur at about 700–1000 °C [8,11,12]. For steels containing Ti and Nb, Xie, You et al. [13] believed that fine Ti, Nb (C, N) particles can be precipitated near the austenite grain boundary at temperatures above 1000 °C, and the precipitates may always exist stably. These fine precipitates pinned the grain boundaries and prevented the occurrence of dynamic recrystallization [9,14,15], which reduced the ductility of the steels ultimately. On the other hand, the addition of Sn and Sb can significantly improve the corrosion resistance of microalloyed steels [16–18]. Ahn, SooHoon, et al. [19] found that SnO_2 and Sb_2O_5, the protective corrosion products produced by microalloyed steels containing Sn and Sb in acid chloride corrosion media, can act as inhibitors of anode reaction. Moreover, the Sn and Sb elements can also work synergistically with Cu to form a continuous and dense Sn-Sb-Cu oxide layer to protect the substrate [16,18,20]. However, it was reported that the additional content of Sn and Sb should be strictly controlled, or else it may be converted into the unfavorable elements for the toughness of high-strength steel [21].

This paper aims to study the effect of Sn and Sb on the hot ductility of a typical Nb+Ti microalloyed steel (Q345B). Results fully revealed the high-temperature mechanical properties of the designed steels, and related discussion on the fracture morphology and the precipitation status can provide important guidance for the improvement of the actual production process.

2. Experimental

2.1. Experimental Materials

The samples taken in this research are ingots smelted in a laboratory hollow induction furnace. The chemical composition of the experimental steel (Sb+Sn steel and Sb steel) and Q345B steel [22] are as shown in Table 1. After forging, specimens (Φ4 mm × 10 mm) were manufactured to conduct the thermal expansion test in the dilatometer-DIL 805 (BÄHR THERMOANALYSE GMBH, Hüllhorst, Germany), and specimens Φ10 mm × 120 mm in size were manufactured for the high-temperature tensile test. The processing scheme is presented in Figure 1.

Table 1. Chemical composition of two experimental steels and Q345B (wt.%).

Steel	C	Si	Mn	S	P	Cu+Cr+Ni	Sn	Sb	Al	Nb	Ti	N
Sb+Sn steel	0.062	0.25	0.89	0.003	0.003	1.5–1.9	0.043	0.050		0.039	0.026	0.004
Sb steel	0.083	0.26	0.91	0.006	0.003	1.5–1.9		0.050		0.044	0.025	0.004
Q345B	0.17	0.193	1.438	0.015	0.018				0.028	0.025	0.012	0.004

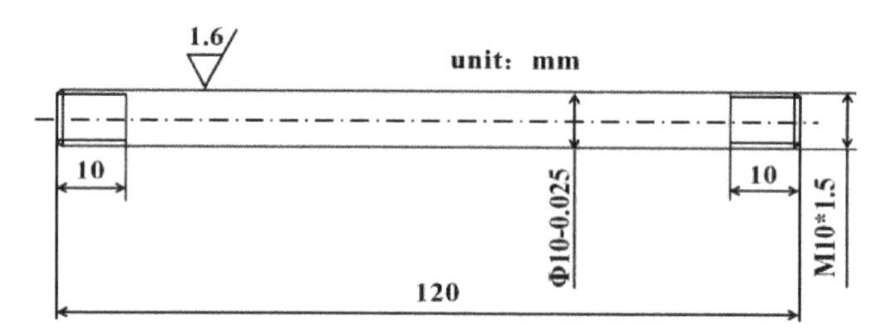

Figure 1. Shape of specimens for hot tensile tests.

The schematic diagram of the specimens after the tensile test is shown in Figure 2a. The cylindrical samples were cut transversely to obtain samples with a length of 10 mm, as shown in Figure 2b, and then cut transversely at a distance of 3 mm from the fracture, as shown in Figure 2c. Finally, the specimens were mechanically polished and etched using 4% nitric acid ethanol solution to observe microstructures.

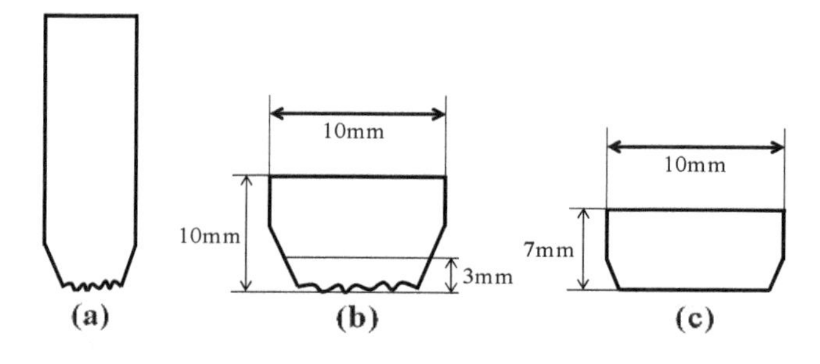

Figure 2. Schematic diagrams of specimen processing showing (**a**) The specimen after fracture; (**b**) the specimen near the fracture surface; (**c**) the final specimen used for microstructure observation.

2.2. Experimental Methods

2.2.1. Thermal Expansion Test

The critical phase transition point of the experimental steels was measured and evaluated by a dilatometer-DIL 805. Specimens were firstly heated to 500 °C at a rate of 10 °C/min and then to 1000 °C at a rate of 3 °C/min, followed by an isothermal process of 3 min, and finally cooled to room temperature at a rate of 3 °C/s. The thermal schedule was depicted in Figure 3a.

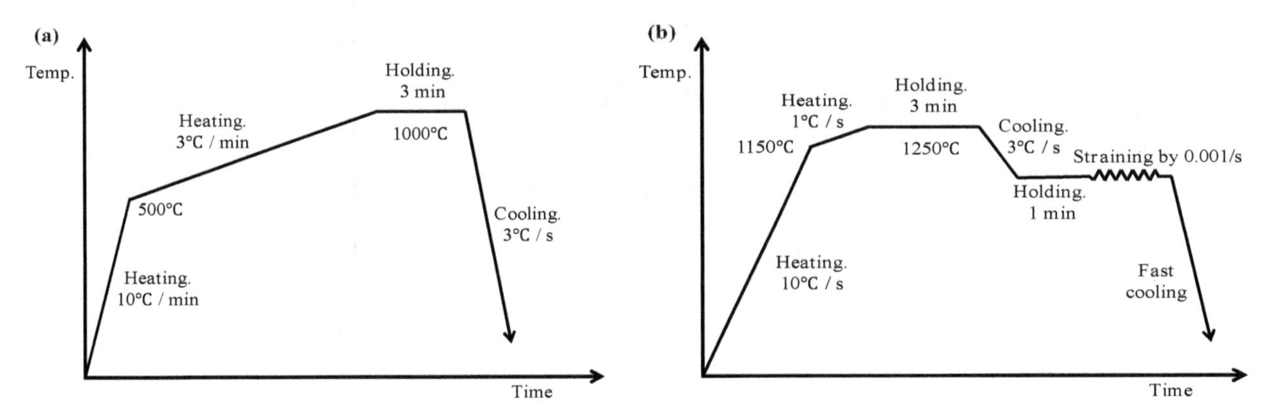

Figure 3. Thermal schedule used to study the critical phase transition and hot ductility. (**a**) Thermal schedule of the dilatometer test; (**b**) thermal schedule of the high-temperature tensile test.

2.2.2. High-Temperature Tensile Test

The hot ductility of the experimental steels was measured and evaluated by a high-temperature tensile test, and the experimental equipment was Gleeble-3500 (Dynamic Systems Inc., San Francisco, CA, USA). The sample was firstly heated to 1150 °C at a rate of 10 °C/s, and then the temperature was increased to 1250 °C at a rate of 1 °C/s. Specimens were held at 1250 °C for 3 min to keep the uniformity of the temperature and compositions and also promote the dissolution of precipitates. The specimens were cooled to the testing temperature at a cooling rate of 3 °C/s, and then the heat was held for 1 min before stretching at the strain rate of 10^{-3}/s. The testing temperature range was 650–1250 °C, and the step length was 50 °C. After tension, the specimens were quickly cooled to room temperature to maintain the fracture morphology and microstructures at the testing temperature. The corresponding thermal schedule is shown in Figure 3b.

3. Results

3.1. Critical Transformation Point of Experimental Steels

According to the results of the dilatometer test, we obtained the critical phase transition temperature (Ac1, Ac3, Ar1, Ar3) [23] of Sb steel Ac1 = 762 °C, Ac3 = 865 °C, Ar1 = 709 °C, Ar3 = 764 °C; and of Sb+Sn steel Ac1 = 741 °C, Ac3 = 868 °C, Ar1 = 705 °C, Ar3 = 759 °C.

3.2. Hot Ductility and Strength

Previous research shows that the continuous casting slab does not crack when the percentage reduction of the area measured by the high-temperature tensile test is above 60% (Z > 60%), and when Z < 60%, on the contrary, the crack sensitivity of the casting slab increases [24,25]. Thus, the 60% percent reduction of the area can be used as a threshold value to divide the high plastic and low plastic zones of steels in this research.

In order to explain the hot ductility characteristics of the two experimental steels more precisely, we compared the hot ductility data of Sb or Sb+Sn modified Nb+Ti microalloyed steels with original steels (Q345B) [22]. Figure 4 shows the relationship between the percentage reduction of area and the testing temperature of two experimental steels, and Q345B, Ar1, Ac1, Ar3, and Ac3 are the critical transformation points of Sb+Sn steel. It indicates that Q345B has two low plasticity zones (1300–1350 °C and 650–900 °C) in the range of 650 °C to 1350 °C, and a high plasticity zone (900–1300 °C). Obviously, Sb steel and Sb+Sn steel exhibited outstanding plasticity at 1000–1250 °C. In addition, the plasticity of Sb steel and Sb+Sn steel showed a tendency of decreasing first and then increasing at the range from 650 °C to 950 °C, and the lowest valley of the percentage reduction of the area appeared at 750 °C. In general, the hot ductility of the two test steels is similar to that of Q345B.

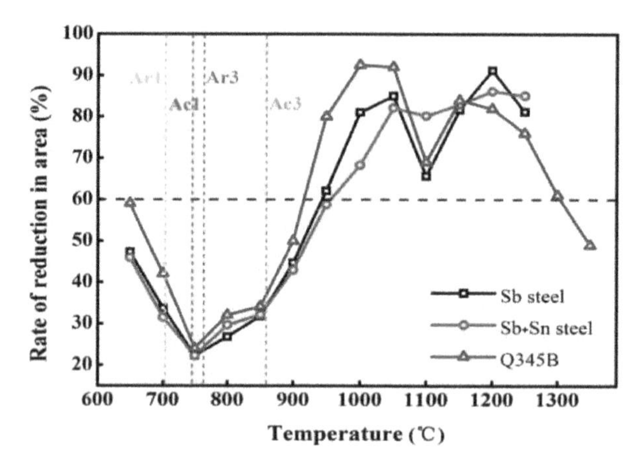

Figure 4. Hot ductility curves.

There are three high-temperature ductility troughs as the temperature rises from 600 °C to the melting points of steel [6,24,26]. The first ductility trough is attributed to S, P, and O elements enriched at the grain boundaries above 1300 °C, and the ductility is reduced due to the inclusion of (Fe, Mn, Si)(S, O) in the dendritic intergranular region. The second ductility trough (which appears at high temperature (800–1300 °C) when the strain rate (higher than 10^{-2}/s) is higher) is related to the strain rate, which usually appears at higher strain rates [27]. The experimental steels did not show the first and second ductility trough in the experimental conditions and temperature range. However, the third ductility trough that generally exists in the range of 700–1000 °C [8,28,29] appeared in the temperature range from 650–950 °C, which can be induced by numerous reasons. It is also strengthened in this research and discussed in detail in the following sections.

Figure 5 shows the relationship between tensile strength and the tensile temperature of three steels. When the maximum stress of the specimens against plastic deformation exceeds the tensile strength, there will be microcracks that affect the quality of the specimens [26]. It can be seen that the tensile strength of Q345B is much lower than that of the two experimental steels below 900 °C, and the performance of steels tends to coincidence when the temperature above 900 °C. Moreover, both of the modified steels present an increase of strength around 1000 °C as compared with Q345B.

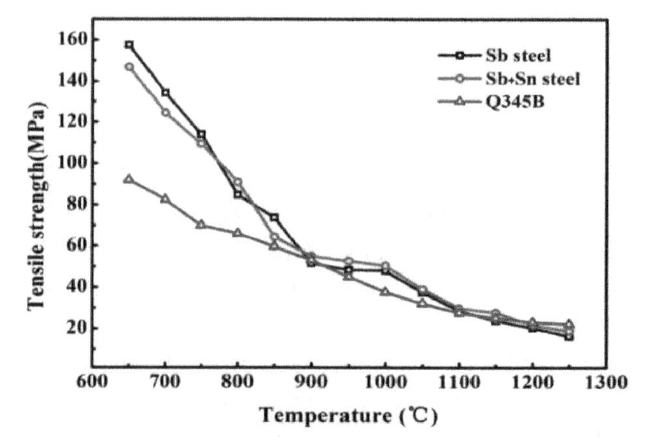

Figure 5. Thermal strength curve.

4. Discussion

4.1. Stress–Strain Behavior

Dynamic recovery and dynamic recrystallization are the two main mechanisms that lead to material softening during high-temperature deformation, and the latter can bring about a more significant softening effect, generally showing obvious stress fluctuations on the high-temperature tensile curve [30,31]. In the process of high-temperature and low-strain-rate deformation, there is also work hardening, and dynamic precipitation strengthening processes, and the softening caused by dynamic recrystallization alternate as the main deformation mechanism will show multipeak stress on the tensile curve status [32].

Ryan [33] observed that the presence of a stress peak in a constant strain rate flow curve leads to an inflection in the stress dependence of the strain hardening rate, $\theta = (\partial\sigma/\partial\varepsilon)_{\dot{\varepsilon}}$. Later, on the basis of considerations of irreversible thermodynamics, the inflections in θ–ε plots in austenitic stainless steels were shown to be due to the initiation of dynamic recrystallization [34,35]. This was subsequently confirmed by the observations in other materials [36]. We find the inflection point from θ-ε plods in the upper right corner of Figure 6a,b, and designate it as the occurrence point of dynamic recrystallization.

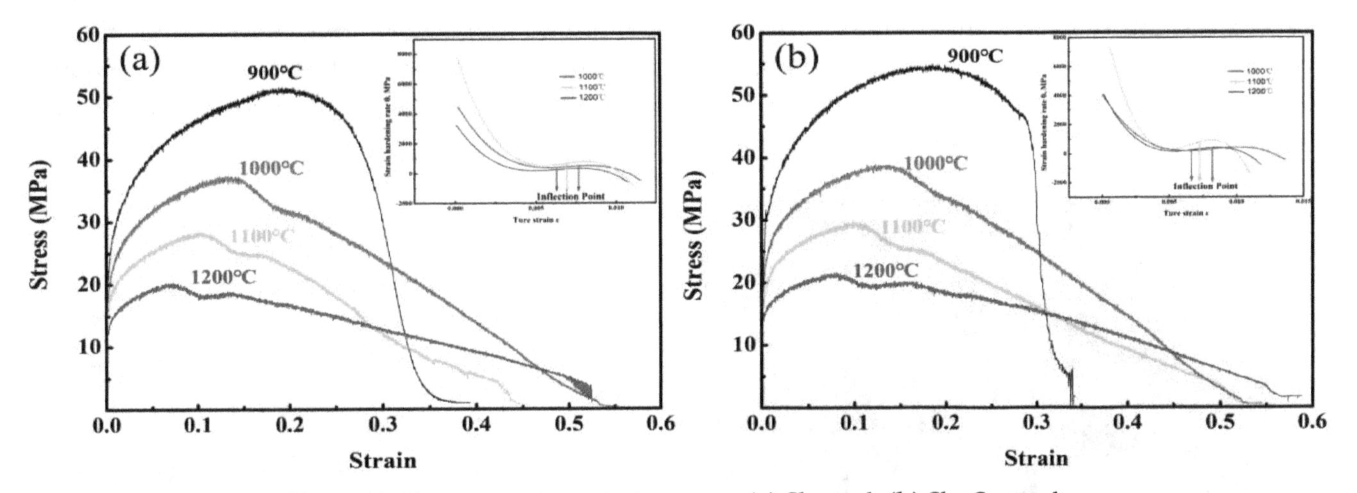

Figure 6. True stress-true strain curve. (**a**) Sb steel; (**b**) Sb+Sn steel.

It can be seen from Figure 6 that, as the deformation temperature increases, the strain required for recrystallization softening to occur becomes smaller. The 900 °C tensile curve shows a unimodal shape, indicating that work hardening always occupies the main deformation mechanism at this temperature. The percentage reduction of the area of Sb steel and Sb+Sn steel is higher than 60% in the lower-higher temperature of 950–1250 °C, which is caused by the occurrence of dynamic recrystallization process.

4.2. Influence of Precipitation on Hot Ductility

Matveev et al. [37] believed that the precipitates, inclusions, interstitial impurities, and other factors on the initial grain boundaries were the reasons for the ductility decrease. As shown in Figure 7, the Thermo-Calc (TCFE7 database) software (Stockholm, Sweden) was used to calculate constituents of precipitates and related precipitation order in Sb+Sn steel. The main precipitation elements of the experimental steel were Ti and Nb. Precipitation of Ti appears between 1050 and 1487 °C, and the main precipitates are in the form of TiN. While the precipitation temperature range of Nb is 800–1120 °C, the main precipitates are in the form of Nb (C, N). Therefore, the Nb (C, N) in the experimental steel can be completely dissolved while the original TiN cannot, due to the influence of the experimental heating temperature. When the testing temperature decreases from 950 to 850 °C, the increased content of precipitates (TiN, Nb (C, N)) resulted in increasingly severe grain boundary pinning, which inhibited boundary migration and hindered the occurrence of dynamic recrystallization.

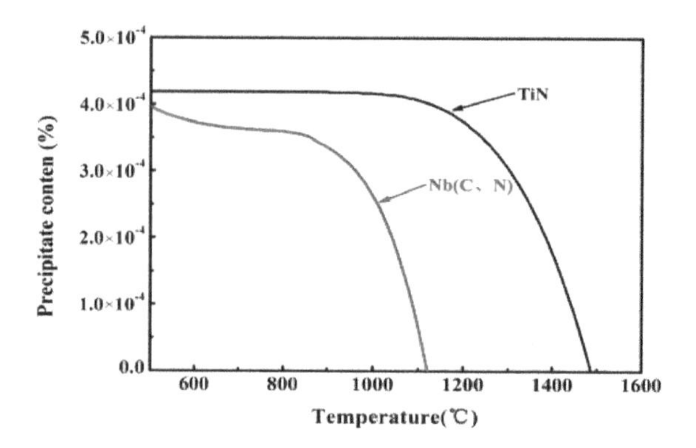

Figure 7. Thermo-Calc software calculates the precipitation order and amount of Sb+Sn steel.

In order to analyze the cause of the worst hot ductility at 750 °C, an energy spectrum analysis is performed on the fracture morphology of Sb+Sn steel. As shown in Figure 8a, there is an empty hole on the fracture surface and a precipitate inside, and the precipitate is determined to be a Ti+Nb carbide, as shown in Figure 8b. However, the diffraction peaks of Sb and Sn are not detected, so the Sn and Sb elements do not produce precipitates, and the phenomenon of pinning grain boundaries is not aggravated. Studies have shown that when the content of Sb and Sn elements is high, they tend to aggregate at the grain boundaries [21]. Figure 9a is an SEM photograph of Sb+Sn steel with obvious grain boundaries. EDS microarea surface scan results show that Sb and Sn elements are evenly distributed in the solid solution in Figure 9b,c.

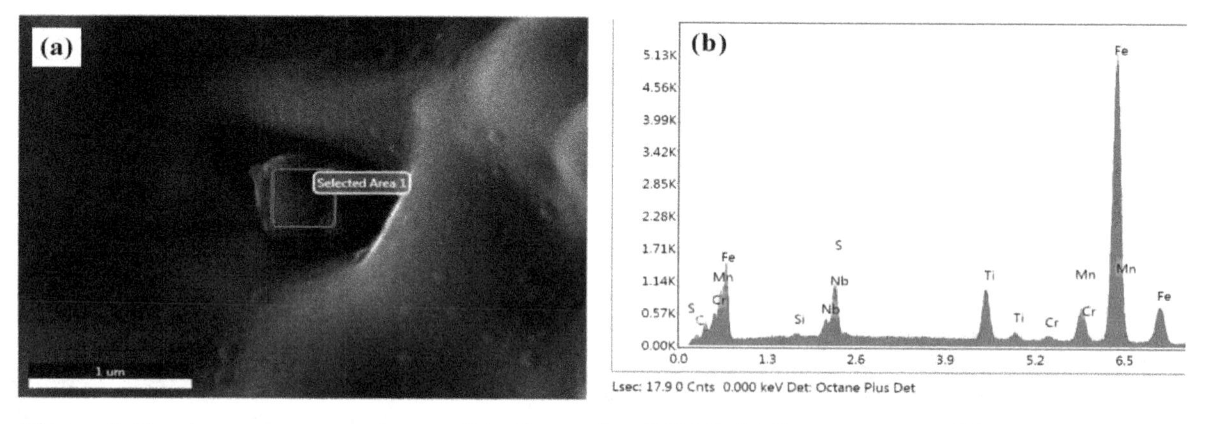

Figure 8. Fracture energy spectrum of Sb+Sn steel at 750 °C. (**a**) Scanning area; (**b**) energy spectrum.

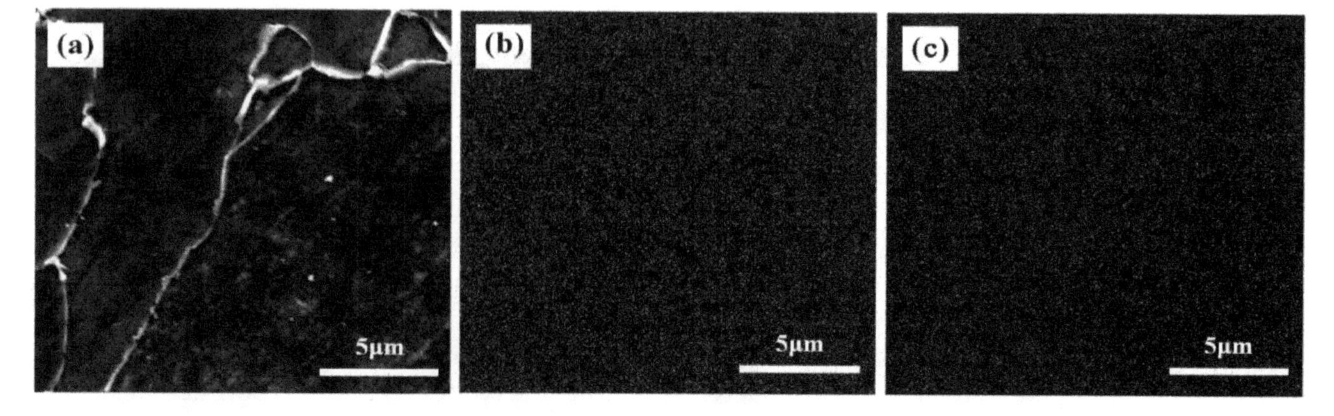

Figure 9. SEM photograph and energy spectrum. (**a**) Scanning area; (**b**) Sb energy spectrum; (**c**) Sn energy spectrum.

However, as can be seen from Figure 4, the percentage reduction of area of the experimental steel is generally lower than that of Q345B especially in the low plastic zone, which is mainly attributed to the precipitate content. The contents of Ti and Nb in Q345B are 0.012 (wt.%) and 0.025 (wt.%), respectively, and the contents of Ti and Nb in experimental steel are 0.025 (wt.%) and 0.04 (wt.%), respectively. The content of Ti and Nb precipitates in the experimental steel is higher, this weakens the ductility of the steel. It can be seen from Figure 5 that the strength of the experimental steels is significantly higher than Q345B at 1000 °C, which may be due to the precipitation strengthening of Ti+Nb. Based on the above analysis, we believe that adding an appropriate amount of Sb and Sn elements to the experimental steel does not affect the hot ductility.

4.3. Fracture Morphology Analysis and Microstructure

The fracture morphologies of Sb steel and Sb+Sn steel at different tensile temperatures are shown in Figure 10. At a test temperature of 1100 °C, there are a large number of dimples that existed on the fracture surface, as can be seen in Figure 10a,e. Under high-temperature strain, dynamic recrystallization allows the grain boundary to obtain sufficient driving force to migrate. At this time, the migration speed of the grain boundary is higher than the slip speed of the grain boundary, which wraps the already formed grain boundary crack into the grain and reduces the cracking speed at grain boundaries. Interconnection of the cracks in the grain requires a greater external force to cut the whole grain and eventually rupture. Therefore, the occurrence of dynamic recrystallization is conducive to the improvement of hot ductility. As the experimental temperature decreased to 850 °C, the fracture morphology showed obvious cleavage fracture characteristics, as shown in Figure 10b,f. The loss of ductility from 1100 °C to 850 °C can be explained by the precipitates, which provide a pinning effect on the austenite grain boundaries and prevent the recrystallization phenomenon of steel. As shown in Figure 10c,g, the fracture morphology under 750 °C shows a typical brittle fractured smooth block structure. At this temperature, the fine grains at the austenite boundary provide nucleation points, and the ferrite film grows the grain boundary to the intragranular region. As the temperature decreases to 650 °C, the fracture morphology can be seen in Figure 10d,h, and the dimples increase significantly.

Figure 10. Fracture morphology of Sb steel and Sb+Sn steel at different temperatures. (**a**) Sb steel-1100 °C; (**b**) Sb steel-850 °C; (**c**) Sb steel-750 °C; (**d**) Sb steel-650 °C; (**e**) Sb+Sn steel-1100 °C; (**f**) Sb+Sn steel-850 °C; (**g**) Sb+Sn steel-750 °C; (**h**) Sb+Sn steel-650 °C.

Microstructures of the specimens of Sb+Sn steel at different temperatures (1100 °C, 850 °C, 750 °C, and 650 °C) are displayed in Figure 11. The microstructures under 1100 °C and 850 °C are acicular bainite and granular bainite, respectively. At 750 °C, ferrite is formed at the austenite boundary, and the remaining austenite is transformed into bainite. As the temperature decreases to 650 °C, the hot ductility increases slightly, which is related to the ferrite and pearlite that have undergone phase transformation.

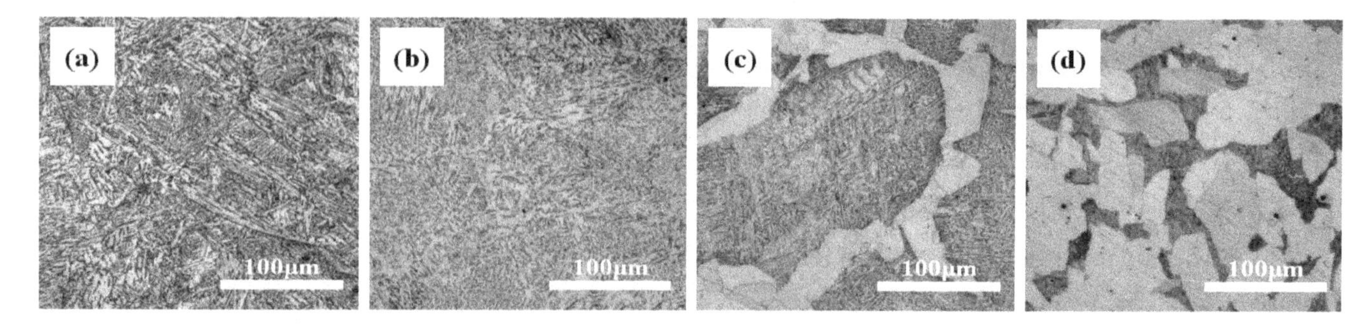

Figure 11. Microstructures of the specimens of Sb+Sn steel at different temperatures. (**a**) 1100 °C; (**b**) 850 °C; (**c**) 750 °C; (**d**) 650 °C.

4.4. Fracture Mechanism of the Lowest Brittle Valley

A fracture mechanism at 750 °C is involved to further illustrate the fracture mechanism. Primary ferrite has already existed at 750 °C and it is distributed along the austenite grain boundaries. Although both austenite and ferrite have good ductility, when the two phases are present together, ferrite can be considerably softer than austenite, so that most of the strain, therefore, concentrates in these bands, giving low ductility [26]. Then, the mixture of precipitates in the ferrite becomes the core of empty holes formation, and the empty holes grow and accumulate to form microcracks with the increase of strain. Finally, the completely cracked microcracks are the tiny dimples observed on the fracture morphology. Taking the cross-section in the direction of parallel tensile force, the mechanism of microcrack formation is shown in Figure 12.

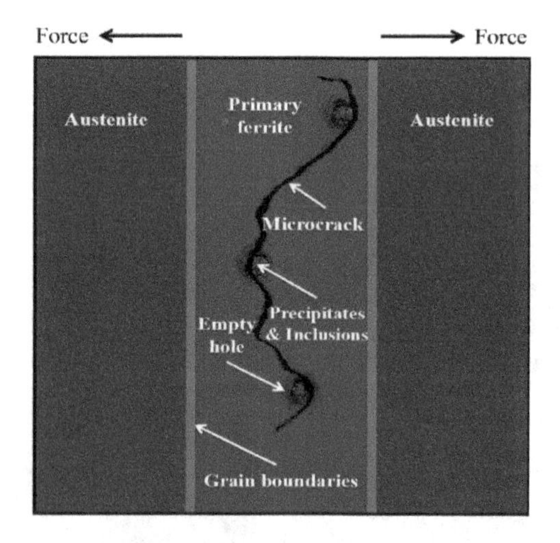

Figure 12. Schematic diagram of the microcrack formation mechanism.

In short, the experimental steel has a ductility trough in the temperature range of 650–950 °C, and the reasons can be concluded as follows: one is the fine precipitates of Ti and Nb pinned to the grain boundary; the other is an austenite-ferrite phase transition. However, the addition of a small number of Sb and Sn microalloying elements does not affect the hot ductility.

5. Conclusions

(1) Sb steel and Sb+Sn steel did not show the first and second ductility trough in the experimental conditions and temperature range, and the third ductility trough is about 650–950 °C.

(2) In the high-plasticity zone with the percentage reduction of area of $Z > 60\%$, high temperature and deformation promote the occurrence of dynamic recrystallization, which is beneficial to the

improvement of hot ductility. The third ductility trough, the pinning of the grain boundaries by fine precipitates and the precipitation of primary ferrite along the grain boundaries are responsible for the decrease in hot ductility.

(3) The $\gamma \rightarrow \alpha$ phase transition temperatures of Sb steel and Sb+Sn steel are 759 °C and 764 °C, respectively. The precipitates of the experimental steel are mainly TiN, Nb (C, N) not Sn and Sb. Adding a small amount of microalloying elements Sb and Sn does not affect the hot plasticity.

Author Contributions: Conceptualization, W.Y. and Y.C.; methodology, C.H. And D.T.; software, J.W.; validation, J.W.; writing—original draft preparation, J.W.; writing—review and editing, C.H., J.W.; funding acquisition, W.Y. All authors have read and agreed to the published version of the manuscript.

References

1. Sahoo, G.; Singh, B.; Saxena, A. Effect of strain rate, soaking time and alloying elements on hot ductility and hot shortness of low alloy steels. *Mater. Sci. Eng. A* **2018**, *718*, 292–300. [CrossRef]
2. Ma, F.; Wen, G.; Wang, W. Effect of Cooling Rates on the Second-Phase Precipitation and Proeutectoid Phase Transformation of a Nb–Ti Microalloyed Steel Slab. *Steel Res. Int.* **2013**, *84*, 370–376. [CrossRef]
3. Faccoli, M.; Roberti, R. Study of hot deformation behaviour of 2205 duplex stainless steel through hot tension tests. *J. Mater. Sci.* **2013**, *48*, 5196–5203. [CrossRef]
4. Zhang, H.J.; Zhang, L.F.; Wang, Y.D. Effect of Sampling Locations on Hot Ductility of Low Carbon Alloyed Steels. *Steel Res. Int.* **2018**, *89*, 1800052. [CrossRef]
5. Wang, Z.; Ma, W.; Wang, C. Effect of Strain Rate on Hot Ductility of a Duplex Stainless Steel. *Adv. Mater. Sci. Eng.* **2019**, *2019*, 6810326. [CrossRef]
6. Yamamoto, K.; Yamamura, H.; Suwa, Y. Behavior of Non-metallic Inclusions in Steel during Hot Deformation and the Effects of Deformed Inclusions on Local Ductility. *ISIJ Int.* **2011**, *51*, 1987–1994. [CrossRef]
7. Du, C.; Zhang, J.; Wen, J.; Li, Y.; Lan, P. Hot ductility trough elimination through single cycle of intense cooling and reheating for microalloyed steel casting. *Ironmak. Steelmak.* **2016**, *43*, 331–339. [CrossRef]
8. Jang, J.H.; Heo, Y.U.; Lee, C.H.; Bhadeshia, H.K.D.H.; Suh, D.W. Interphase precipitation in Ti–Nb and Ti–Nb–Mo bearing steel. *Mater. Sci. Technol.* **2013**, *29*, 309–313. [CrossRef]
9. Liu, Y.; Du, L.X.; Wu, H.Y.; Misra, R.D.K. Hot Ductility and Fracture Phenomena of Low-Carbon V–N–Cr Microalloyed Steels. *Steel Res. Int.* **2019**, *91*, 1900265. [CrossRef]
10. Zhou, Q.; Li, Z.; Wei, Z.S.; Wu, D.; Li, J.Y.; Shao, Z.Y. Microstructural features and precipitation behavior of Ti, Nb and V microalloyed steel during isothermal processing. *J. Iron Steel Res. Int.* **2019**, *26*, 102–111. [CrossRef]
11. Qian, G.; Cheng, G.; Hou, Z. Effect of the Induced Ferrite and Precipitates of Nb–Ti Bearing Steel on the Ductility of Continuous Casting Slab. *ISIJ Int.* **2014**, *54*, 1611–1620. [CrossRef]
12. Lan, P.; Tang, H.; Zhang, J. Hot ductility of high alloy Fe–Mn–C austenite TWIP steel. *Mater. Sci. Eng. A* **2016**, *660*, 127–138. [CrossRef]
13. Xie, Y.; Cheng, G.; Chen, L.; Zhang, Y.; Yan, Q. Characteristics and Generating Mechanism of Large Precipitates in Nb–Ti-microalloyed H13 Tool Steel. *ISIJ Int.* **2016**, *56*, 995–1002. [CrossRef]
14. Vervynckt, S.; Verbeken, K.; Thibaux, P.; Houbaert, Y. Recrystallization–precipitation interaction during austenite hot deformation of a Nb microalloyed steel. *Mater. Sci. Eng. A* **2011**, *528*, 5519–5528. [CrossRef]
15. Carpenter, K.R.; Dippenaar, R.; Killmore, C.R. Hot Ductility of Nb- and Ti-Bearing Microalloyed Steels and the Influence of Thermal History. *Metall. Mater. Trans. A* **2009**, *40*, 573–580. [CrossRef]
16. Nam, N.D.; Kim, M.J.; Jang, Y.W.; Kim, J.G. Effect of tin on the corrosion behavior of low-alloy steel in an acid chloride solution. *Corros. Sci.* **2010**, *52*, 14–20. [CrossRef]
17. Pardo, A.; Merino, M.C.; Carboneras, M.; Coy, A.E.; Arrabal, R. Pitting corrosion behaviour of austenitic stainless steels with Cu and Sn additions. *Corros. Sci.* **2007**, *49*, 510–525. [CrossRef]
18. Park, S.A.; Kim, S.H.; Yoo, Y.H.; Kim, J.G. Effect of Chloride Ions on the Corrosion Behavior of Low-Alloy Steel Containing Copper and Antimony in Sulfuric Acid Solution. *Met. Mater. Int.* **2015**, *21*, 470–478. [CrossRef]
19. Ahn, S.; Park, K.J.; Oh, K.; Hwang, S.; Park, B.; Kwon, H.; Shon, M. Effects of Sn and Sb on the Corrosion Resistance of AH 32 Steel in a Cargo Oil Tank Environment. *Met. Mater. Int.* **2015**, *21*, 865–873. [CrossRef]

20. Pardo, A.; Merino, M.C.; Carboneras, M.; Viejo, F.; Arrabal, R.; Muñoz, J. Influence of Cu and Sn content in the corrosion of AISI 304 and 316 stainless steels in H_2SO_4. *Corros. Sci.* **2006**, *48*, 1075–1092. [CrossRef]
21. Kameda, J.; Mcmahon, C.J. The effects of Sb, Sn, and P on the strength of grain boundaries in a Ni-Cr Steel. *Metall. Trans. A* **1981**, *12*, 31–37. [CrossRef]
22. Liu, J.W.; Wang, C.M.; Yu, L.; Qi, Y.F. Hot Ductility of Q345B low carbon high strength steel. *Nonferrous. Met. Sci. Eng.* **2015**, *6*, 61–67.
23. Song, W.X. *Metal Science*, 2nd ed.; Metallurgical Industry Press: Beijing, China, 1989; pp. 318–319.
24. Chen, D.; Cui, H.; Wang, R. High-Temperature Mechanical Properties of 4.5%Al δ-TRIP Steel. *Appl. Sci.* **2019**, *9*, 5094. [CrossRef]
25. Fu, J.; Wang, F.M.; Hao, F.; Jin, G.X. High-temperature mechanical properties of near-eutectoid steel. *Int. J. Miner. Metall. Mater.* **2013**, *20*, 829–834. [CrossRef]
26. Zeng, Y.N.; Sun, Y.H.; Cai, K.K.; Ma, Z.F.; Xi, A. Failure mode and hot ductility of Ti-bearing steel in the brittle zone. *Rev. Métall.* **2013**, *110*, 315–323. [CrossRef]
27. Mintz, B.; Jonas, J.J. Influence of strain rate on production of deformation induced ferrite and hot ductility of steels. *Mater. Sci. Technol.* **1994**, *10*, 721–727. [CrossRef]
28. Lanjewar, H.A.; Tripathi, P.; Singhai, M.; Patra, P.K. Hot Ductility and Deformation Behavior of C-Mn/Nb-Microalloyed Steel Related to Cracking during Continuous Casting. *J. Mater. Eng. Perform.* **2014**, *23*, 3600–3609. [CrossRef]
29. Xie, S.S.; Lee, J.D.; Yoon, U.S.; Yim, C.H. Compression Test to Reveal Surface Crack Sensitivity between 700 and 1100 °C of Nb-bearing and High Ni Continuous Casting Slabs. *ISIJ Int.* **2002**, *42*, 708–716. [CrossRef]
30. Mejía, I.; Bedolla-Jacuinde, A.; Maldonado, C.; Cabrera, J.M. Hot ductility behavior of a low carbon advanced high strength steel (AHSS) microalloyed with boron. *Mater. Sci. Eng. A* **2011**, *528*, 4468–4474. [CrossRef]
31. Ghosh, S.; Mula, S. Fracture toughness characteristics of ultrafine grained Nb–Ti stabilized microalloyed and interstitial free steels processed by advanced multiphase control rolling. *Mater. Charact.* **2020**, *159*, 110003. [CrossRef]
32. Sakai, T.; Jonas, J.J. Dynamic recrystallization: Mechanical and microstructural considerations. *Acta. Metal.* **1984**, *32*, 189–209. [CrossRef]
33. Ryan, N.D.; Mcqueen, H.J. Flow stress, dynamic restoration, strain hardening and ductility in hot working of 316 steel. *J. Mater. Process. Technol.* **1990**, *21*, 177–199. [CrossRef]
34. Mirzadeh, H.; Najafizadeh, A. Prediction of the critical conditions for initiation of dynamic recrystallization. *Mater. Des.* **2010**, *31*, 1174–1179. [CrossRef]
35. Poliak, E.I.; Jonas, J.J. Initiation of Dynamic Recrystallization in Constant Strain Rate Hot Deformation. *ISIJ Int.* **2007**, *43*, 684–691. [CrossRef]
36. Kim, S.I.; Lee, Y.; Byon, S.M. Study on constitutive relation of AISI 4140 steel subject to large strain at elevated temperatures. *J. Mater. Process. Technol.* **2003**, *140*, 84–89. [CrossRef]
37. Matveev, M.A.; Kolbasnikov, N.G.; Kononov, A.A. Causes of High Temperature Ductility Trough of Microalloyed Steels. *Trans. Indian Inst. Met.* **2017**, *70*, 2193–2204. [CrossRef]

Modeling of Precipitation Hardening during Coiling of Nb–Mo Steels

Jean-Yves Maetz [1], Matthias Militzer [1,*], Yu Wen Chen [2], Jer-Ren Yang [2], Nam Hoon Goo [3], Soo Jin Kim [3], Bian Jian [4] and Hardy Mohrbacher [5]

1 Centre for Metallurgical Process Engineering, The University of British Columbia, Vancouver, BC V6T 1Z4, Canada; jymaetz@hotmail.com
2 Department of Materials Science and Engineering, National Taiwan University, Taipei 10617, Taiwan; f01527051@ntu.edu.tw (Y.W.C.); jryang@ntu.edu.tw (J.-R.Y.)
3 Technical Research Center, Hyundai-Steel Company, Dangjin 167-32, Korea; namhgoo@hyundai-steel.com (N.H.G.); soojkim@hyunday-steel.com (S.J.K.)
4 Niobium Tech Asia, Singapore 068898, Singapore; jian.bian@niobiumtech.com
5 NiobelCon bvba, 2970 Schilde, Belgium; hm@niobelcon.net
* Correspondence: matthias.militzer@ubc.ca

Abstract: Nb–Mo low-alloyed steels are promising advanced high strength steels (AHSS) because of the highly dislocated bainitic ferrite microstructure conferring an excellent combination of strength and toughness. In this study, the potential of precipitation strengthening during coiling for hot-strip Nb–Mo-bearing low-carbon steels has been investigated using hot-torsion and aging tests to simulate the hot-rolling process including coiling. The obtained microstructures were characterized using electron backscatter diffraction (EBSD), highlighting the effects of Nb and Mo additions on formation and tempering of the bainitic ferrite microstructures. Further, the evolution of nanometer-sized precipitates was quantified with high-resolution transmission electron microscopy (HR-TEM). The resulting age hardening kinetics have been modelled by combining a phenomenological precipitation strengthening model with a tempering model. Analysis of the model suggests a narrower coiling temperature window to maximize the precipitation strengthening potential in bainite/ferrite high strength low-alloyed (HSLA) steels than that for conventional HSLA steels with polygonal ferrite/pearlite microstructures.

Keywords: advanced high strength steels; HSLA steels; precipitation strengthening; tempering; bainitic ferrite; EBSD; austenite-to-ferrite transformation; hot-torsion test; coiling simulation

---

1. Introduction

High strength low-alloyed (HSLA) and similar microalloyed low-carbon steels have been developed since the 1960s and their production continuously increases, especially for automotive and pipeline industries, because of their high strength, excellent ductility, and good weldability [1–3]. Originally the high strength of these steels was obtained by a combination of refining the ferrite-pearlite microstructure and precipitation hardening due to adding microalloying elements such as Nb, V, or Ti [1,4]. Optimizing the coil cooling process is critical to realize the precipitation strengthening potential in hot band. Models have been developed to predict the precipitation behavior for these conventional HSLA steels [4,5].

Nowadays there is a strong driving force to develop new advanced high strength steels (AHSS) to meet the future challenges in the automotive, energy and construction sectors [2,6]. In particular, Nb-Mo-bearing HSLA steels offer a tremendous opportunity to develop high-performance steels with increased strength [7–10]. Nb is known to be a powerful microalloying element to accumulate strain in austenite by delaying recrystallization because of solute drag and/or strain-induced precipitation [11–13]. Mo addition affects considerably the austenite-to-ferrite transformation by delaying polygonal ferrite formation, inhibiting pearlite and promoting bainite formation [7,8,14]. The high dislocation density and the resulting fine substructure significantly increases the strength of these steels. In addition to transformation hardening, the precipitation strengthening potential of Nb–Mo HSLA steels have been observed to be substantial [8,14]. Indeed, in highly dislocated ferrite or bainite, precipitation occurs on dislocations [9,15,16]. The precipitation strengthening is thus expected to be promoted as compared to polygonal ferrite by a higher precipitate density as the dislocation density increases [8,14]. In addition, Mo has a supplementary potential for precipitation strengthening by increasing the volume fraction of precipitates as (Nb_x, Mo_{1-x}) (CN) mixed carbides or carbo-nitrides have been observed [8,9].

The goal of this study is to develop a precipitation strengthening model for hot-rolled Nb–Mo HSLA steels applicable to industrial coiling conditions. Four low-carbon steels with various levels of Nb and Mo content were investigated. Laboratory hot-torsion tests were performed to simulate the hot-rolling process. The hardening potential was assessed performing aging at typical coiling temperatures. Systematic electron backscatter diffraction (EBSD) and high-resolution transmission electron microscopy (HR-TEM) studies were conducted to capture the microstructure and precipitation changes. A first version of the precipitation strengthening model using the approach used for conventional HSLA steels [5] has been presented previously [17], in which tempering of the bainitic-ferrite microstructure during aging was not considered even though it appears to have a non-negligible effect on the resulting mechanical properties [18]. The present paper provides the detailed microstructure analyses for tempering and coiling simulations at various temperatures to assess the microstructure contribution to mechanical properties as well as a revised age hardening model that considers both precipitation and transformation hardening contributions. The proposed modeling approach is validated with laboratory coil cooling simulations of the investigated Nb–Mo HSLA steels.

2. Materials and Methods

The steel compositions are similar to those of typical industrially hot-rolled HSLA steels but with systematic variations in the Nb and Mn contents, as listed in Table 1. The steels are laboratory steels that were vacuum melted to produce 50 kg ingots from which torsion specimens were machined with a gage section of 12.7 mm in length and a diameter of 10.3 mm.

Table 1. Chemical composition (wt.%) of the four investigated steels.

Steel	C	Mn	Si	Nb	Mo
Nb1	0.08	1.5	0.2	0.05	0
Nb1Mo1	0.08	1.5	0.2	0.05	0.1
Nb1Mo2	0.08	1.5	0.2	0.05	0.2
Nb2Mo2	0.08	1.5	0.2	0.1	0.2

The thermo-mechanical simulations are summarized in Figure 1. The as-received specimens were subjected to hot-torsion tests using a DSI HTS 100 hot torsion tester (Dynamic Systems Inc., Poestenkill, NY, USA) in order to simulate hot-rolling where the nominal deformation occurs within approximately a layer of 1 mm from the sample surface [19]. The as-received samples were soaked at 1250 °C for 15 min to dissolve the Nb and Mo containing precipitates, before undergoing 10 torsion passes at a strain rate of 1/s between 1100 and 900 °C, including three roughing passes (R1–R3) and seven finishing passes (F1–F7) for a total strain of 3.2. The details of the hot-torsion test schedule are shown in Table 2. Subsequently, different cooling strategies were employed to produce (i) as-quenched (AQ), i.e. under-aged, specimens and (ii) coil cooling simulated specimens. In the first test series (path 1 in

Figure 1), the primary goal of the hot-torsion tests had been to create as-quenched specimens without Nb and Mo containing precipitates but with a typical hot-strip rolling microstructure for subsequent age hardening tests (path 2). The specimens were quenched with pressurized He-gas from 900 °C to 400 °C at an initial cooling rate of 40 °C/s followed by air cooling to room temperature. In the second series of hot torsion testing (path 3 in Figure 1), the specimens were cooled from 900 °C to a designated coiling temperature, T_c, in the range of 650 °C to 500 °C at a cooling rate of 40 °C/s followed by cooling at 0.5 °C/min for 3 h to simulate coil cooling.

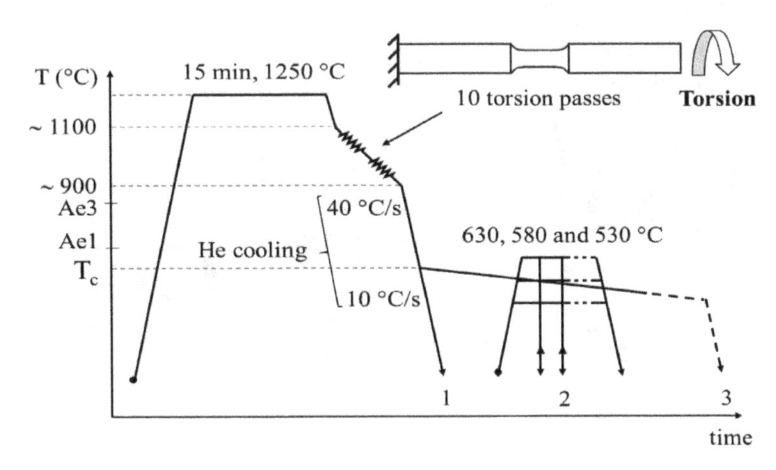

Figure 1. Summary of the various thermo-mechanical treatments employed in this study: path 1 for as-quenched condition, path 2 for isothermal age hardening treatments, and path 3 for coiling simulation.

Table 2. Schedule for hot-torsion tests.

Pass	R1–R3	F1	F2	F3	F4	F5	F6	F7
Temperature (°C)	1100	1024	982	958	950	922	910	900
Strain (1 s^{-1} rate)	0.33 each	0.35	0.5	0.4	0.3	0.3	0.3	0.1
Interpass time (s)	10	4	2.4	1.6	1.2	0.8	0.6	-

The as-quenched specimens were then subjected to isothermal aging tests (see path 2 in Figure 1) at three temperatures, i.e., 530 °C, 580 °C and 630 °C, to cover the range of typical coiling temperatures. Heat treatments were carried out in a tube furnace under argon atmosphere. The aging tests were interrupted at pre-selected times to perform microhardness measurements at room temperature after natural air cooling. The same sample is thus used to follow the entire aging kinetics for a given temperature. The reported aging time, t, represents the cumulative effective time, not counting the heating time, i.e. the time before the temperature of the sample reaches 20 °C below the investigated temperature. For each aging time, ten hardness measurements were made in the cross-section of the specimen at 500 μm from the surface to avoid edge effects but to remain in the fully deformed area. A load of 1 kgf was applied for 15 s, leading to an indent diagonal of order of 100 μm.

Electron backscatter diffraction (EBSD) mapping was conducted to characterize the microstructure at 500 ± 200 μm from the sample surface in different conditions, i.e. as-quenched (AQ), peak-aged (PA), over-aged (OA) and coil cooling simulated. SiC grinding papers and diamond polishing down to 1 μm were used for sample preparation, followed by electro-polishing performed at room temperature using an electrolyte of 95% acetic and 5% perchloric acids, applying a voltage of 15 V for 30 s. EBSD acquisition was performed in a Zeiss Sigma field emission scanning electron microscope (SEM) using an EDAX DigiView EBSD Camera (EDAX/Ametek, Draper, UT, USA) and TSL Orientation Imaging Microscopy (OIM) software (Version 6.2, EDAX/Ametek, Mahwah, NJ, USA). A 20 kV accelerating voltage and 60 μm aperture were set for the SEM parameters and a square grid with a step size of 0.2 μm for the EBSD scan. The possible phases selected for EBSD analysis were body-centered-cubic

(BCC) and face-centered-cubic (FCC) structures corresponding to ferrite and austenite, respectively. For data processing, high angle grain boundaries (HAGB) are defined by disorientations higher than 15°, whereas low angle grain boundaries (LAGB) have disorientations between 2° and 15° [20]. Kernel average misorientation (KAM) is used to measure lattice disorientation below 2° using all Kernel points up to the second nearest neighbors. The martensite areas were not properly indexed because martensite laths are highly dislocated and relatively small with respect to the applied step size. Martensite islands appear on the inverse pole figure (IPF) maps as clusters of randomly oriented points [21] (see Figure 2a) with low index quality (IQ) values [22] (see Figure 2b). A combination of both criteria was used to determine potential martensite constituents (see Figure 2c). Clusters of low IQ (typically <0.35) with a size of at least 0.2 μm^2 (i.e., 5 pixels) are considered as martensite islands (see Figure 2d).

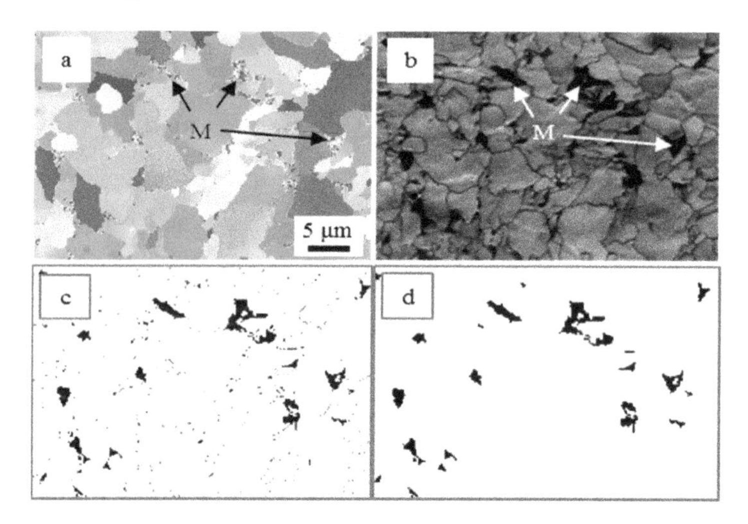

Figure 2. Example of electron backscatter diffraction (EBSD) analysis of martensite islands: (**a**) inverse pole figure (IPF) map; (**b**) index quality (IQ) map; (**c**) clusters of randomly-oriented points with low IQ values; and (**d**) clusters considered as martensite islands (>0.2 μm^2) excluding grain boundary points.

The mean grain size of the microstructure is determined as the average equivalent area diameter (EQAD), accounting for all microstructure constituents (i.e., irregular ferrite, martensite and austenite), with a minimum disorientation between grains of 2°, as given by Equation (1):

$$EQAD_{2^\circ} = 2\sqrt{\frac{A}{\pi(n_{IF} + n_M + n_{RA})}} \tag{1}$$

where A is the investigated area and the n_i (i = IF, M, RA) are the numbers of irregular ferrite (IF), martensite (M) and retained austenite (RA) grains. No significant difference of the EQAD values was observed including or excluding the grains cut at the border of the map, because of a sufficiently large number of grains in the investigated maps (>2000).

Further, selected high-resolution transmission electron microscopy (HR-TEM) studies were conducted to characterize the precipitates at various aging stages. TEM specimens were prepared by cutting discs at 0.5–1 mm from the edge of the torsion specimens, thinned mechanically down to 0.06 mm, and then twin-jet electro-polished to perforation using a mixture of 5% perchloric acid, 15% glycerol, and 80% ethanol at a potential of 35 V and a temperature of −2 °C. TEM observations were conducted using a Tecnai G^2 F20 field emission gun transmission electron microscope (FEG-TEM) equipped with an energy dispersive spectrometer (EDS). The size measurement of carbides was conducted by HR-TEM and at least 50 carbide particles were measured for each condition. The specimens were tilted to the <100> zone axis to obtain Moiré fringes from the BCC matrix and the (Nb,Mo)C particle. The crystallographic orientation relationship between the matrix and the (Nb,Mo)C was determined by fast Fourier transformed (FFT) image and identified as the Baker-Nutting orientation relationship (BN-OR), $\{100\}_\alpha//\{110\}_{carbide}$, $<100>_\alpha//<110>_{carbide}$, as usually observed for these carbides [8,9]. Based on the BN-OR, the (Nb,Mo)C carbides could be detected precisely and their sizes could be measured accurately. The number density of carbides (the number of carbides per unit volume of ferrite matrix, μm^{-3}) for each steel was determined from 10 HR-TEM frames with a total examined area of 10 μm^2. The thickness of the observed area was measured by using electron energy-loss spectroscopy (EELS). EELS is an analysis of the energy distribution of electrons which involves the inelastical scattering events. The inelastically scattered electrons with different energy loss are categorized into the spectrum and the thin foil thickness, η, is obtained as $\eta = \lambda \ln(I_t/I_0)$ where I_0 is the integrated intensity of the zero-loss peak, I_t is the total intensity and λ is the inelastic mean free path of electrons.

3. Results and Discussion

3.1. Age Hardening Results

3.1.1. Hardness Kinetics

The age hardening curves at the three investigated temperatures are shown for the four steels in Figure 3. Age hardening peaks are observed for all curves, due to the precipitation strengthening effect, well-known in microalloyed steels [1]. The magnitude of the hardness peaks appears to be independent of temperature for a given steel grade. The as-quenched and the peak hardness increase with alloying content, as shown in Figure 4. The difference of peak and as-quenched hardness indicates the apparent peak precipitation hardening increment, ΔH_p, that increases from 22 HV for the Nb1 and Nb1Mo1 steels, to 29 HV for the Nb1Mo2 steel and 30 HV for the Nb2Mo2 steel. It is worth noting that no hardening was observed for an Nb-free 0.2 wt.% Mo steel initially included in the study, meaning that Mo itself does not produce significant precipitation strength.

Figure 5 provides a comparison of the age hardening kinetics for the four Nb-bearing steels by presenting the normalized hardness change, i.e., $\Delta H/\Delta H_p$ where $\Delta H = H - H(\textit{as-quenched})$, for each temperature. The kinetics are similar for the four steels, with peak aging times of approximately 20, 120, and 1200 min at 630, 580, and 530 °C, respectively. Nevertheless, the aging curve of the Nb1 steel that does not contain Mo is slightly shifted to shorter times compared to the other steels.

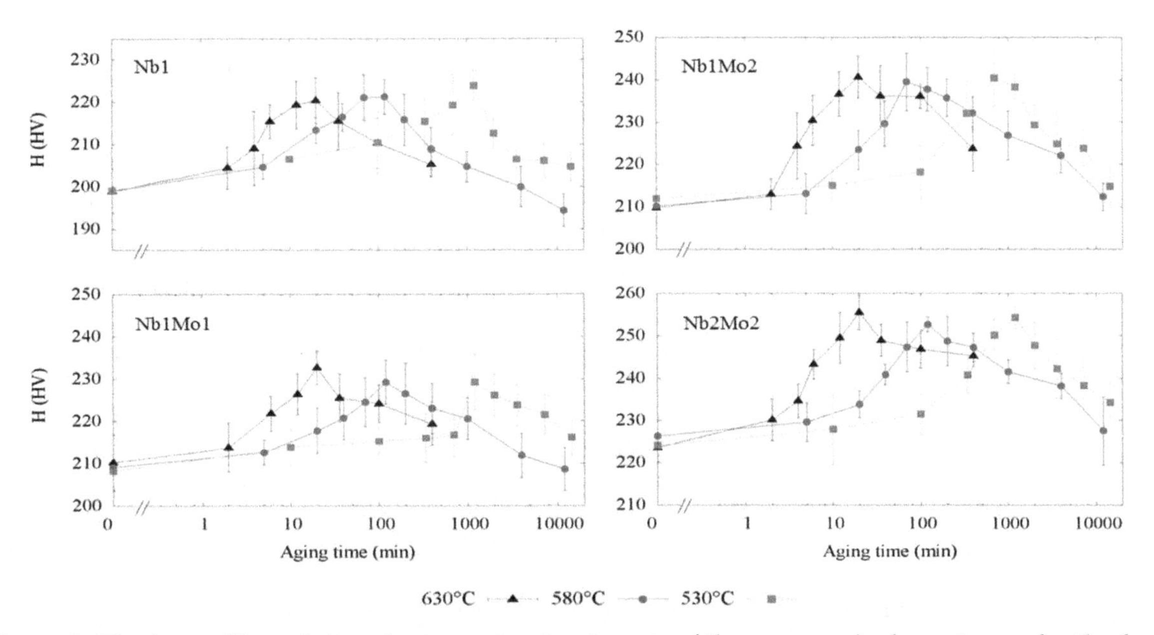

Figure 3. Hardness, H, evolution during aging treatments of the as-quenched specimens for the four investigated steels: Nb1, Nb1Mo1, Nb1Mo2, Nb2Mo2. The error bars indicate the standard deviation from the ten hardness measurements.

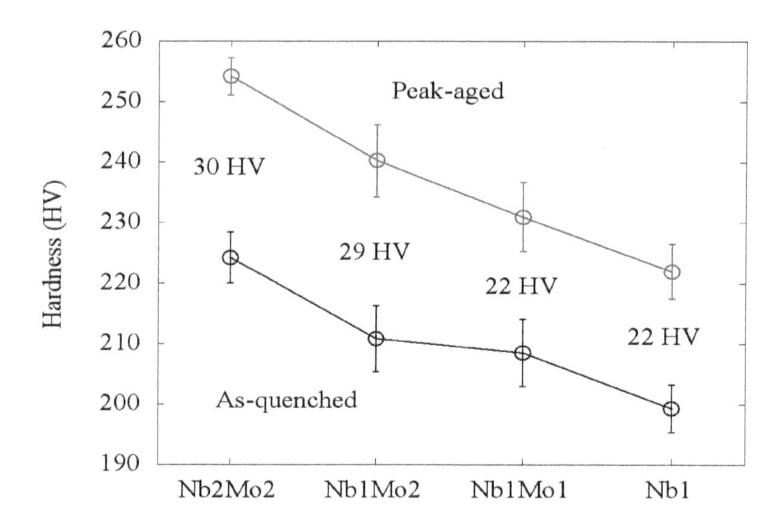

Figure 4. As-quenched, peak-aged, and associated precipitation hardness for the four investigated steels.

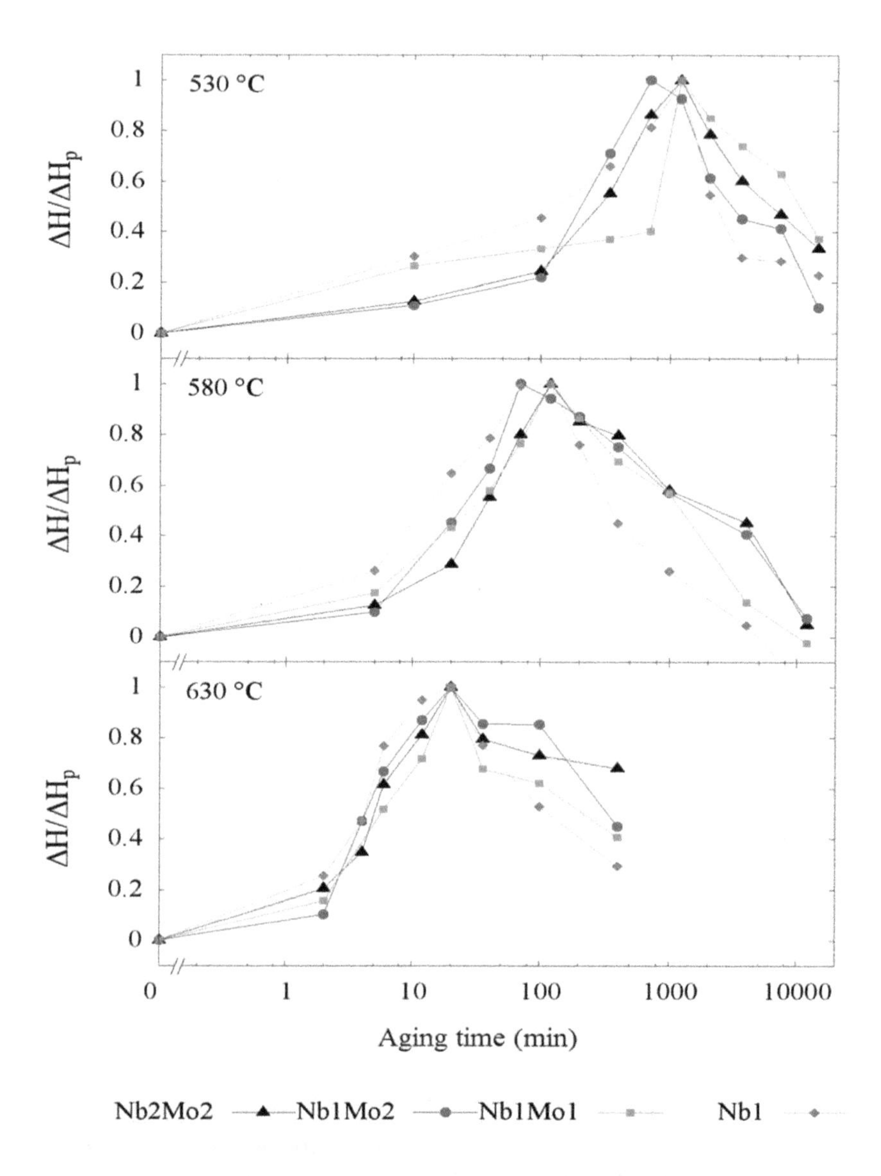

Figure 5. Comparison of the age hardening kinetics for the four Nb-bearing steels at the three aging temperatures of 530 °C, 580 °C and 630 °C.

3.1.2. Microstructure

Figure 6 shows the as-quenched microstructures of the four investigated steels. All of them are composed of a mixture of ferrite and bainitic ferrite grains, denoted here as irregular ferrite (IF) microstructure, with small fractions of martensite and retained austenite (M/A) islands (see Table 3). No pearlite was observed in the as-quenched conditions. The bainitic ferrite is characterized by large grains delimited by HAGB with high KAM and a high density of LAGB, corresponding to bainite laths or highly dislocated substructures. The ferrite grains are small, more polygonal, and less dislocated. However, the ferrite and bainitic ferrite grains cannot be strictly separated by EBSD features (KAM, grain size, etc.) because of the continuity of the IF. Nevertheless, the microstructure appears to be more ferritic for the Nb1 steel and to become more bainitic with increasing Mo content. Secondly, the bainitic ferrite is more and more dislocated with increasing Nb and Mo contents. Consequently, the mean KAM values of IF increases with the alloying content (see Table 3). A similar trend is obtained when considering the LAGB densities. Further, the HAGB have predominantly disorientation angles between 54° and 60° that are characteristics of the FCC-BCC transformation according to specific orientation relationships, such as Kurdjumov-Sachs or Nishiyama-Wasserman. The resulting bimodal disorientation frequency distribution is typical for a bainitic ferrite microstructure, whereas a random, i.e., MacKenzie, distribution is observed for polygonal ferrite microstructures [20]. The frequency and density of LAGB increases with increasing Nb and Mo contents, as listed in Table 3 indicating an increased bainitic ferrite fraction with increasing Mo and/or Nb content.

Finally, the overall grain size ($EQAD_{2°}$), taking into account the substructure of the bainitic ferrite grains, decreases with increasing Nb and Mo content (Table 3). The refinement of the as-quenched IF microstructure with increasing Nb content is consistent with the known effect of Nb on ferrite grain refinement through increasing austenite pancaking by inhibiting recrystallization [11,12]. Further, Mo is known to considerably delay the austenite-to-ferrite transformation and to promote bainite formation [7,8,14]. These microstructure changes lead to an increase of the as-quenched hardness with the Nb and Mo contents, as shown in Figures 3 and 4. Note that the relatively wide standard deviation observed (approx. 5 HV) is the result of local microstructure heterogeneities after the hot-rolling simulation process because of the complex irregular ferrite microstructures with areas of fine ferrite grains and large bainitic ferrite grains.

Table 3. As-quenched microstructure features as determined by EBSD analysis.

Steel	Nb1	Nb1Mo1	Nb1Mo2	Nb2Mo2
EQAD (>2°) (μm)	2.6	2.4	2.2	1.9
LAGB density (mm^{-1})	191	282	335	392
Mean KAM (<2°)	0.48	0.55	0.60	0.62
M/A fraction (%)	3.1	3.5	3.5	3.9

The evolution of the microstructure during aging has been investigated for two steels, the least and the most alloyed, at two aging conditions. Figure 7 shows the EBSD maps of the Nb1 and Nb2Mo2 steels after 15 min (peak-aged condition) and 690 min (over-aged condition) holding at 630 °C. Tempering of the microstructure during aging is obvious for both steels compared to the as-quenched conditions shown in Figure 6. After 15 min of aging no M/A is observed, and after over-aging the microstructure appears to be slightly recovered, with the ferrite microstructure appearing to be less irregular. The tempering leads to coarsening of the microstructure with an increase of the overall grain size for the Nb1 steel from 2.6 μm in the AQ condition to 3.2 μm in the over-aged (OA) condition and for the Nb2Mo2 steel from 1.9 to 2.4 μm, respectively. The recovery effect on the mean KAM is more challenging to capture because of statistical reasons, but a significant decrease is observed for the Nb1 steel from 0.48 in the AQ condition to 0.38 for the OA condition.

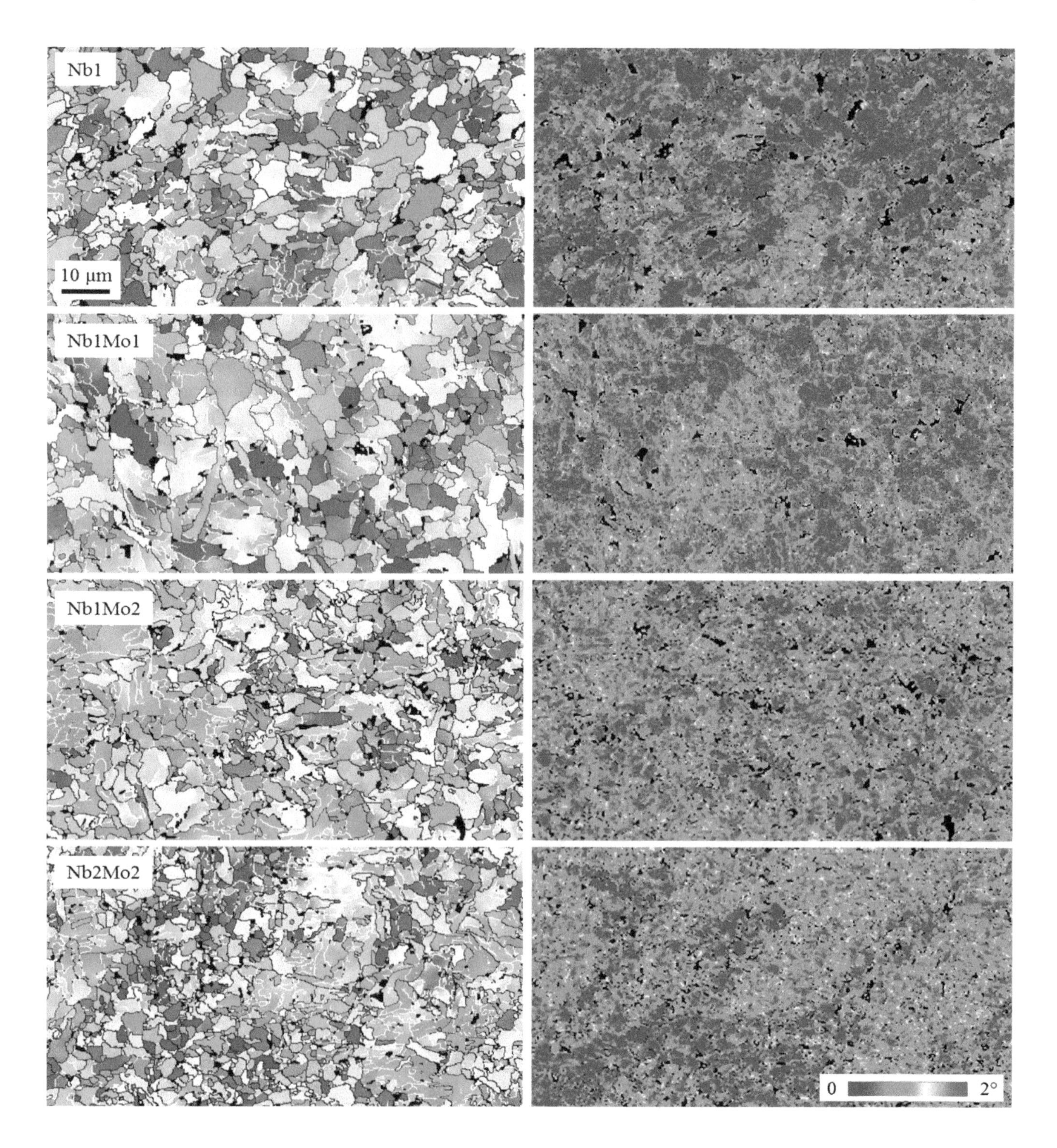

Figure 6. EBSD IPF maps (left row) and kernel average misorientation (KAM) maps (right row) of the investigated steels Nb1, Nb1Mo1, Nb1Mo2, Nb2Mo2 (top to bottom) for the as-quenched conditions (high angle grain boundaries (HAGB) in dark grey and low angle grain boundaries (LAGB) in white, martensite and retained austenite (M/A) in black).

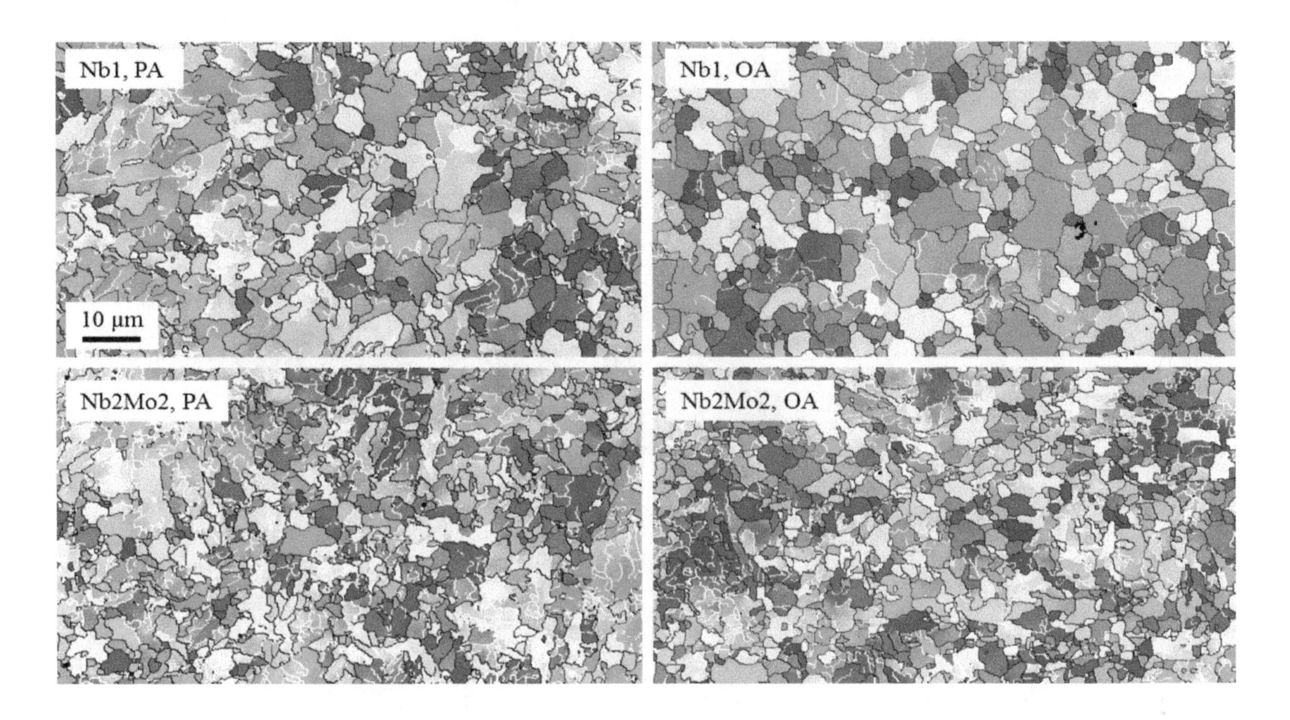

Figure 7. EBSD IPF maps of the Nb1 and Nb2Mo2 steels peak-aged (PA), 15 min at 630 °C, (Nb1, PA and Nb2Mo2, PA) and over-aged (OA), 690 min at 630 °C (Nb1, OA and Nb2Mo2, OA).

Further, the HR-TEM investigation confirmed the presence of fine precipitates, as shown in Figure 8 for the Nb1 and Nb2Mo2 steels. Two types of carbides were observed in the AQ samples, i.e., coarse and monolayer platelet carbides. The coarse carbides were likely strain-induced carbides formed in austenite at higher temperatures [9,11], while the tiny platelets, of approximately 2 nm in length, precipitated in the irregular ferrite. However, the ultra-fine carbides are rarely observed in the as-quenched condition and it is suggested that the strength of the present steels is not markedly influenced by these tiny and rare carbides. The compositions of these ultra-fine carbides have not been characterized because of their small size. In the samples with peak-aged condition (aged for 25 min at 630 °C), a large number of nano-sized precipitates is observed in the irregular ferrite for all four samples. The carbide number densities are 4×10^3, 5×10^3, 5.8×10^3, and 9.2×10^3 μm^{-3} for the Nb1, Nb1Mo1, Nb1Mo2, and Nb2Mo2 steels, respectively. The increase in carbide number density with increasing Mo and Nb additions is consistent with the observations made in a recent study by Zhang et al. [23]. The fast Fourier transform diffractograms are obtained from the HR-TEM Moiré fringe images and confirm the Baker-Nutting orientation relationship (BN-OR) between carbides and ferrite. On the other hand, the carbide sizes, i.e., their lengths, for the four investigated steels in the peak-aged condition are very similar with 3.9 ± 1.3, 4.1 ± 1.6, 4.2 ± 1.4, and 3.7 ± 1 nm, respectively. For the over-aged samples (aged for 400 min at 630 °C), the size of these precipitates is larger, i.e., 5.5–6 nm in length with a few atomic layers wide in all investigated steels, as summarized in Figure 9. In the over-aged Nb-Mo-bearing steels, the chemical compositions of the nanometer-sized carbides, $(Nb_x, Mo_{x-1})C$, were determined using EDX spectroscopy. The Nb/Mo ratio was estimated to be 1:1 for Nb1Mo2 and Nb2Mo2, while 3:2 for Nb1Mo1, indicating an increase of the Mo content in the precipitate composition with increase of the nominal Mo bulk composition, in accordance with other studies [8,9]. These results provide, however, only an approximate composition analysis because the nanometer sized carbides are embedded in the ferrite matrix and the nanoprobe of the electron beam may have spread from the nanometer-sized carbide to the ferrite matrix. Thus, a more advanced and precise analysis would be required to quantify the three-dimensional distribution of chemical compositions in the corresponding $(Nb_x, Mo_{x-1})C$ carbides.

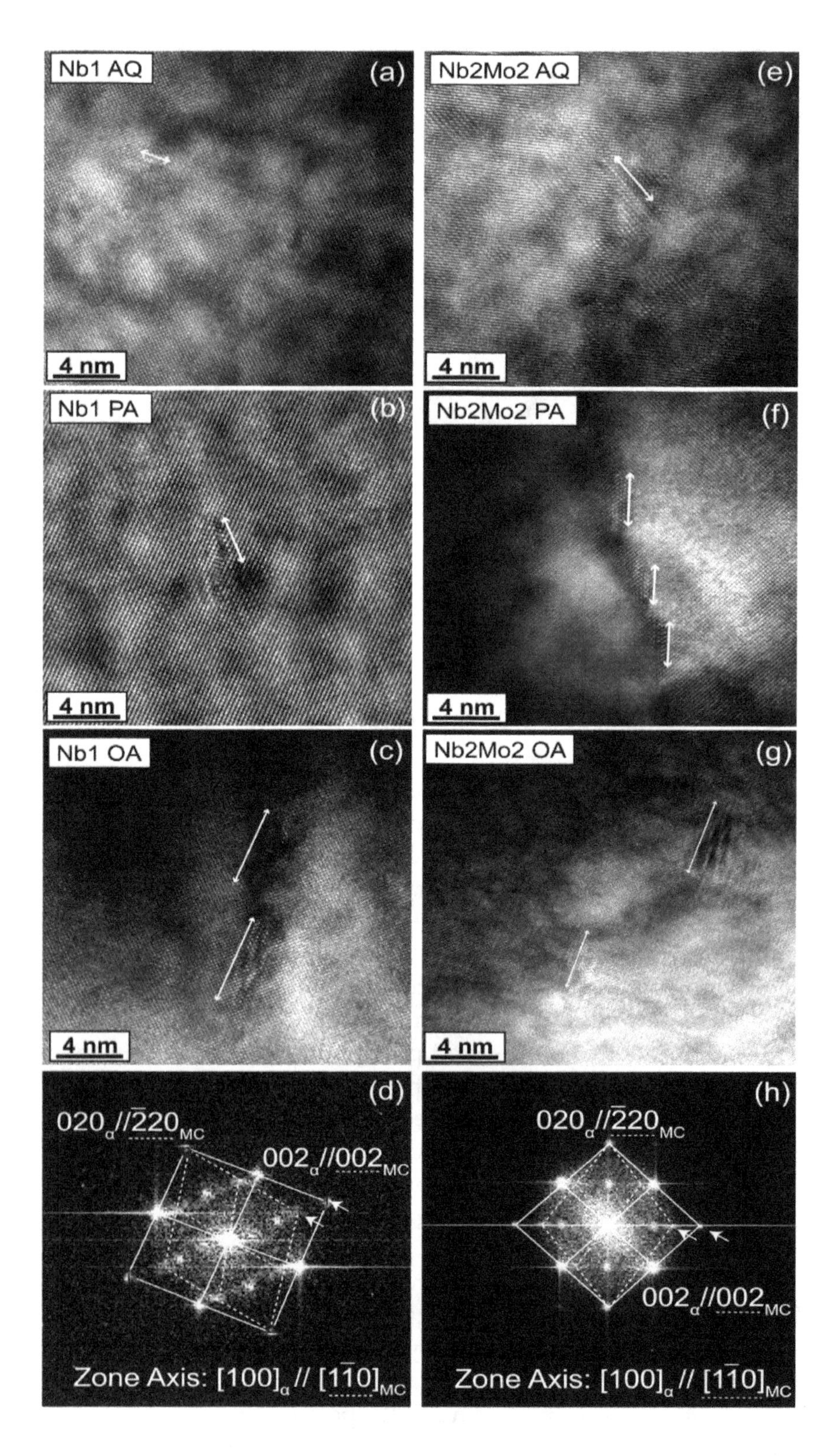

Figure 8. High-resolution transmission electron microscopy (HR-TEM) showing the precipitates in the Nb1 and Nb2Mo2 steels: (**a**) as-quenched Nb1; (**b**) peak-aged Nb1; (**c**) over-aged Nb1; (**d**) fast Fourier transform diffractogram for the carbide and ferrite of peak-aged Nb1; (**e**) as-quenched Nb2Mo2; (**f**) peak-aged Nb2Mo2; (**g**) over-aged Nb2Mo2; (**h**) fast Fourier transform diffractogram for the carbide and ferrite in peak-aged Nb2Mo2. The arrows in the FFT images indicated the corresponding index of diffraction spots which belong to matrix and carbides.

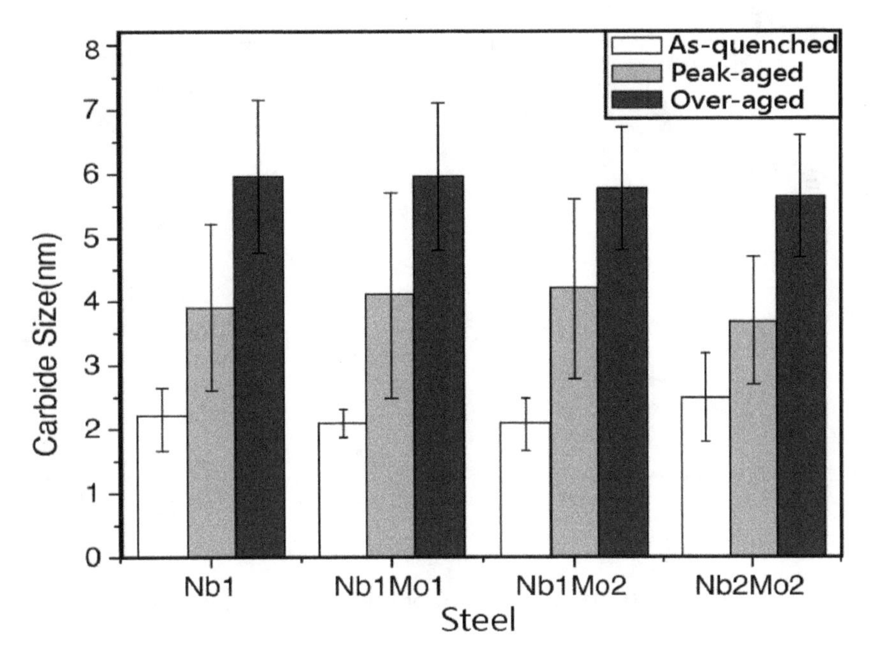

Figure 9. The average carbide size in as-quenched, peak-aged, and over-aged conditions for the four studied steels.

3.2. Modeling

3.2.1. Time-Temperature Equivalence

The hardening peak observed during aging is due to precipitation of carbides but microstructure analysis shows that it is also affected by softening of the as-quenched microstructure because of tempering. To model the true precipitation strengthening effect this softening contribution is determined from the microstructure parameters obtained by EBSD using a general yield strength model [1].

The temperature independence of the peak hardness suggests that the tempering state is similar and the same fraction of alloying elements have precipitated at the peak. This is consistent with the solubility of Nb in ferrite being much lower than the Nb content of the steels such that the equilibrium fraction precipitated is essentially independent of temperature.

The timescale of the age hardening kinetics shifts to longer times with decreasing temperature. A temperature-corrected time, P, is introduced in Equation (2) based on the Arrhenius law such that:

$$P = t\exp\left(-\frac{Q}{RT}\right) \tag{2}$$

where Q (kJ/mol) is an effective activation energy, T is the aging temperature (K), and R the universal gas constant (J/mol/K). At the peak aging time, t_p, the precipitation and tempering states are the same for the three temperatures. The relationship between the peak aging time and the temperature is then given by Equation (3):

$$\ln(t_p) = \frac{Q}{RT} + \ln(P_P) \tag{3}$$

where $\ln(P_P)$ is the intercept of the Arrhenius-plot. The peak aging time is determined for each age hardening curve with a polynomial fit to consider all the hardness measurements near the peak. The analysis of the peak aging times permits to determine Q, as shown in Figure 10. For the four investigated steels, Q is approximately 230 kJ/mol. The value of P_P is the same for the Nb–Mo-bearing steels but is slightly smaller for the Nb1 steel, i.e., 10^{-12} min vs 7×10^{-13} min. The slightly faster age hardening behavior in the Nb1 steel may be rationalized with the faster diffusion of Nb compared to Mo [24].

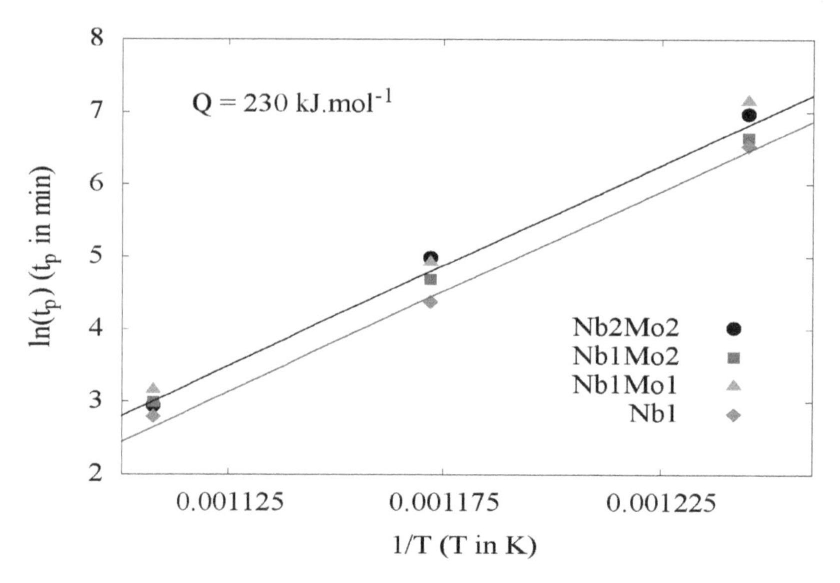

Figure 10. Arrhenius plot of the peak aging times.

3.2.2. Tempering

To determine the microstructure softening during aging, a general yield stress model is used with Equation (4), considering a linear sum of the various yield stress contributions [1]:

$$\sigma_y = \sigma_0 + \sigma_{ss} + \sigma_{gs} + \sigma_\rho + \sigma_{ppt} \tag{4}$$

where σ_0 is the friction stress, σ_{ss} accounts for solid solution strength, σ_{gs} is the Hall-Petch relation to account for the role of grain size, σ_ρ is the contribution from dislocations and σ_{ppt} is the precipitation strength. Except σ_{ppt} all these strengthening contributions may be assessed from the literature and/or from the microstructure analyses previously detailed. For convenience, all of these contributions are referred to as the yield stress of the bulk, i.e., $\sigma_{bulk} = \sigma_0 + \sigma_{ss} + \sigma_{gs} + \sigma_\rho$.

The friction stress and the solid solution strength from alloying elements are estimated to be 108 MPa for the four steels according to Equation (5), as proposed by Choquet et al. [25]:

$$\sigma_0 + \sigma_{ss} = 63 + 23\ Mn + 53\ Si + 700\ P \tag{5}$$

where the stress is in MPa and concentrations are in wt.%. This relationship does not account for potential solute solution strength of Nb and Mo, which, however, appears to be rather low even in the as-quenched condition of the present steels. For example, the solution strength for 0.2 wt.% Mo is estimated to be 2 MPa [26]. The grain size contribution is determined using the Hall-Petch relationship implemented by Iza-Mendia and Gutiérrez [20]. Here, a lower strengthening contribution is taken into account for LAGB than for HAGB that is proportional to the square root of the disorientation angle between grains [27]. This model appears to be a realistic approach to describe irregular ferrite microstructures with various disorientation profiles and LAGB densities. After implementation of the model to include M/A, the grain boundary strengthening contribution is described by Equation (6) as:

$$\sigma_{gs} = k_y \left[\sum_{2^\circ \le \theta < 15^\circ} f_{IF_\theta} \sqrt{\frac{\theta}{15}} + \sum_{\theta \ge 15^\circ} f_{IF_\theta} + f_{M/A} \right] EQAD_{2^\circ}{}^{-1/2} \tag{6}$$

where f_{IF_θ} is the grain boundary frequency of disorientation θ (°) in the irregular ferrite and $f_{M/A}$ is the M/A phase boundary frequency with the sum of the boundary fraction $\sum f_{IF_\theta} + f_{M/A} = \sum_{2^\circ \le \theta < 15^\circ} f_{IF_\theta} + \sum_{\theta \ge 15^\circ} f_{IF_\theta} + f_{M/A} = 1$. The parameter k_y is the Hall-Petch coefficient that had been obtained after calibration of the model with ferrite-pearlite microstructures [20], i.e., $k_y = 19.4$ MPa.mm$^{0.5}$.

The dislocation strengthening contribution is determined using the Taylor relationship, i.e. Equation (7), considering the geometrically necessary dislocation density ρ_{GND} [28] only:

$$\sigma_{\rho} = \alpha M G b \sqrt{\rho_{GND}} \tag{7}$$

where $\alpha = 0.3$ is a constant, $M = 3$ the Taylor factor, $\mu = 80\ GPa$ the shear modulus, and $b = 0.25$ nm the magnitude of the Burgers vector. Here ρ_{GND} is assessed from $\overline{KAM}$ (in rad), the mean KAM value of the irregular ferrite, using the approach of Calcagnotto et al. [22]. However, overestimations of the dislocation density using this law were reported [10,20,22]. As shown in Equation (8), a noise term, KAM_{noise}, to account for the relative angular resolution of the EBSD measurement was thus introduced:

$$\rho_{GND} = 2\frac{(\overline{KAM} - KAM_{noise})}{ub} \tag{8}$$

where u is the unit length, corresponding to the mean distance between the center of the reference pixel and the centers of all the pixels included in the KAM calculation. $KAM_{noise} = 0.3°$ has been determined as the y-intercept of the mean KAM as a function of the unit length measured in several irregular ferrite grains with various distortion states. Finally the dislocations densities in the investigated conditions range from between 5×10^{13} m^{-2} for the over-aged Nb1 to 2.2×10^{14} m^{-2} for the as-quenched Nb2Mo2. It is worth noting that part of the geometrically necessary dislocations are not included in this calculation because the maximum disorientation considered in the KAM calculation is 2°. However, the larger disorientations are incorporated into the grain size strengthening contribution term by having different strengthening coefficients for LAGB.

The validity of the general yield stress model in Equation (4) is tested on the as-quenched conditions because the precipitation strengthening contribution is expected to be negligible, considering the small size and low density of the precipitates. The modeled yield stress values are compared to the hardness measurements, using a basic relationship $H = \kappa\sigma$ [29,30] where κ is a conversion factor that here was determined to be 3.7 HV/MPa from the average of the conversion factors obtained for each steel. This conversion factor provides a consistent description of the experimental hardness measurements with an increase of the grain size and dislocation strengthening contributions with increasing Nb–Mo content (see Figure 11). However, the conversion factor is higher than expected [29] as the present analysis considers a lower limit for the dislocation density and does not account for any solute solution and/or precipitation strengthening contributions from Nb and Mo.

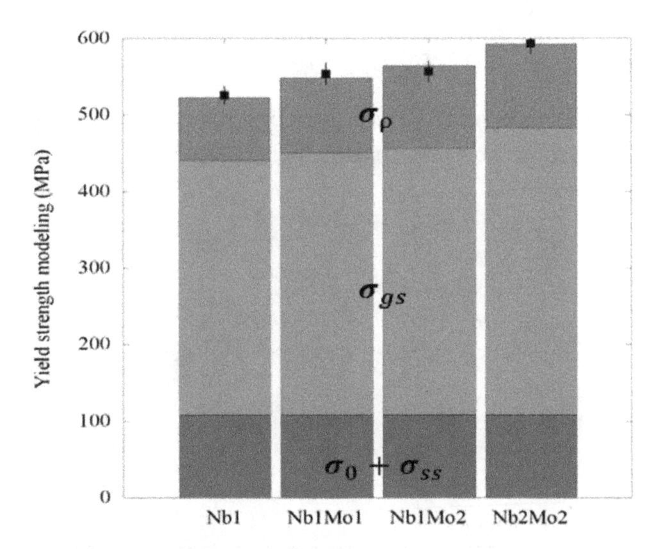

Figure 11. Comparison of the predicted yield stresses with hardness measurements for as-quenched conditions. The square symbols represent the hardness measurements after conversion using the relationship $\sigma = H/\kappa$ with the error bar indicating the standard deviation.

The softening due to tempering of the microstructure during aging is then estimated from Equation (4) using the EBSD data obtained for tempered Nb1 and Nb2Mo2 steels (see Figure 7) and considering $\sigma_{ppt} = 0$. Figure 12 shows the softening fraction plotted as a function of aging time at 630 °C. In a first approximation, the normalized tempering behavior is assumed to be the same for both steels. Equation (9) is employed to describe the softening fraction because of tempering [31]:

$$\tau(t) = 1 - \frac{\sigma_{bulk}(t)}{\sigma_{bulk}^{AQ}} = 1 - e^{(-D_1 * t^{m_1})} \tag{9}$$

where the predicted yield stress of the bulk as a function of the aging time (in min) is normalized to the as-quenched conditions for each steel. The adjustable parameters in Equation (9) are taken to be $D_1 = 0.045$ and $m_1 = 0.13$ in order to reasonably well describe the trend of the tempering behavior at 630 °C for the six microstructures investigated. Considering that the timescale for tempering is shifted with the temperature according to the Arrhenius law, the model can then be extrapolated to other aging temperatures using the temperature-corrected time of Equation (2) with the Q and P_p determined from the age hardening curves, see Figure 10.

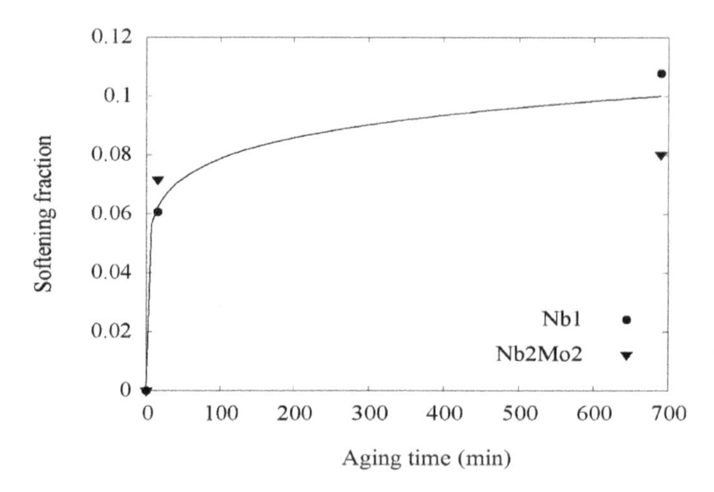

Figure 12. Softening fraction due to tempering of the bulk microstructure during aging at 630 °C for Nb1 and Nb2Mo2 steels, calculated using Equations (4)–(8) with the EBSD data. Solid line depicts the proposed tempering model according to Equation (9).

3.2.3. Precipitation Strengthening Model

The age hardening curve is the result of a combination of precipitation hardening and bulk softening due to tempering, as schematically presented in Figure 13.

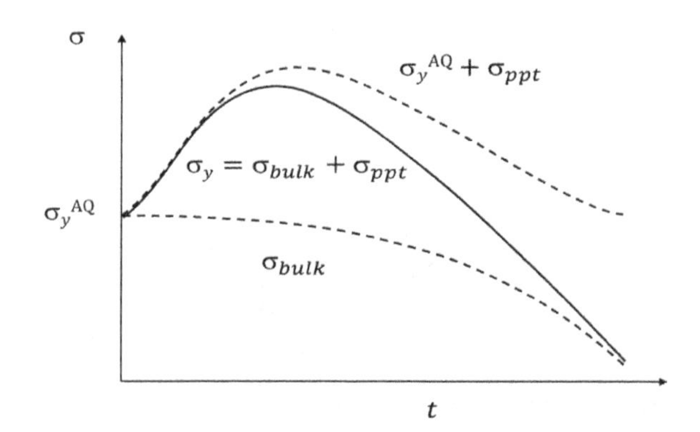

Figure 13. Schematic representation of the effects of precipitation and tempering on yield stress.

The true precipitation strengthening curves are thus obtained by subtracting the microstructure softening from the age hardening curves. A precipitation hardening model has been developed to describe the normalized master curves. The model is based on the Shercliff-Ashby approach [32] with Equations (10)–(12), assuming that the total precipitation strengthening is the harmonic mean of precipitate shearing (Δ_{sh}, also known as Friedel effect) and bypassing (Δ_{by}, also known as Orowan looping), i.e.,:

$$\Delta\sigma_{ppt} = \frac{2}{\frac{1}{\Delta\sigma_{sh}} + \frac{1}{\Delta\sigma_{by}}} \tag{10}$$

$$\Delta\sigma_{sh} = K_1 f^{1/2} r^{1/2} \tag{11}$$

$$\sigma_{by} = K_2 \frac{f^{1/2}}{r} \tag{12}$$

here f and r are the volume fraction and mean radius of the precipitates, respectively. K_1 and K_2 are constants. In the Shercliff-Ashby approach originally applied to aluminum alloys [32] and adopted in previous studies for HSLA steels containing Ti, V, and Nb [4,5], a precipitate coarsening law according to the Lifshitz-Slyozov-Wagner (LSW) theory is assumed, i.e., $r(t) \propto t^{1/3}$, such that Equation (13) provides the hardening model:

$$\frac{\Delta H_{ppt}}{\Delta H_p} = \frac{2C_1(1 - \exp(-P^*/\tau_1))^{1/2} P^{*1/6}}{1 + P^{*1/2}} \tag{13}$$

where $\tau_1 = 0.35$ and $C_1 = 1.03$ are normalization constants. This model appears to be suitable to describe the over-aging behavior for Nb–Mo-bearing steels. However, it overestimates the under-aged strengthening behavior.

For shorter aging times a growth law is assumed instead, i.e., r(t) $\propto t^{1/2}$. For the evolution of the precipitation fraction, a Johnson-Mehl-Avrami-Kolmogorov (JMAK) law given by Equation (14) is adopted, i.e.,:

$$f(t) = 1 - \exp\left(-\frac{t^n}{\beta}\right) \tag{14}$$

where β is a temperature-dependent rate parameter. The exponent n is taken to be 1.35 from Perrard et al. [16], who studied the precipitation of NbC in highly dislocated HSLA Nb-bearing steels. Assuming that $\Delta\sigma_{sh}$ and $\Delta\sigma_{by}$ are equal at the aging peak and combining Equations (10) and (11), Equation (15) is obtained as hardening model:

$$\frac{\Delta H_{ppt}}{\Delta H_p} = \frac{2C_2(1 - \exp(-P^{*n}/\tau_2))^{1/2} P^{*1/4}}{1 + P^{*3/4}} \tag{15}$$

with $\tau_2 = 0.36$ and $C_2 = 1.03$ as normalization parameters. This model describes adequately the under-aged hardening kinetics. Therefore, the precipitation hardening model used in this study is a combination of these two models, with Equation (15) for $P^* < 1$ and Equation (13) for $P^* > 1$, i.e., the transition from a growth to a coarsening law occurs at the aging peak. Such a transition is reasonable as the decrease in precipitation strength is associated with an increase in inter-particle spacing due to coarsening.

The contributions of tempering and precipitation on age hardening are shown in Figure 14 for the four investigated steels. The master curve obtained from the above-described analysis to describe the hardness measurements performed at the three aging temperatures is plotted using Equation (2) as a function of the normalized temperature-corrected time $P^* = P/P_p$. The red curve shows the microstructure softening because of tempering described in Equation (9) and the blue curve shows the age hardening behavior after adding the precipitation strengthening contribution from the models according to Equations (13) and (15). The simulated hardening evolution is in good agreement with the experimental measurements except for short times at lower aging temperatures (580 °C and mainly

530 °C). A more detailed investigation of the microstructure softening because of tempering, especially at lower aging temperatures, may explain this discrepancy.

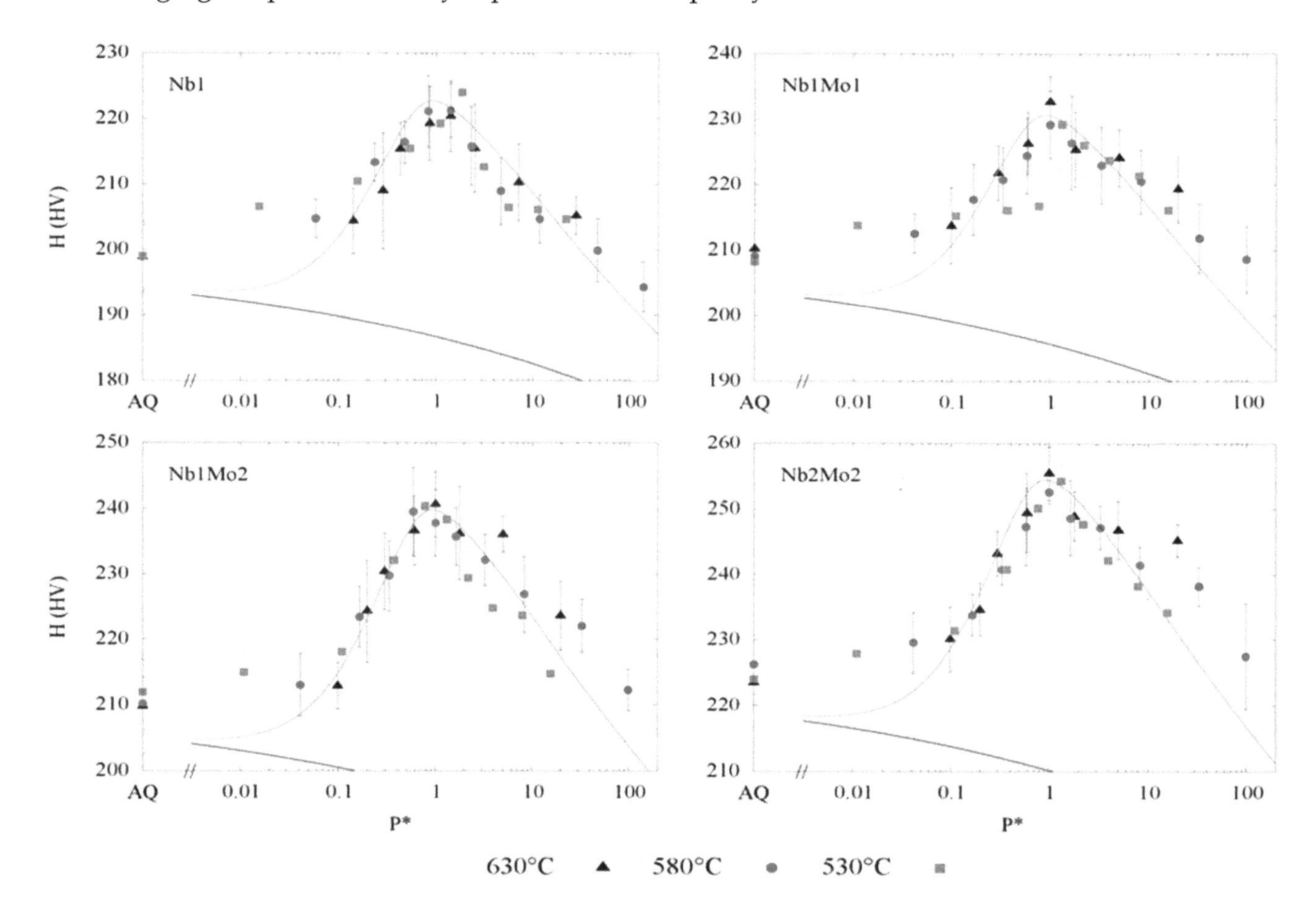

Figure 14. Comparison of modelled and measured hardness for the four investigated steels, Nb1, Nb1Mo1, Nb1Mo2 and Nb2Mo2, at the three aging temperatures. The red curve represents the microstructure softening and the blue curve is obtained by adding the precipitation strengthening contribution.

3.3. *Coiling Simulations*

3.3.1. Microstructure

The transformation hardening is assessed by microstructure investigation after coiling simulations (see path 3 in Figure 1) using the general yield stress model previously described with Equations (4)–(8). Coiling temperatures of 500 °C, 550 °C, 600 °C, and 650 °C were studied for Nb1 and Nb1Mo2 steels to quantify the effect of Mo on transformation hardening and for Nb1Mo1 and Nb2Mo2 steels only the coiling temperature of 600 °C was considered.

As an example, Figure 15 shows the Nb1 and Nb1Mo2 microstructures after coiling simulations. For Nb1, no M/A is observed for any of the coiling temperatures compared to the as-quenched condition (see Figure 6). However, at the coiling temperature of 500 °C the irregular ferrite microstructure is otherwise very similar to that of the as-quenched condition, e.g., in terms of the bainitic ferrite fraction, while with the increase of coiling temperature this fraction seems to decrease and ferrite grains appear to be coarser. Similarly, the decrease of bainitic ferrite in the irregular ferrite microstructure with increasing coiling temperature is also observed for Nb1Mo2. However, M/A is observed after coiling simulation in Nb1Mo2. The fraction is rather low at 550 °C and 650 °C but significant at 600 °C. Figure 16a,b show that M/A islands are coarser and localized in the bainitic ferrite for a coiling temperature of 600 °C compared to the as-quenched condition where M/A was in the form of fine islands randomly distributed in the microstructure. For coiling simulation at 650 °C a small portion of pearlite is also observed (see Figure 16c,d). This illustrates the microstructure changes from an irregular ferrite microstructure containing significant bainitic ferrite fraction at low coiling temperature to a ferrite-pearlite microstructure with increasing coiling temperature. These observations confirm the delaying effect of Mo on the austenite-to-ferrite transformation [7,8,14].

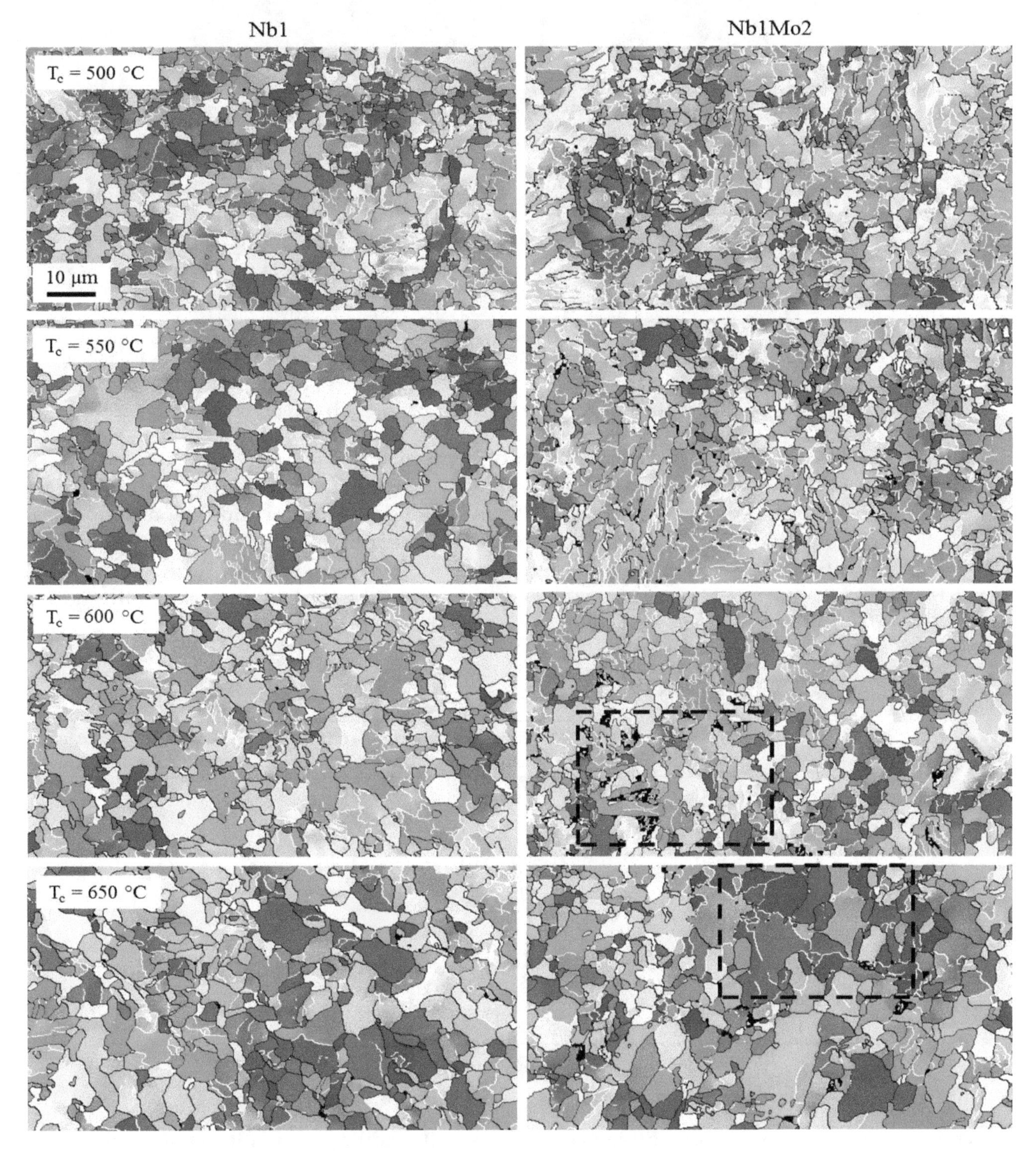

Figure 15. EBSD IPF maps for Nb1 steel (**left**) and Nb1Mo2 steel (**right**) after coiling simulation at four coiling temperatures, i.e., 500 °C, 550 °C, 600 °C and 650 °C, with a cooling rate of 0.5 °C/min for 3 h (HAGB in dark grey and LAGB in white, M/A in black). The areas indicated by dashed lines are shown in more detail in Figure 16.

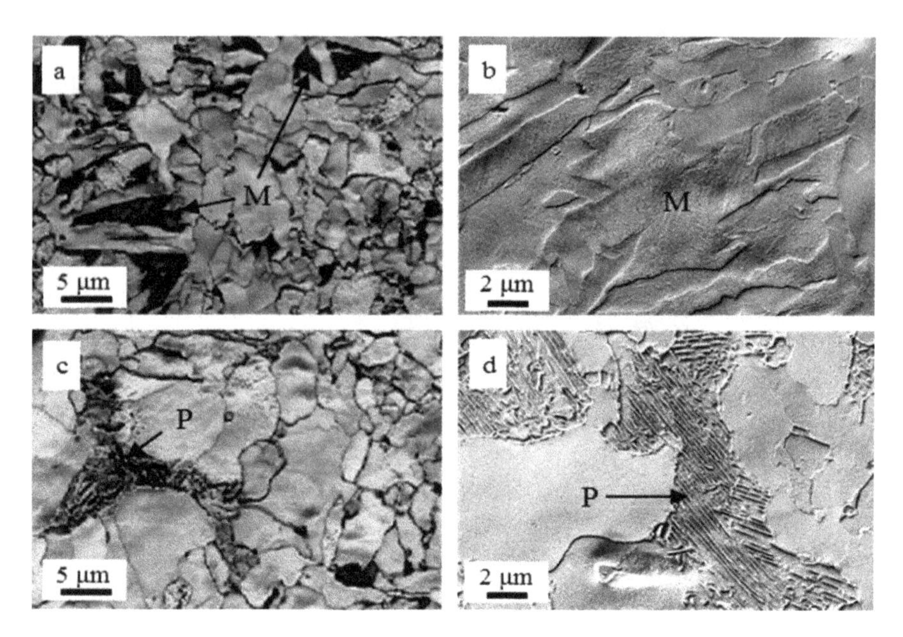

Figure 16. Microstructure investigation of Nb1Mo2 after coiling simulation at 600 °C (**a**,**b**) and at 650 °C (**c**,**d**): (**a**,**c**) are IQ EBSD maps corresponding to the selected area in Figure 15; (**b**,**d**) are SEM images to identify the structures with low IQ values, i.e., martensite (M) and pearlite (P), respectively.

Similar observations were made for Nb1Mo1 and Nb2Mo2 coiling simulated at 600 °C. The microstructure features for coiling simulation are reported in Table 4 for the four investigated steels. The increase of the overall grain size with increasing the coiling simulation temperature is clearly observed, with an $EQAD_{2°}$ approximately 15% larger after coiling at 600 °C than in the as-quenched condition (see Table 3). The measurements of the irregular ferrite mean KAM indicate that the bainitic ferrite fraction is approximately the same for the as-quenched microstructure and after coiling simulation at 500 °C and decrease with increasing the coiling temperature.

Table 4. Microstructure features in coiling simulated samples as determined by EBSD analysis.

Steel	Coiling Temperature (°C)	M/A Fraction (%, ±0.5)	Mean KAM (°, ±0.05)	$EQAD_{2°}$ (µm, ±0.1)
Nb1	500	0	0.51	2.7
	550	0	0.41	2.9
	600	0.3	0.42	3.0
	650	0	0.40	3.4
Nb1Mo1	600	0.3	0.47	2.9
Nb1Mo2	500	0.4	0.59	2.4
	550	1.0	0.55	2.4
	600	2.1	0.48	2.5
	650	0.7	0.40	3.2
Nb2Mo2	600	2.1	0.56	2.3

3.3.2. Transformation Hardening

The significant microstructure differences observed after coiling simulations have a clear effect on the mechanical properties. Using the EBSD data presented in Table 4 combined with the general yield stress model described in Equations (4)–(8), the softening fraction of the resulting transformation products from the coiling simulation with respect to the as-quenched condition is calculated and shown in Figure 17. Although the microstructures are different between the four steels, no significant differences in the relative coil cooling softening fraction is obtained. Similar to the softening because of tempering, see Equation (9), Equation (16) describes the softening fraction as a function of coiling temperature such that:

$$\tau\,(T_c) = 1 - \frac{\sigma_{bulk}(T_c)}{\sigma_{bulk}^{AQ}} = 1 - e^{(-D_2 * T_c{}^{m_2})} \tag{16}$$

where T_c is the coiling temperature (in °C), whereas $D_2 = 5.0 \times 10^{-19}$ and $m_2 = 6.2$ are fitting constants.

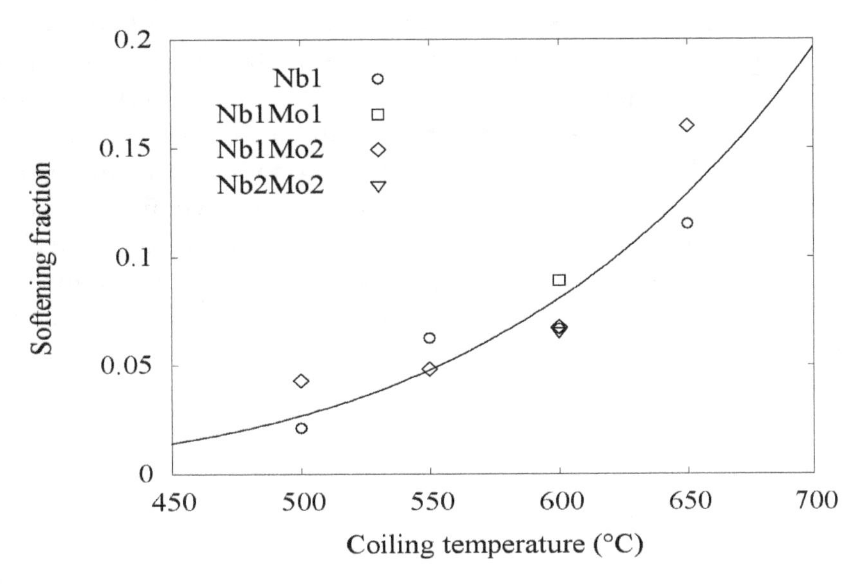

Figure 17. Evolution of the microstructure softening for coiling simulation at various temperatures, calculated from the EBSD data presented in Table 4. Softening model according to Equation (16) is shown with the solid line.

3.3.3. Model Validation

The generalization of the precipitation strengthening model for non-isothermal treatment paths is given by Equation (17) obtained by integrating P over the time-temperature path, i.e., [4]:

$$P = \int_{t_{T_c}}^{t_{450^\circ C}} \exp\left(-\frac{Q}{RT(t)}\right) dt \tag{17}$$

where $T(t)$ is the temperature as a function of time during coiling. The limit of 450 °C is taken here because precipitation rates become insignificant below this temperature for coil cooling conditions.

Figure 18 shows the hardening obtained after coiling as a function of temperature for the coil cooling rate of 0.5 °C/min employed in the tests. The red curve is the microstructure strength without precipitation as obtained with the softening analysis (see Figure 17) and the black curve is the overall strength including precipitation strength contribution, calculated from Equations (13), (15), and (17). The model predictions are in reasonable agreement with the measured hardness values on the coil cooled samples considering the complex microstructural changes occurring during coiling. At low coiling temperature, the strength is similar to the as-quenched conditions with a predominantly bainitic ferrite microstructure with little or no precipitation hardening. When the coiling temperature increases, the microstructure is less bainitic but the loss of strength is compensated by precipitation hardening, until a maximum at 600 °C is reached. Above that temperature, precipitate coarsening is expected at the detriment of precipitation hardening. Finally, for higher coiling temperatures, a lower strength than at the as-quenched conditions is expected, with a ferrite-pearlite microstructure and reduced precipitation hardening.

The conclusions regarding the precipitation hardening as a function of coiling simulation have been confirmed with TEM observations. The largest carbide number density is observed for coiling simulations at 600 °C with values of 8.6×10^2, 9×10^2, and 2.9×10^3 μm^{-3} for the Nb1Mo1, Nb1Mo2, and Nb2Mo2 steels, respectively. The carbide sizes for this coiling condition falls independent of steel chemistry in the range of 4.2 to 4.5 nm that is comparable to the peak-aged condition (see Figure 9). The carbide density is about a factor 3 larger in the Nb2Mo2 steel as compared to the other steels with lower Nb content. For 550 °C coiling simulation, the number density of carbides is significantly lower and their sizes are smaller than at 600 °C confirming an under-aged condition. In detail, the carbide number density is 1.1×10^2, 3.3×10^2, and 8.1×10^2 μm^{-3} for the Nb1Mo1, Nb1Mo2 and Nb2Mo2 steels, respectively, and the average carbide size is 3 nm independent of steel chemistry. At 650 °C coiling temperature, precipitates are only recorded in the Nb2Mo2 specimens with a carbide number density of 1.1×10^3 μm^{-3} that is comparable to that observed at 550 °C and an average carbide size of 5.3 nm which is consistent with an over-aged condition. In the steels with a lower Nb content, however, the carbide density is remarkably low for this coiling condition where the microstructure is primarily polygonal ferrite with a low dislocation density reducing the effective nucleation density for carbide formation. Thus, the reduced precipitation strength for these steels appears to also be affected by a lower nuclei density rather than by the otherwise proposed coarsening behavior. Further studies will be required to quantify in more detail the precipitation behavior at higher coiling temperatures (>650 °C) in the steels with lower Nb contents (<0.1 wt.%).

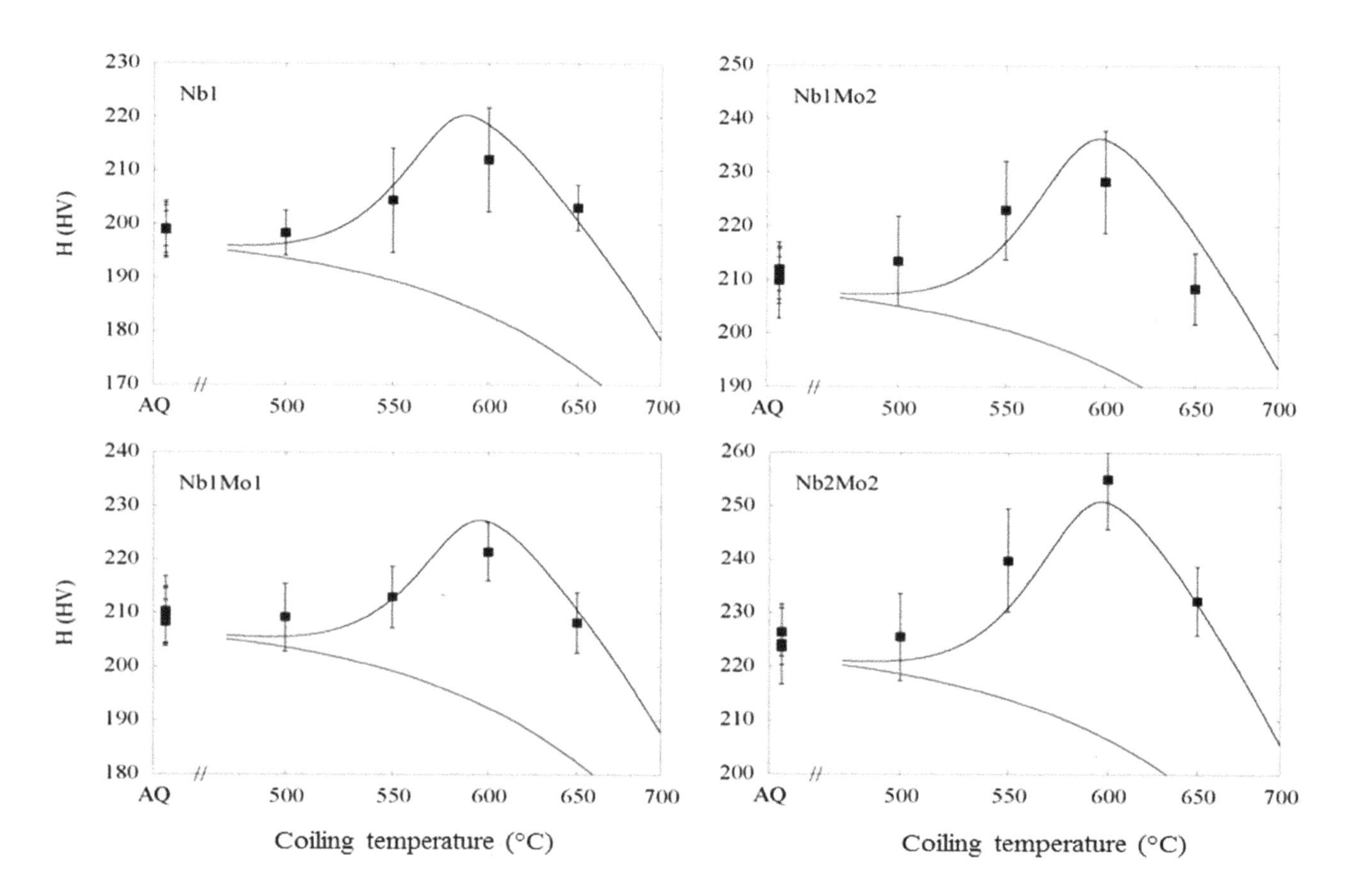

Figure 18. Comparison of measured (symbols) and predicted (solid lines) hardness of hardening after coiling as a function of coiling temperature for the four investigated steels, i.e., Nb1, Nb1Mo1, Nb1Mo2 and Nb2Mo2. The red curves represent the calculated microstructure hardness without precipitation strengthening contribution.

Optimum precipitation strengthening windows were determined in previous studies for a complex-phase Nb–Mo steel (0.05C-1.9Mn-0.05Nb-0.5Mo-0.004N-0.04Si-0.05Al (wt.%)) by

Sarkar et al. [33] and conventional HSLA-Nb steels with a polygonal ferrite/pearlite microstructure by Militzer et al. [4]. As shown elsewhere [18], the predicted precipitation strengthening contribution as a function of coiling temperature is similar for the Nb–Mo complex-phase steel as that obtained for the Nb–Mo steels investigated here. The optimum coiling temperature window to have at least 90% of precipitation peak strength is much narrower in the Nb–Mo steels as compared to that of 100 °C (570–670°C) in conventional Nb-microalloyed steels with ferrite-pearlite microstructures where transformation hardening is negligible such that only precipitation hardening is affected during coiling. In the Nb–Mo steels, on the other hand, transformation hardening is another strength contribution that depends in detail on the coiling condition. This aspect was not quantified by Sarkar et al. but the current study suggests that the coiling temperature window to have overall strength levels within 5% of the absolute peak strength is, as shown in Figure 17, about 50 °C, i.e., 575 °C to 625 °C for the investigated Nb–Mo steels (565–615 °C for Nb1).

4. Conclusions

The potential of precipitation strengthening during coiling for hot-strip Nb–Mo-bearing low-carbon steels has been investigated with laboratory studies simulating the hot-rolling process. A hardening model predicting strength changes as a function of coil cooling conditions has been developed. The model is based on a standard additive approach of the strengthening contributions, where the microstructure strengthening contribution is determined based on microstructure parameters obtained from EBSD analyses and the precipitation strengthening contribution is assessed from age hardening kinetics using a modified Shercliff-Ashby approach.

In the investigated conditions, the microstructures of the four studied steels ranges from ferrite-pearlite to irregular bainitic ferrite. The potential of Nb–Mo bearing microalloyed steels was verified as Nb and Mo have a clear strengthening effect on the microstructure, decreasing the overall grain size and increasing the dislocation density after direct quenching or coiling simulations. Increasing Nb and Mo contents also increases the precipitation strengthening potential but in the investigated composition ranges the microstructure strengthening is more significant. The microstructure strengthening potential by decreasing the coiling temperature for those steels was also verified, going from a ferrite-pearlite microstructure at higher coiling temperature to a more bainitic ferrite microstructure for lower coiling temperature. Increasing Nb and Mo content also appears to increase the M/A fraction.

The key point of the developed model is that it accounts for both precipitation age hardening and tempering of highly dislocated bainitic and/or irregular ferrite. The model provides insight into optimizing the hardening after coiling by finding the best balance between precipitation strengthening and transformation hardening due to a highly dislocated bainitic ferrite microstructure. The model predictions are in reasonable agreement with experimental hardness measurement in coil cooling simulated samples. An optimum coiling temperature window of 50 °C between 575 °C and 625 °C is concluded from this analysis for the Nb–Mo bearing steel. To extend the proposed model to predict property variations in industrial coils it is suggested to integrate it with a coil cooling model that will have to be developed and benchmarked with temperature measurements at different coil positions.

Author Contributions: J.-Y.M. and M.M. conceived and designed the hot-torsion tests and developed the model. J.-Y.M. performed the hot-torsion tests, characterized the microstructures with EBSD and analyzed the data. J.-R.Y. and Y.W.C. conducted the HR-TEM studies. N.H.G. and S.J.K. organized the laboratory steel making and conducted the hardness measurements. B.J. and H.M. coordinated the work and contributed to the interpretation of the data. J.-Y.M., M.M. and H.M. wrote the paper with the input of all co-authors.

Acknowledgments: The fruitful discussions with W.J. Poole, J.D. Embury, and A. Deschamps as well as F. de Geuser are acknowledged with gratitude.

References

1. Gladman, T. *The Physical Metallurgy of Microalloyed Steels*, 1st ed.; Institute of Materials: London, UK, 1997.
2. Bouaziz, O.; Zurob, H.; Huang, M. Driving force and logic of development of advanced high strength steels for automotive applications. *Steel Res. Int.* **2013**, *84*, 937–947. [CrossRef]
3. Kim, Y.M.; Kim, S.K.; Lim, Y.J.; Kim, N.J. Effect of microstructure on the yield ratio and low temperature toughness of linepipe steels. *ISIJ Int.* **2002**, *42*, 1571–1577. [CrossRef]
4. Militzer, M.; Hawbolt, E.B.; Meadowcroft, T.R. Microstructural model for hot strip rolling of high-strength low-alloy steels. *Metall. Mater. Trans.* **2000**, *31*, 1247–1259. [CrossRef]
5. Militzer, M.; Poole, W.J.; Sun, W. Precipitation hardening of HSLA steels. *Steel Res.* **1998**, *69*, 279–285. [CrossRef]
6. Matlock, D.K.; Speer, J.G.; De Moor, E.; Gibbs, P.J. Recent developments in advanced high strength sheet steels for automotive applications: An overview. *JEStech* **2012**, *15*, 1–12.
7. Mohrbacher, H.; Sun, X.; Yong, Q.; Dong, H. MoNb-based alloying concepts for low-carbon bainitic steels. In *Advanced Steels: The Recent Scenario in Steel Science and Technology*; Weng, Y., Dong, H., Gan, Y., Eds.; Springer: Berlin/Heidelberg, Germany, 2011; pp. 289–302.
8. Lee, W.-B.; Hong, S.-G.; Park, C.-G.; Park, S.-H. Carbide precipitation and high-temperature strength of hot-rolled high-strength low-alloy steels containing Nb and Mo. *Metall. Mater. Trans. A* **2002**, *33*, 1689–1698. [CrossRef]
9. Huang, B.M.; Yang, J.R.; Yen, H.W.; Hsu, C.H.; Huang, C.Y.; Mohrbacher, H. Secondary hardened bainite. *Mater. Sci. Technol.* **2014**, *30*, 1014–1023. [CrossRef]
10. Isasti, N.; Jorge-Badiola, D.; Taheri, M.L.; Uranga, P. Microstructural features controlling mechanical properties in Nb–Momicroalloyed steels. Part I: yield strength. *Metall. Mater. Trans. A* **2014**, *45*, 4960–4971. [CrossRef]
11. Dutta, B.; Palmiere, E.J.; Sellars, C.M. Modelling the kinetics of strain-induced precipitation in Nb microalloyed steels. *Acta. Mater* **2001**, *49*, 785–794. [CrossRef]
12. Fernandez, A.I.; Uranga, P.; Lopez, B.; Rodriguez-Ibabe, J.M. Dynamic recrystallization behavior covering a wide austenite grain size range in Nb and Nb–Ti microalloyed steels. *Mater. Sci. Eng. A* **2003**, *361*, 367–376. [CrossRef]
13. Hutchinson, C.R.; Zurob, H.S.; Sinclair, C.W.; Brechet, Y.J.M. The comparative effectiveness of Nb solute and NbC precipitates at impeding grain-boundary motion in Nb steels. *Scr. Mater.* **2008**, *59*, 635–637. [CrossRef]
14. Uemori, R.; Chijiiwa, R.; Tamehiro, H.; Morikawa, H. AP-FIM study on the effect of Mo addition on microstructure in Ti-Nb steel. *Appl. Surf. Sci.* **1994**, *76–77*, 255–260. [CrossRef]
15. Charleux, M.; Poole, W.J.; Militzer, M.; Deschamps, A. Precipitation behavior and its effect on strengthening of an HSLA-Nb/Ti steel. *Metall. Mater. Trans. A* **2001**, *32*, 1635–1647. [CrossRef]
16. Perrard, F.; Deschamps, A.; Maugis, P. Modelling the precipitation of NbC on dislocations in α-Fe. *Acta Mater.* **2007**, *55*, 1255–1266. [CrossRef]
17. Maetz, J.-Y.; Militzer, M.; Goo, N.H.; Kim, S.J.; Jian, B.; Mohrbacher, H. Modelling of precipitation hardening in Nb–Molow carbon steels. In Proceedings of the Thermomechanical Processing (TMP) Conference, Milan, Italy, 26–28 October 2016.
18. Maetz, J.-Y.; Militzer, M.; Chen, H.Y.W.; Yang, J.R.; Goo, N.H.; Kim, S.J.; Jian, B.; Mohrbacher, H. Modeling of age hardening kinetics during coiling of high performance Nb–Mosteels. In Proceedings of the International Symposium on New Developments in Advanced High-Strength Sheets steels, Warrendale, PA, USA, 30 May–2 June 2017.
19. Hall, D.; Worobec, J. Torsion simulation of the hot strip rolling process. In *Phase Transformations during the Thermal/Mechanical Processing of Steel, Proceedings of 34th Annual Conference of Metallurgists of CIM, Vancouver, BC, Canada, 20–24 August 1995*; Canadian Institute of Mining, Metallurgy and Petroleum: Westmount, QC, Canada, 1995; pp. 305–316.
20. Iza-Mendia, A.; Gutiérrez, I. Generalization of the existing relations between microstructure and yield stress from ferrite-pearlite to high strength steels. *Mater. Sci. Eng. A* **2013**, *561*, 40–51. [CrossRef]
21. Reichert, J. Structure and properties of complex transformation products in Nb/Mo-microalloyed steels. Ph.D. Thesis, The University of British Columbia, Vancouver, BC, Canada, April 2016.

22. Calcagnotto, M.; Ponge, D.; Demir, E.; Raabe, D. Orientation gradients and geometrically necessary dislocations in ultrafine grained dual-phase steels studied by 2D and 3D EBSD. *Mater. Sci. Eng. A* **2010**, *527*, 2738–2746. [CrossRef]
23. Zhang, Z.Y.; Sun, X.J.; Yong, Q.L.; Li, Z.D.; Wang, Z.Q.; Wang, G.D. Precipitation behavior of nanometer-sized carbides in Nb–Momicroalloyed high strength steel and its strengthening mechanism. *Acta Metall. Sin.* **2016**, *52*, 410.
24. Mehrer, H. *Landolt-Börnstein—Group III Condensed Matter*; Springer: Berlin/Heidelberg, Germany, 1990.
25. Choquet, P.; Fabregue, P.; Guisti, J.; Chamont, B.; Pezant, J.N.; Blancet, F. Modelling of forces, structure and final properties during the hot rolling process on the hot strip mill. In *Mathematical Modelling of Hot Rolling of Steel*; Canadian Institute of Mining and Metallurgy: Montreal, QC, Canada, August 1990; pp. 34–43.
26. Pickering, F.B.; Gladman, T. *Metallurgical Developments in Carbon Steels*; Iron and Steel Institute: London, UK, 1963.
27. Hansen, N. Hall-Petch relation and boundary strengthening. *Scr. Mater.* **2004**, *51*, 801–806. [CrossRef]
28. Brewer, L.N.; Field, D.P.; Merriman, C.C. Mapping and assessing plastic deformation using EBSD. In *Electron Backscatter Diffraction in Materials Science*, 2nd ed.; Schwartz, A.J., Kumar, M., Adams, B.L., Field, D.P., Eds.; Springer: Berlin/Heidelberg, Germany, 2009; p. 251.
29. Ashby, M.F.; Jones, D.R.H. *Engineering Materials 2*, 3rd ed.; Pergamon: Oxford, UK; Burlington, MA, USA, 1980.
30. Zhang, P.; Li, S.X.; Zhang, Z.F. General relationship between strength and hardness. *Mater. Sci. Eng. A* **2011**, *529*, 62–73. [CrossRef]
31. Zhang, Z.; Delagnes, D.; Bernhart, G. Microstructure evolution of hot-work tool steels during tempering and definition of a kinetic law based on hardness measurements. *Mater. Sci. Eng. A* **2004**, *380*, 222–230. [CrossRef]
32. Shercliff, H.R.; Ashby, M.F. A process model for age hardening of aluminium alloys—I. The model. *Acta Metall. Mater.* **1990**, *38*, 1789–1802. [CrossRef]
33. Sarkar, S.; Militzer, M. Microstructure evolution model for hot strip rolling of Nb–Mo microalloyed complex phase steel. *Mater. Sci. Technol.* **2009**, *25*, 1134–1146. [CrossRef]

5

Microalloyed Steels through History until 2018: Review of Chemical Composition, Processing and Hydrogen Service

Julio C. Villalobos [1], Adrian Del-Pozo [2], Bernardo Campillo [3,4], Jan Mayen [5] and Sergio Serna [2,*]

[1] Instituto Tecnológico de Morelia, Avenida Tecnológico No. 1500, Col. Lomas de Santiaguito, Morelia 58120, México; julio.villalobos@uaem.mx
[2] CIICAp, Universidad Autónoma del Estado de Morelos, Av. Universidad 1001, Col. Chamilpa, Cuernavaca 62609, Mexico; adrian.delpozo@alumnos.uaem.mx
[3] Instituto de Ciencias Físicas-UNAM, Av. Universidad 1001, Col. Chamilpa, Cuernavaca 62609, Mexico; bci@fis.unam.mx
[4] Facultad de Química-UNAM, Circuito de la Investigación Científica S/N, Mexico City 04510, Mexico
[5] CONACYT, CIATEQ, Unidad San Luis Potosí, Eje 126 No. 225, Zona Industrial, San Luis Potosí 78395, Mexico; jan.mayen@ciateq.mx
* Correspondence: aserna@uaem.mx

Abstract: Microalloyed steels have evolved in terms of their chemical composition, processing, and metallurgical characteristics since the beginning of the 20th century in the function of fabrication costs and mechanical properties required to obtain high-performance materials needed to accommodate for the growing demands of gas and hydrocarbons transport. As a result of this, microalloyed steels present a good combination of high strength and ductility obtained through the addition of microalloying elements, thermomechanical processing, and controlled cooling, processes capable of producing complex microstructures that improve the mechanical properties of steels. These controlled microstructures can be severely affected and result in catastrophic failures, due to the atomic hydrogen diffusion that occurs during the corrosion process of pipeline steel. Recently, a martensite–bainite microstructure with acicular ferrite has been chosen as a viable candidate to be used in environments with the presence of hydrogen. The aim of this review is to summarize the main changes of chemical composition, processing techniques, and the evolution of the mechanical properties throughout recent history on the use of microalloying in high strength low alloy steels, as well as the effects of hydrogen in newly created pipelines, examining the causes behind the mechanisms of hydrogen embrittlement in these steels.

Keywords: microalloyed steels; mechanical properties; processing; microstructural and chemical composition; hydrogen embrittlement

---

1. Introduction

Steel has represented a great advance in the history of the humanity, due to its multiple uses and excellent properties. Throughout history, great discoveries have been made through the knowledge of the phenomena that dominate the behavior of alloys, such as chemical composition, microstructure and thermomechanical processes. Many researchers have contributed to this knowledge and laid the foundations responsible for the continuous developments in the field of metallurgy.

In the last 50 years, the strength of steels has increased progressively thanks to advances in metallurgy and manufacturing techniques in response to the market demand for lighter and stronger steels. The need to achieve these characteristics of higher strength and weldability with sufficient toughness and ductility has led to the development of high strength low alloy (HSLA) steels.

HSLA steels typically contain very low carbon content and small amounts of alloying elements [1], and these are classified by the American Petroleum Institute (API) in order of its strength (X-42, X-46, X-52, X-56, X-60, X-65, X-70, X-80, X-100 and X-120). These properties are achieved by a careful selection of microalloy composition and optimization of thermomechanical processing (TMP) and accelerated cooling conditions subsequent to the TMP.

HSLA steels have been developed to obtain improved mechanical properties compared to normal carbon steels, as well as superior corrosion resistance properties. Alloying elements include 0.05–0.25% carbon, manganese content up to 2%, and small amounts of chromium, nickel, molybdenum, copper, nitrogen, vanadium, niobium, titanium and zirconium that can be used in different proportions [2].

There are many types of HSLA steels, such as weathering steels, microalloyed ferrite-pearlite steels, as-rolled pearlitic steels, acicular ferrite steels, dual phase steels, and inclusion shape-controlled steels. Commonly weathering steels have compositions that include small quantities of alloying elements to improve corrosion resistance; additions of copper are often used in atmospheres where the levels of phosphorus and sulfur are low in the air. In general, microalloyed ferrite-pearlite steels present compositions with small percentages of carbide formers like vanadium and titanium, which strengthen the steel via precipitation hardening and grain refinement. As-rolled pearlitic steels typically only contain carbon and manganese but can have additions of other elements to increase their strength. Dual phase steels use microalloying to create a ferrite matrix with dispersed martensite [2].

The importance of these steels lies in their use in a wide variety of commercial and industrial applications; as a result of this, they have served as crucial materials to achieve great advances in many fields, such as: construction, heavy duty vehicles, storage tanks, railroad cars, oil rigs, pipes for hydrocarbon transport and many more.

2. Chemical Composition of Microalloyed Steels

High Strength Low Alloy (HSLA) steels were introduced at the beginning of the 20th century and the main elements used in a low wt % combination are Nb, V, and Ti. The microalloyed steels (as later were named) use a small amount of microalloying elements ranging from 0.10 up to 0.15 wt %, for single and multi-elements [3], and are either Nb, V or Ti alloyed [4].

The use of microalloyed elements was first reported in the literature around the years 1938–1939, with the use of Nb as a strengthener of hot rolled C-Mn steels [5]. Subsequently, in the year 1945, the V effects were used to increase the strength of normalized steels, but these new microalloyed steels had low impact toughness and ductility [6,7]. It was well known the advantage of grain refining effects due to precipitation of microalloyed elements in the presence of N and C, promoting the precipitation of carbides, nitrides and carbonitrides (Nb, V) (C, N), but the strengthening mechanism was not clear.

The increase of Mn content has been correlated to a reduction of the harmful effect on toughness caused by the precipitation of V and Nb carbonitrides [8], and it was later discovered that Nb could retard the austenite recrystallization [9], which was the introduction of controlled rolling and thermomechanical controlled processing [10]. Mo and Al were used to retard the austenite recrystallization, but it was determined that Nb was the most effective [11]. Later, the use of other elements Zr, B and rare earth metals were used for the control of inclusions, and the controlled additions of S were added to improve the machinability [12]. The effect of microalloying content (wt %) on austenite recrystallization is shown in Figure 1.

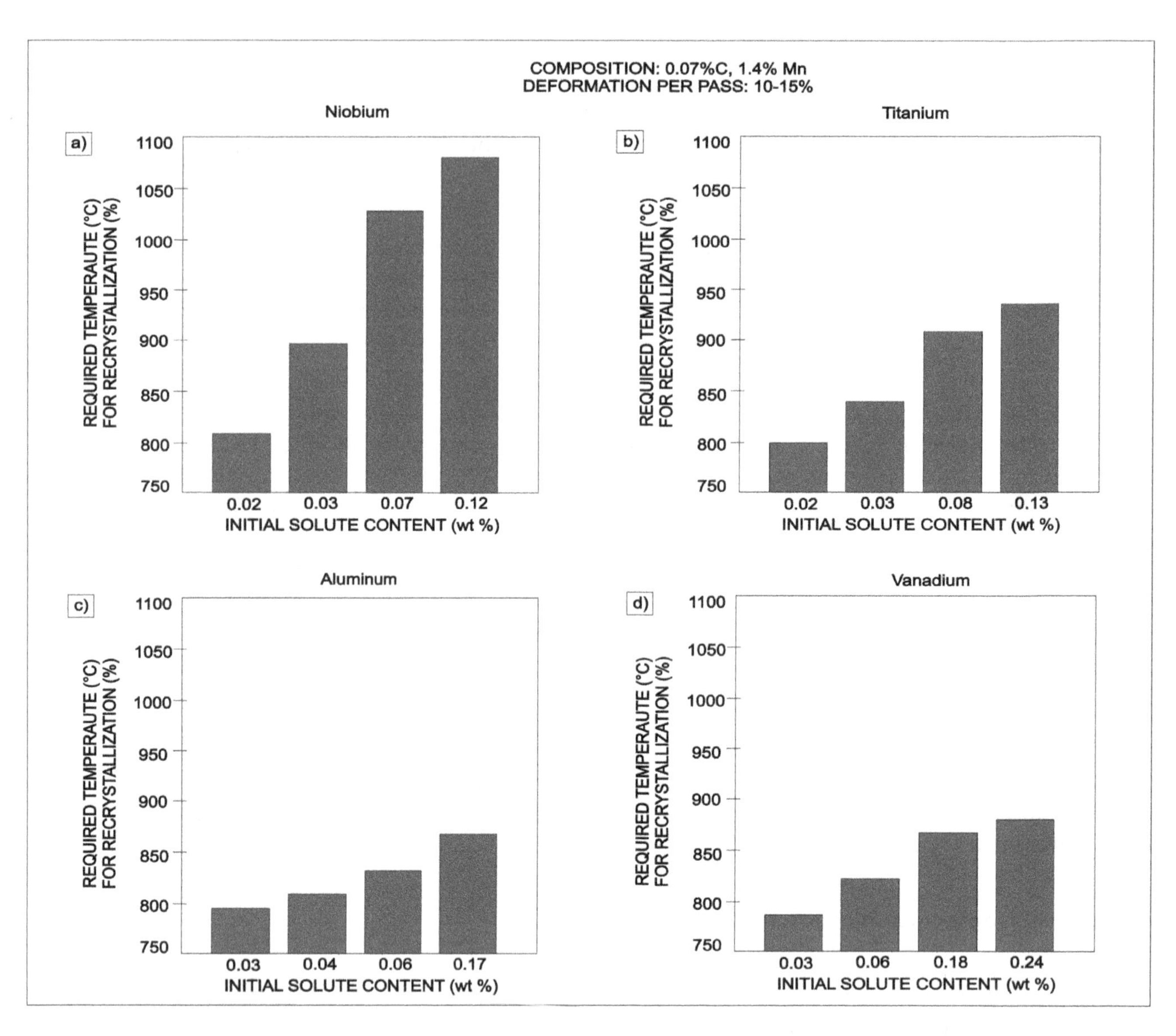

Figure 1. Austenite recrystallization as a function of microalloying content, (**a**) niobium, (**b**) titanium, (**c**) aluminum and (**d**) vanadium, content [12].

The early work on Nb microalloyed steels was concentrated in the USA and UK; in the year 1958, the American company National Steel Corporation developed an Nb-treated steel with relatively low cost with the addition of 0.005–0.03% of Nb content, obtaining a combined effect on strengthening and good toughness [13], this semikilled C-Mn steel is able achieve a yield strength in a range of 300 up to 415 MPa. Furthermore, V was used in combination of Cr-Mo to improve the creep resistance at high temperatures, which was attributed to their high temperature stability [14], with a maximum working temperature of 400 °C.

In the 1970s, the strengthening mechanism was developed by the optimization of chemical composition and processing. The principal mechanisms that even modern high strength steels still use are grain size refinement, solid solution by Mn, Mo, Si, Cu, Ni, Cr addition, precipitation hardening due to the formation of Nb, V, Ti carbonitrides, increased dislocation density and fine grain microstructure generated by a thermomechanical controlled processing and controlled cooling that enhance the mechanical properties [15–17], and decreases the transformation temperature. In accordance with Vervynct et al. [18], the solid solution strengthening is directly related to the microalloyed content, while grain size refinement and precipitation hardening is a function of the interaction between microalloying elements and strain percent during thermomechanical controlled processes (TMCPs).

The C content used in HSLA steels before 1980 was 0.07–0.12%; meanwhile, up to 2% of Mn content was commonly employed together with different additions and combinations of V, Nb and Ti (max. 0.1%) [19]. Reducing the C content could improve the weldability maintaining strength, and it was equal to mild steels, but the principal problem of reducing the C content was that ductility and toughness were not as good as quenched and tempered steels [19].

Controlling the critical temperatures of austenite, microalloying elements can be adjusted to achieve the final mechanical properties required for some applications. These critical temperatures are: the grain coarsening temperature during reheating, recrystallization temperature during hot rolling and transformation temperature during cooling [20,21]. The principal effects of microalloying elements are summarized in Table 1.

Table 1. Principal effect of microalloyed elements [18].

Element	wt %	Effect
C	<0.25	Strengthener
Mn	0.5–2.0	Retards the austenite decomposition during accelerated cooling Decreases ductile to brittle transitions temperature Strong sulfide former
Si	0.1–0.5	Deoxidizer in molten steel Solid solution strengthener
Al	<0.02	Deoxidizer Limits grain growths as aluminum nitride
Nb	0.02–0.06	Very strong ferrite strengthener as niobium carbides/nitrides Delays austenite-ferrite transformation
Ti	0–0.06	Austenite grain control by titanium nitrides Strong ferrite strengthener
V	0–0.10	Strong ferrite strengthener by vanadium carbonitrides
Zr	0.002–0.05	Austenite grain size control Strong sulfide former
N	<0.012	Strong former of nitrides and carbonitrides with microalloyed elements
Mo	0–0.3	Promotes bainite formations Ferrite strengthener
Ni	0–0.5	Increase fracture toughness
Cu	0–0.55	Improves corrosion resistance Ferrite strengthener
Cr	0–1.25	In the presence of copper, increase atmospheric corrosion resistance
B	0.0005	Promotes bainite formation

The evolution of the chemical composition of microalloyed steels from 1959 to the present is shown in Table 2. Microalloyed steels present a large variety of complex combinations in their chemical composition (Nb-V, Nb-Mo, Nb-Cr) depending on the mechanical properties required.

The precipitation of carbides and nitrides occurs in three distinct stages during the microalloyed steels processing as is described in Table 3 [22,23]. These precipitates could be homogeneous, coherent, semi-coherent or incoherent with the crystalline lattice of steel. They can even precipitate in grain boundaries and dislocations, where they can later promote strain induced precipitation.

Modern studies have primarily focused on nanoprecipitates since they make a significant contribution to the yield strength [24]. The presence of precipitates (Ti, V, Nb) (C, N) inhibits the movement of the austenitic grain boundary and retards the grain growth at high temperatures [25];

this process impedes the movement of the dislocations and causes an increase in strength of the steel [26]. Other studies have determined that precipitates of TiN and Nb (C, N) have been effective in inhibiting the growth of austenitic grain [11]. The austenite grain inhibition effect depends on the forming temperature of carbonitrides during the processing of steels.

The principal effects associated with the formation of Nb, V and Ti carbonitrides are listed in Table 4 and Figure 2 shows the solubility temperature of the most common precipitates in microalloyed steels. These effects depend strongly on the stability and solubility as a function of temperature processing. According to this figure, the VN, NbCN and TiC particles are stable at the typical normalizing temperature, around 900 °C, which induces a sufficient volume fraction of fine particles for grain growth controlling [27,28].

Table 2. Evolution of chemical compositions as a function of the API steel grade [21].

C	Mn	S	Si	Cu	Mo	Nb	V	Ti	Al	Cr	Ni	B	Grade
0.041–0.17	0.30–1.68	0.0002–0.03	0.02–1.39	0.02–0.31	0.005–0.14	0.018–0.06	0.042–0.21	0.002–0.01	-	0.02–0.157	0.005–0.8	-	X-65
0.037–0.125	1.44–1.76	0.001–0.015	0.14–0.44	0.006–0.27	0.01–0.3	0.051–0.092	0.001–0.095	0.009–0.03	-	0.007–0.266	0.02–0.23	-	X-70
0.028–0.142	1.52–1.90	0.001–0.009	0.17–0.31	0.015–0.20	0.05–0.3	0.038–0.090	0.002–0.1	0.007–0.024	-	0.015–0.12	0.02–0.75	-	X-80
0.025–0.1	1.56–2.0	0.001–0.0024	0.1–0.25	0.25–0.46	0.19–0.43	0.043–0.089	0.003–0.07	0.011–0.02	0.006–0.030	0.016–0.42	0.13–0.54	-	X-100
0.027–0.05	0.54–2.14	0.001–0.004	0.08–0.31	0.010–0.015	0.001–0.40	0.048–0.1	<0.025	<0.015	0.045–0.04	0.22–0.42	0.017–1.35	0.0013–0.0017	X-120

Table 3. Stages of carbides and nitrides precipitation during microalloyed steels processing [22,23].

Stage	Description
1	Formed during the liquid phase or after solidification process. Very stable precipitates, generally too coarse to influence on austenite recrystallization. Smallest can retard austenite coarsening during reheating.
2	Precipitation induced by strain during controlled rolling, retarding the austenite recrystallization and causing grain refinement.
3	Formed during or after austenite-ferrite transformation, nucleation in austenite-ferrite interfase or ferrite. Fine precipitation is observed.

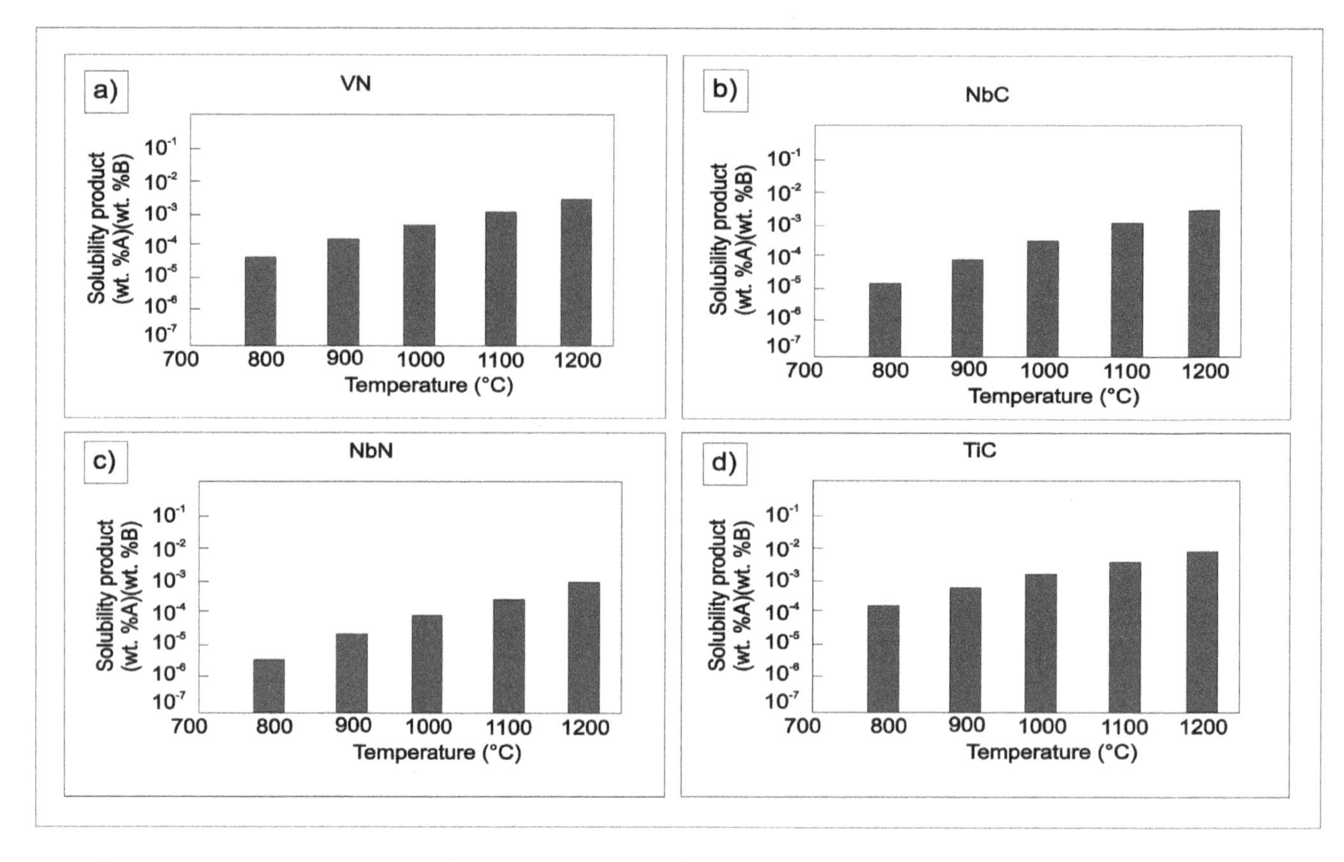

Figure 2. Carbonitrides solubility as a function of temperature, (**a**) vanadium nitride, (**b**) niobium carbide, (**c**) niobium nitride and (**d**) titanium carbide [27,28].

Table 4. The effect of carbides, nitrides and carbonitrides on steels processing [27,28].

Precipitate Type	Effect
Nb	Control austenite transformation during hot rolling processing
TIN	Pin and refine the grain size during high temperature austenitizing
VN, NbCN, TiC	Refines steels microstructure and grain size
Nb, NbCN	Increase recrystallization temperature during hot rolling
VC	Induces precipitation strengthening after normalizing
VN, VC, NbCN, TiC	Induce precipitation strengthening after hot rolling

Nb addition is used for grain refinement and precipitation strengthening, but this strengthening effect is reduced when Nb (C, N) precipitates during hot deformation. These precipitates retard the recovery and recrystallization of deformed austenite and increase the non-recrystallization temperature [29]. The principal role for Nb additions are: NbC and Nb (C, N) precipitation at austenite-ferrite transformation during cooling or cooling treatment that increases strengthening [30]; Nb solid solution and its carbonitrides precipitates retard the austenite grain growth and suppress austenite-ferrite transformation that cause a grain refinement.

On the other hand, small additions of Ti cause a finely dispersed nanoscale nitrides precipitation that restricts austenite grain growth at higher temperatures (1200 °C) [31]. Ti segregates during solidification of the steel and causes a local concentration inducing a precipitation of large TiN particles that cause a pinning effect and the formation of acicular ferrite producing a good heat affected zone (HAZ) toughness [32]. TiC precipitation can also cause strengthening; however, V is the most versatile precipitation strengthening element, and is effective in different compositions of microalloyed steels as well as in those with higher C content.

The Nb addition has a high influence on the transformation temperature of HSLA steels under normal processing conditions [21,33]. Nb in solid solution in austenite reduces the Ar_1 and Ar_3 transformation temperatures of C-Mn steel. This process causes a large austenite grain size that transforms in ferrite or bainite Widmanstädter. The presence of this microstructure conveys a lower fracture toughness than has been observed in Nb-treated steels finish rolled at temperatures around 1000 °C [34]. However, low temperature finish rolled steels containing a fine austenite grain size are able to use the Nb transformation effect in combination with that of bainite-forming elements such as Mo or B to form stronger and tougher steels. The use of controlled cooling reduces the amount of alloying elements required and the steels can reach strengths around 600 MPa [35].

In 2018, microalloyed steels with vanadium were studied, in which an increase in resistance up to 120 MPa was obtained. It was determined that the vanadium atoms in solution delay the bainite reaction at lower transformation temperatures (by 30–40 °C) within cooling rates of 1–50 °C/s [36].

3. Processing of Microalloyed Steels

One of the most efficient and commonly employed methods to improve and control the mechanical properties of steels is the TMCP. This technology has been developed and applied to the industry in a fast manner. A high production volume of steels has been manufactured through this process due to its excellent results. The main principles behind this process are the ability to refine the austenite grain size, increase the defect density, accelerate the precipitation of micro-alloying elements, and control the type of obtained phases [37]. This process has been used more and more, replacing the heat treatment techniques of normalizing, quenching and tempering the heat process. The TMCP helps to make cheaper and simpler the production process of low-carbon bainitic steels, in addition to providing high strength, high tenacity, and good weldability. One of the options to achieve these properties is to increase the microalloying contents, such as Nb, Ti or V; besides affecting austenite microstructural evolution, it contributes greatly to increasing precipitation hardening and dislocation density. One of the problems with high strength low alloy (HSLA) steels is the complex interaction between their hardening mechanisms, making it difficult to optimize parameters for their manufacture.

In order to maintain good control of the TMCP, a good understanding of the two processes involved is required; the complex interaction between strain precipitation of niobium carbonitride in austenite and static recovery [38–41] (Figure 3). The effect of initial austenite grain size and rolling temperature on the critical strain for recrystallization was studied by Tanaka et al. [42] and Kozasu et al. [43], and they reported excellent results in the year 1975, Figures 4 and 5.

Most microalloyed steels are manufactured by controlled rolling and accelerated cooling (TMCP). Severe low temperature rolling is often applied, which may negatively affect fracture toughness (ductile crack arrestability) and resistance to sour environments [44]. Consequently, other technologies have emerged such as the High Temperature Processed Steels [45].

Research by Kozasu et al. [43] showed that equivalent low temperature fracture toughness can be achieved at 1000 °C finishing temperature when 0.10 percent niobium is used compared with 800 °C finish rolling temperature (Figure 6).

In a TMCP, the mechanical properties are greatly affected by different parameters, such as rolling ratio, rolling temperature, cooling pattern, cooling rate and the coiling temperature. Among these factors, one of the most significant is the cooling temperature [46,47]. On the other hand, the temperature of the TMCP contributes to the precipitation of the microalloying elements [48–52].

In the year 2016, Misra and Jansto [53] reported a high strength low alloy (HSLA) steel. At a TMCP temperature of 579 °C, the reported yield strength (YS) was in the range of 701–728 MPa, tensile strength was 996–997 MPa, and elongation was 21–23% (Table 1) [54]. When the TMCP temperature was 621 °C, yield strength, tensile strength, and elongation were in the range of 749–821 MPa, 821–876 MPa, and 19–25%, respectively (Table 5) [54].

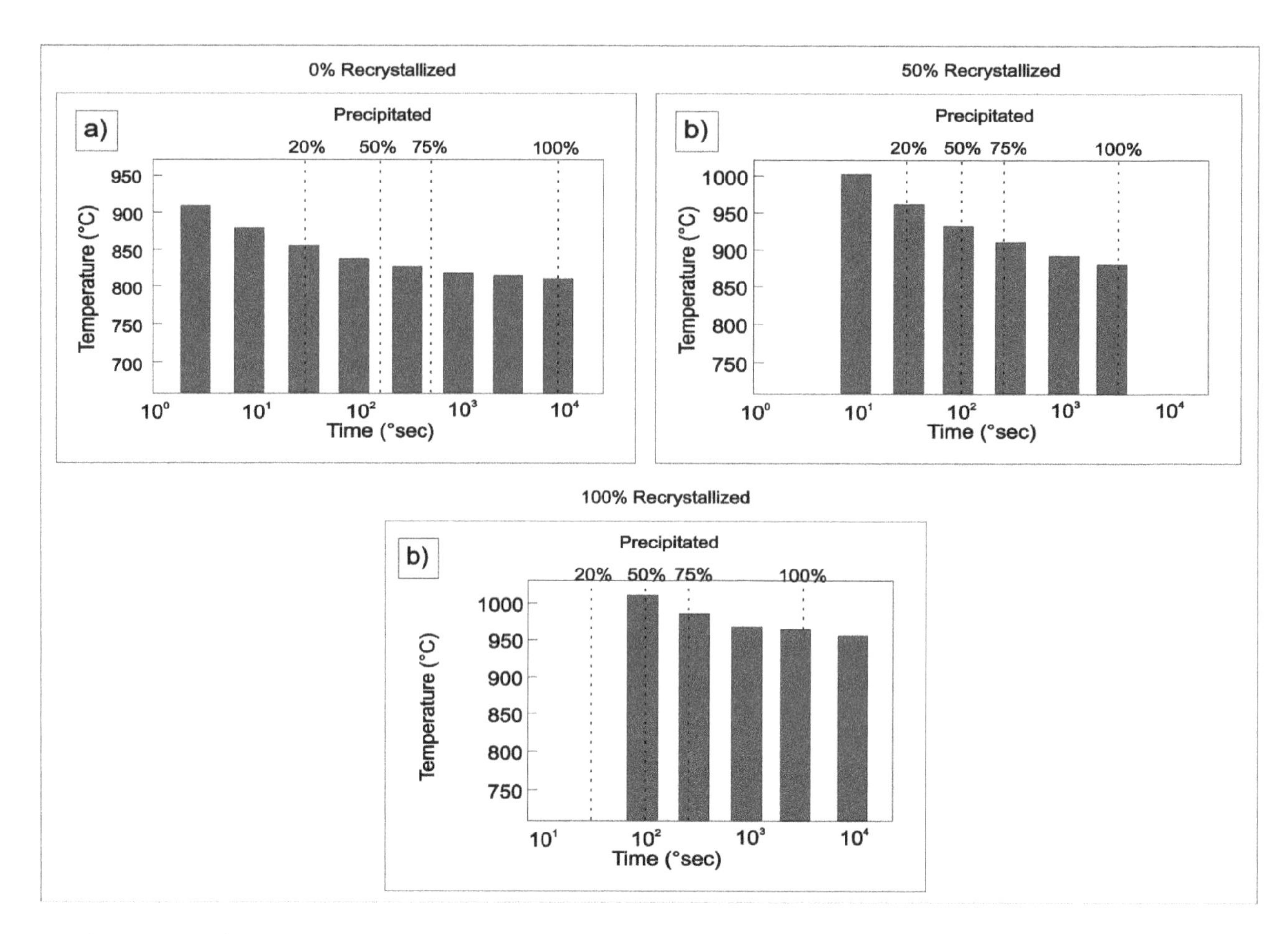

Figure 3. Isothermal recrystallization and precipitation in a columbium steel after 50% deformation. (**a**) 0%, (**b**) 50% and (**c**) 100%, recrystallized. The steel contained 0.10% C, 0.99% Mn, 0.040% Nb, and 0.008% N [38].

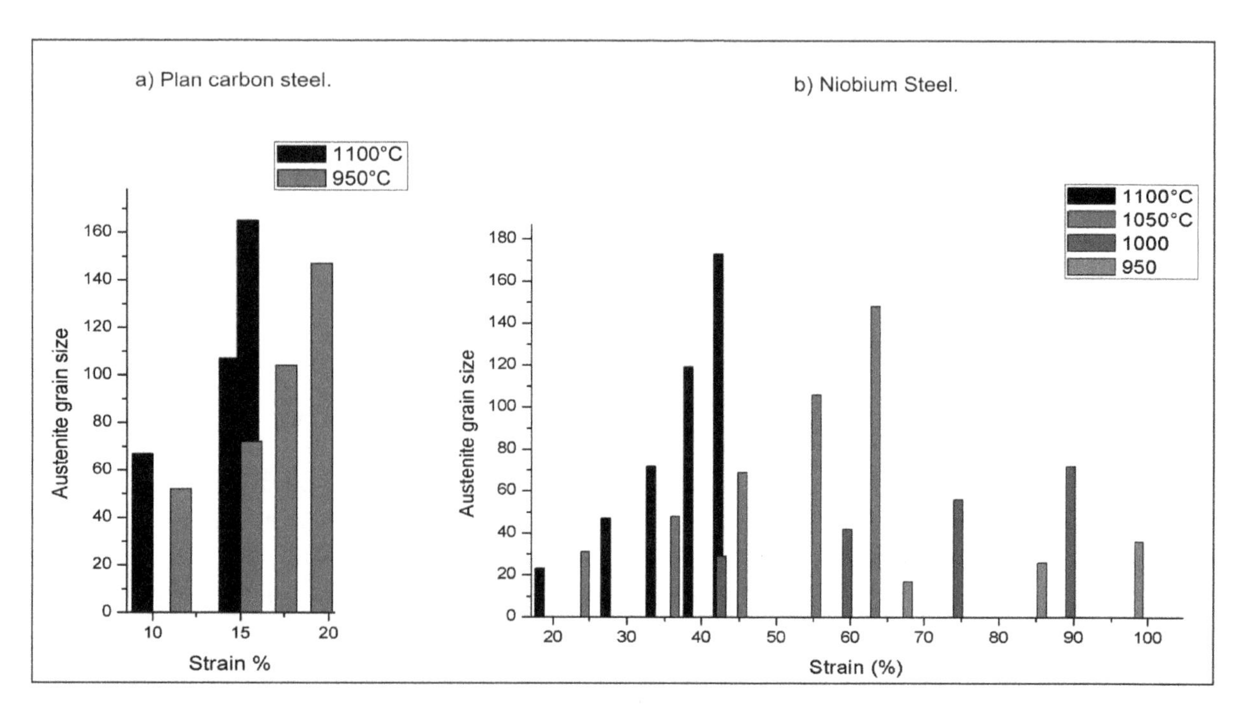

Figure 4. Effect of deformation temperature and initial grain size on critical amount of deformation required for completion of recrystallization in the (**a**) plain-carbon and (**b**) niobium steels [42].

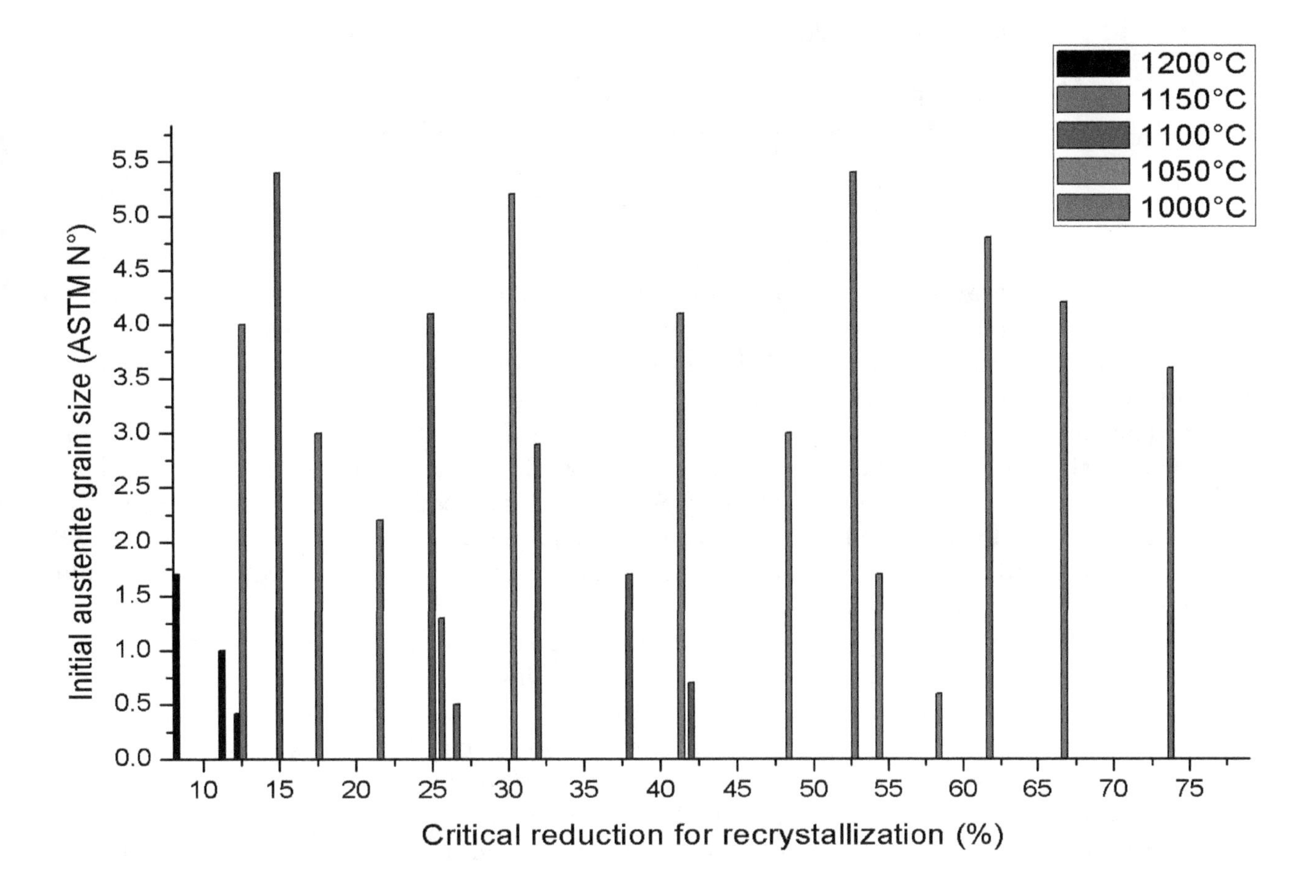

Figure 5. Effect of initial grain size on the critical rolling reduction needed for austenite recrystallization of a 0.03% columbium steel reheated to 1250 °C for 20 min. The initial grain size of the test plates was varied by rolling at higher temperatures [43].

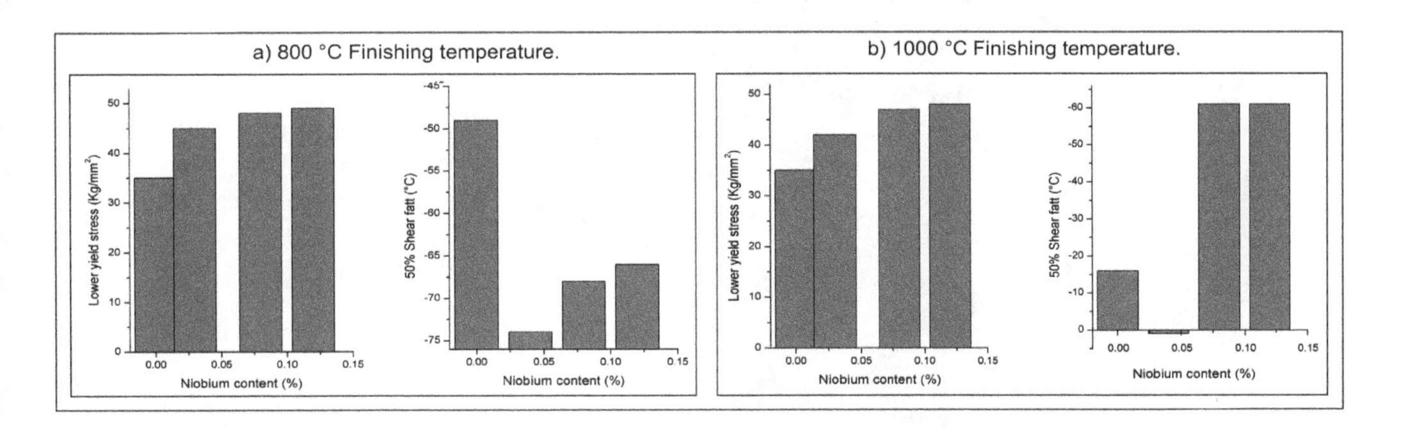

Figure 6. Effect of columbium on lower yield stress and transverse 50% shear fracture appearance transition temperature (FATT) for 20-mm thick steel with a 0.08% carbon, 0.25% silicon, and 1.50% manganese base composition. The reheating temperature was 1250 °C, (**a**) 800 °C finishing temperature, (**b**) 1000 °C finishing temperature [43].

Table 5. Average yield strength, tensile strength, and elongation [54].

Temperature	579 °C	621 °C
Yield strength	701–728 MPa	749–821 MPa
Tensile strength	996–997 MPa	821–876 MPa
Elongation	21–23%	19–25%

Some works have focused on developing equations to determine the optimal parameters in specific compositions, such as Maubane et al. [55], who deduced an equation to calculate the instantaneous flow stress using multiple linear regressions. They observed that the main variables were dislocation density, precipitation hardening in ferrite [56], temperature, solid solution strengthening [57], and grain refinement through the lowering of the Ar_3 [58].

Some researchers [59] have reported that there are several ways to achieve excellent mechanical properties by designing materials incorporated with nanoscale microstructures. Three ways to obtain these properties are the use of finely distributed nanoparticles (based on precipitate strengthening), increase both the strength and ductility of the steels by designing an ultrafinely ferritic lamellae by the cold-rolling process and by altering the morphology, content and distribution, as well as concentration of the retained carbon in the austenite.

Grain size refinement in steels (Figure 7) holds a great deal of promise since it provides improved strength as well as toughness. Currently, ferrite grain refinement is achieved by a combination of controlled rolling and accelerated cooling. It has been found that this process can achieve a refinement of about 50% in steel plates of simple compositions. In the plate rolling process of a plain carbon-manganese (C-Mn) steel, grain size can be refined from 10 μm up to 5 μm when the plate is controlled rolled and accelerated cooled. This refinement in grain size increases the yield strength of the steel by about 80 MPa according to the well-known Hall–Petch relationship [60,61].

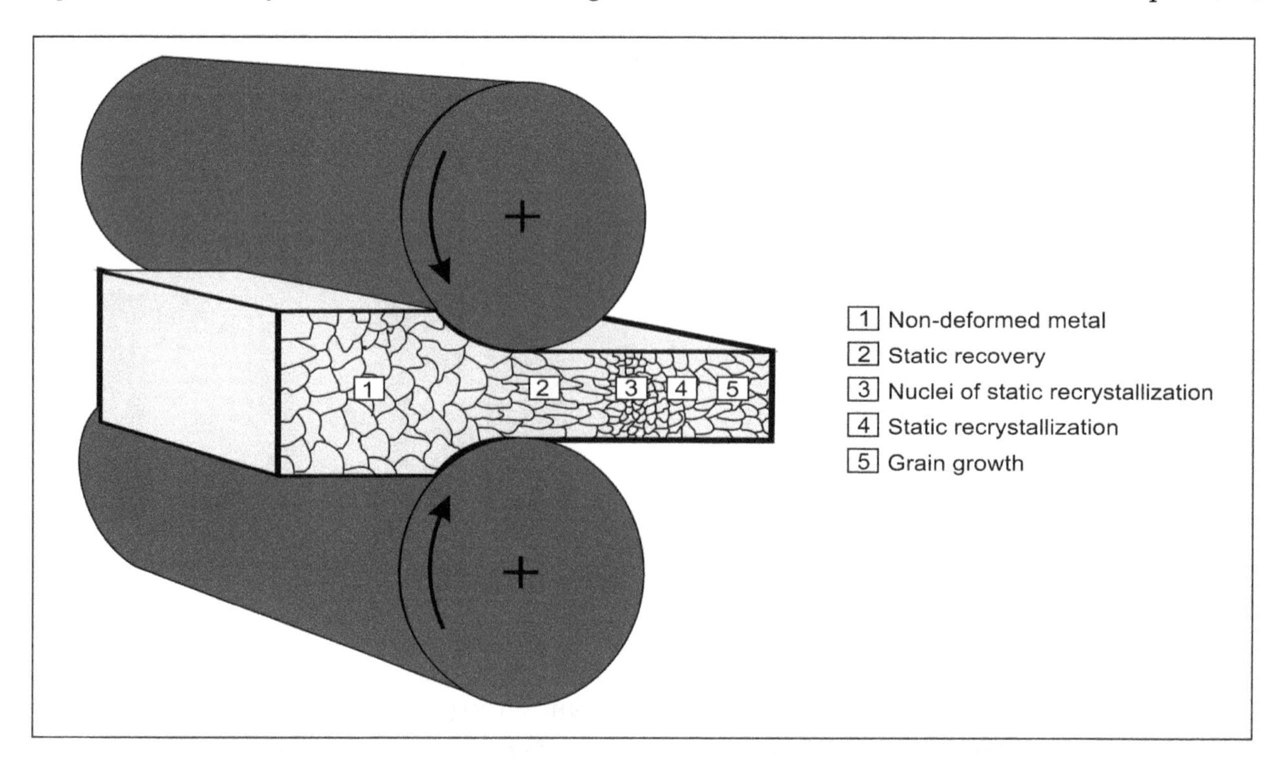

Figure 7. Schematic diagram of grain size refinement in steels.

A further fivefold reduction in grain size to about 1 μm is expected to increase the yield strength by another 260 MPa. Another fivefold decrease in grain size to submicron ranges would raise the strength by a further 586 MPa to levels >1000 MPa.

In view of the enormous advantages that are to be gained by the refinement of the grain size for simple steel compositions, researchers are active all over the world trying to produce ultrafine-grain ferritic steels. In Figure 8, a thermomechanical process is depicted with different strain levels and different cooling rates. It shows that deformation in the biphasic zone and air cooling produces a higher grain size in comparison with that the process with deformation above the Ar_3 transformation temperature and fast cooling.

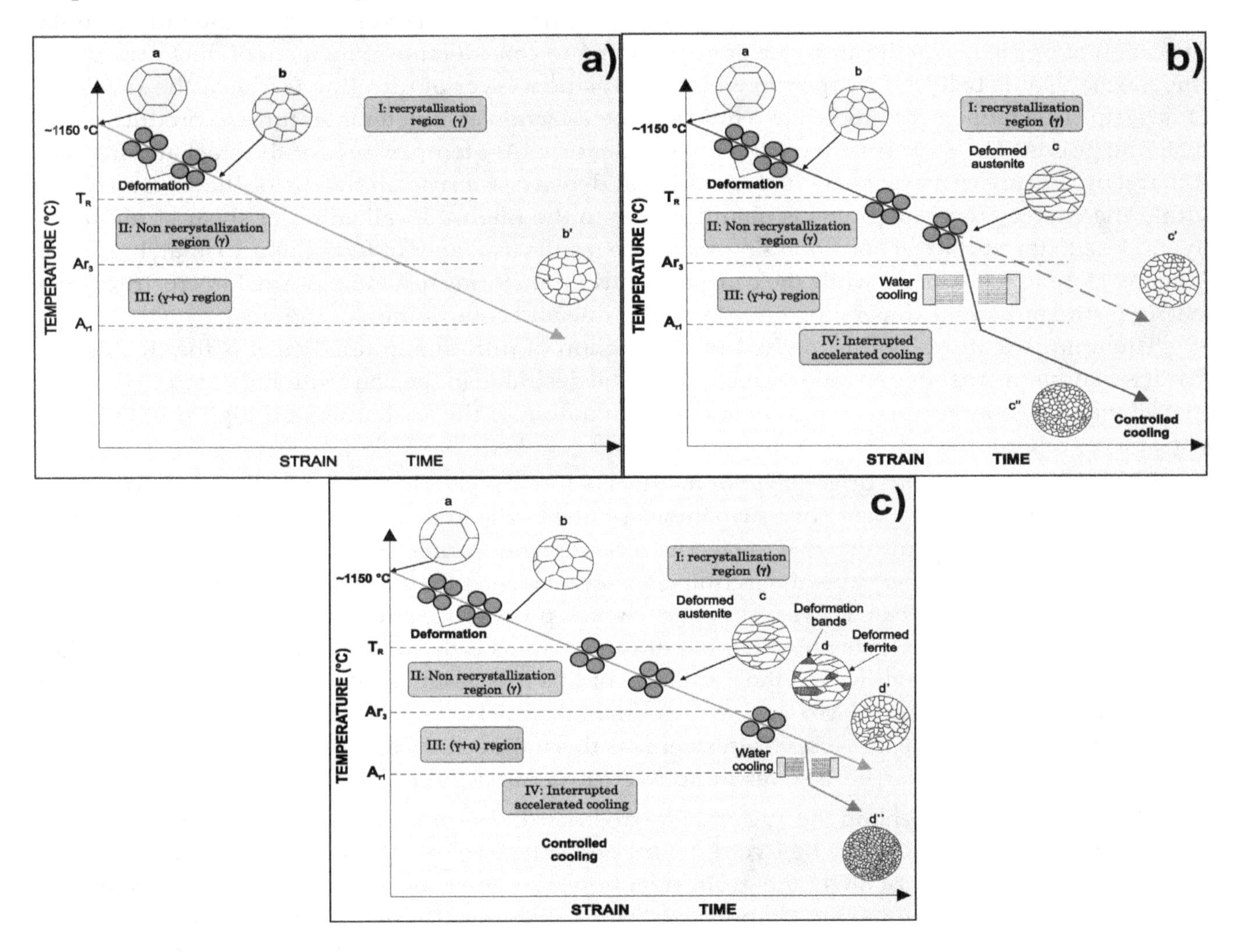

Figure 8. Grain size refinement as a function of rolling temperature and cooling rate. (**a**) Initial austenite grain size, (**b**) Austenite grain after rough rolling in recrystallization zone, (b′) air cooled, (**c**) Deformed austenite grain just above T_r, (c′) air cooled, (c″) water cooled, (**d**) Austenite grain after final rolling in non-recrystallization zone, (d′) water cooled.

Presented below is a brief review of the current status of knowledge of this research field and the developments that have taken place in recent years regarding attempts to produce ultrafine grained ferritic steel. The following mechanisms have been proposed for obtaining ferrite grains as fine as 1–2 μm. These mechanisms are intended to operate during the thermomechanical processing of plain carbon steels and low carbon microalloyed steels.

Austenite is dynamically recrystallized, and the majority of deformation is obtained just above the austenite to ferrite transformation (Ar_3) temperature. This leads to the formation of ferrite grains that are 1–2 μm in size.

High-strain-rate deformation is carried out just below the $\gamma \rightarrow \alpha$ transformation temperature. Due to the heat generated as a result of the high-strain-rate deformation, the ferrite transforms to austenite for a while before transforming back to ferrite. This process, when carried out in a cyclic manner, leads to the formation of very fine ferrite grains.

The severe straining of ferrite is used to initiate dynamic recovery. However, this approach cannot lead to ultrafine grain sizes, but ferrite grain sizes of ~3 μm can be achieved. This occurs because further grain refinement in ferrite is very difficult due to the low strain-hardening exponent and the higher stacking fault energy of ferrite.

The deformation of coarse-grained austenite beyond a critical strain leads to the intragranular nucleation of ferrite inside the austenite grains, leading to considerable refinement of the ferrite grains. This mechanism is believed to operate while producing a layer of ultrafine ferrite at the surface of a thin strip. This refinement procedure requires austenitization at high temperatures to produce coarse austenite grains. The strip is then cooled to very near the Ar_3 temperature of the steel and is rolled at that temperature. An extremely high nucleation density of ferrite on the dislocation substructure within the coarse austenite grains is obtained due to the intense localization of shear strain in the layers close to the surface. This processing sequence results in a layer of ferrite approximately 1 μm in diameter close to the surface, while the core of the strip transforms to a more normal microstructure of coarser ferrite or bainite, depending on the cooling rate and composition of the steel.

The unique features of this method of production of ultrafine-grained steel is that it employs characteristics that are not generally considered to be desirable in the conventional controlled rolling of steel, namely, a large austenite grain size, supercooling of the austenite, and high friction at the work-piece/roll interface.

Water quenching of the surface layer of the steel before the penultimate hot-rolling pass is another way to transform the surface layer of a strip into very fine ferrite grains. Heat from the core of the plate raises the surface temperature, so that the ferrite recrystallizes during the final pass, leading to a grain size of about 2 μm. This mechanism of refining the grain size at the surface is used when producing HiArrest plates (Nippon Steel, Kawasaki, Japan), which possess superior crack-arresting properties.

An altogether different route of obtaining ultrafine grains is the cold rolling and warm annealing of a martensitic structure. However, the limitation of this processing route is the achievement of the martensitic structure in plain C-Mn steel, which limits the thickness of the sheet.

In recent studies, a 0.013% Nb, C-Mn steel was thermomechanically processed in an attempt to obtain very fine grains of ferrite. The resulting microstructure was characterized.

The effect of Nb content on the austenite microstructural evolution during hot deformation has been studied in the past [62], having two different percentages of Nb (0.04 and 0.11%); it is observed that this difference does not affect the grain size; however, their mechanical properties are greatly affected. A difference in the cooling temperature leads to a large difference in the obtained mechanical strength. They observed that, for the alloy with low Nb content at 700 °C, the lowest strength values were obtained, while at a temperature of 600 °C they observed an increase of 90 MPa in the YS and finally for a temperature of 500 °C the YS decreased 20 MPa. For the alloy with high Nb content, the tendency was similar, although the increase in strength was greater, raising 140 MPa as the temperature descended from 700 to 600 °C. This corroborates the great impact that the microalloying elements provide for improving the strength of HSLA steels.

Due to the needs of the industry to manufacture stronger and thinner steels, a new technology called ultra-fast cooling (UFC) has been developed in the last few years to obtain steels with better mechanical properties. In 2004, CRM Group implemented a UFC technology to improve control and efficiency in the manufacturing of high strength steels [63]. By increasing the cooling rate, they were able to produce cheaper high strength steels (up to 850 MPa); this was done with cooling rates of around 300 °C/s. Likewise, Mohapatra et al. [64], applied UFC technology to a hot stationary AISI-304 steel plat by using air atomized spray at different air and water flow rates. They observed that the water flow rate or spray impingement density are found to be an important parameter during high temperature air atomized spray cooling. Other authors [65] have focused their research to find an analytical method, in function of the droplet diameter from a fundamental heat transfer perspective based on the premise that a spray can be considered as a multi-droplet array of liquid at low spray flux density, for a spray evaporative cooling as an alternative to conventional laminar jet impingement cooling.

Techniques such as physical simulation have been broadly employed to optimize parameters such as casting, reheating, rolling, and accelerated cooling conditions. Physical simulation has proven to be a powerful tool in the design and manufacture of new linepipe steels by reducing production costs and improving performance [66].

4. Development of Steels throughout History

Through history, great developments have been made concerning the manufacture processes of microalloyed steels, which have brought great benefits to the development of society, owing to their high strength provided by the alloying elements in their composition. Figures 9 and 10 show a brief summary of the most relevant events spanning from the year 1865, with the first patent of a steel alloy with Chromium (Cr), through the formation of the group for the exchange of information called steel treaters club, which later became the American Society for Metals (ASM), until the decade of the 1970s.

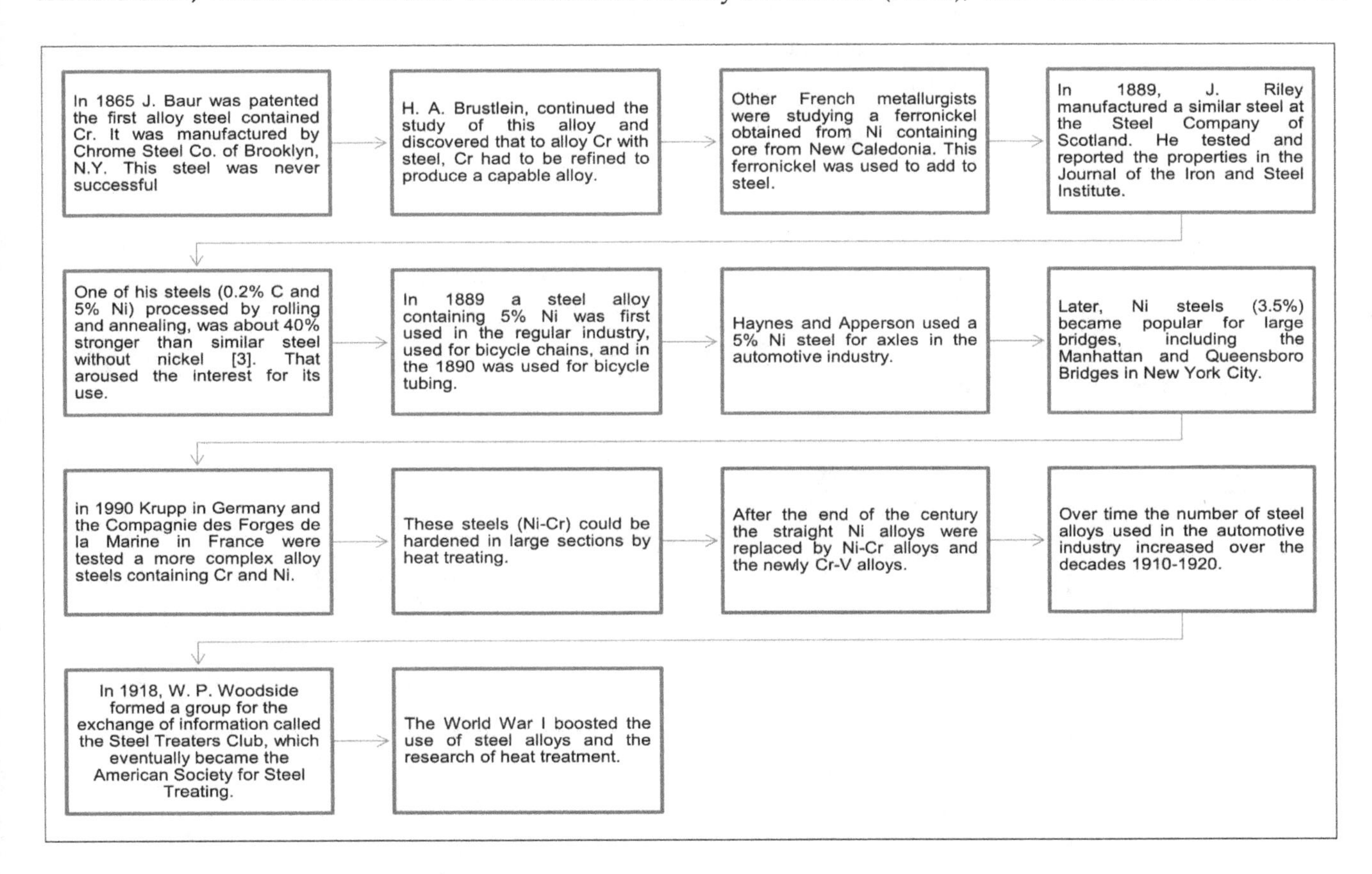

Figure 9. Brief history from 1889 to 1918.

Research in the field of microalloyed steels has increased over the decades, focusing on improving their strength and environmental resistance by precipitates and microstructure controlling. Table 6 presents an analysis of some relevant topics reported in the reviewed literature, ranging from the decade of the 1940s to the present, making a strong emphasis on the development of microalloying elements, control of microstructure, thermal and thermomechanical treatments, as well as the influence they have on steel properties.

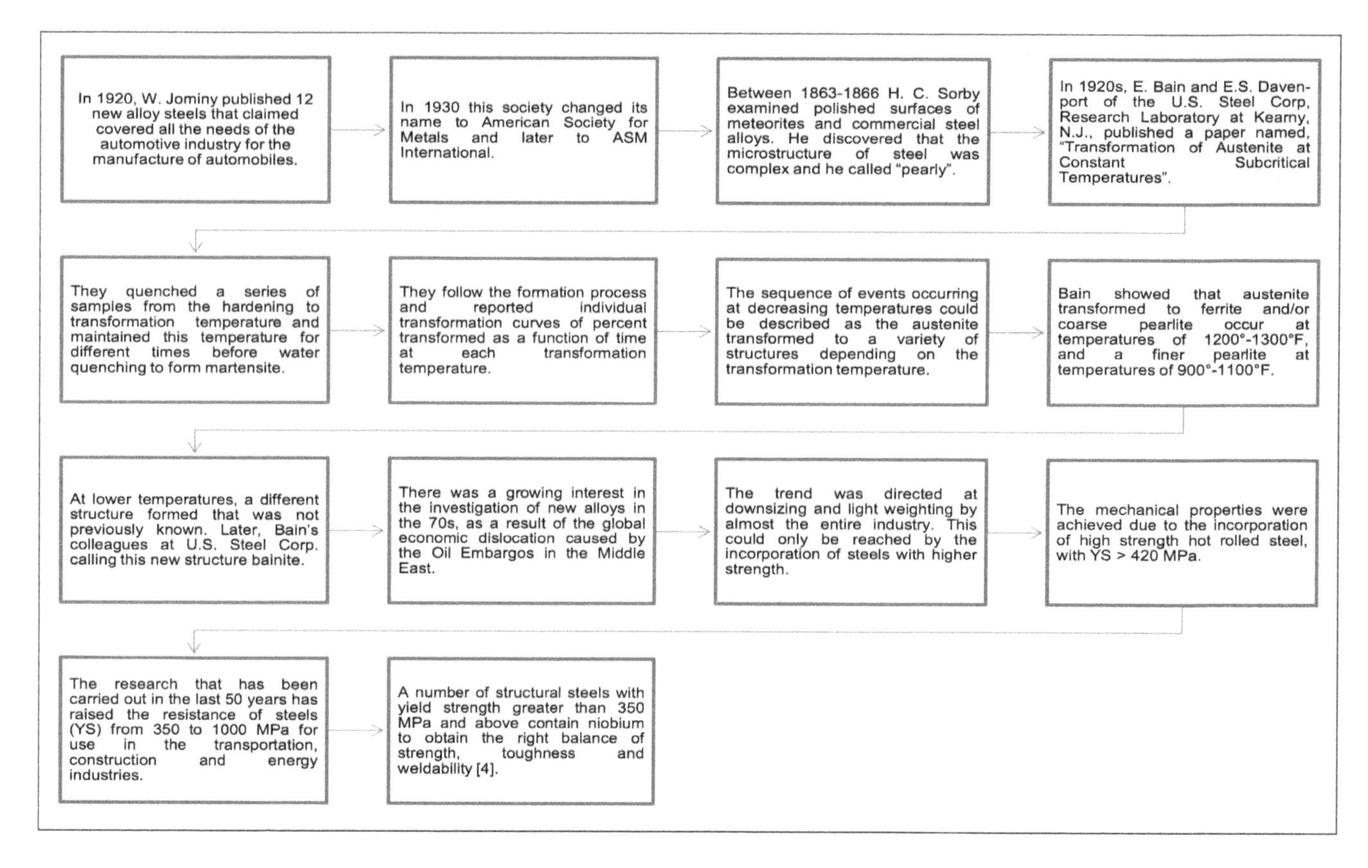

Figure 10. Brief history from 1920 to the 1970s.

Figure 11 shows the development of HSLA steels from 1940 to 2000, the data of which were published and included in the proceedings of international conferences [67–71].

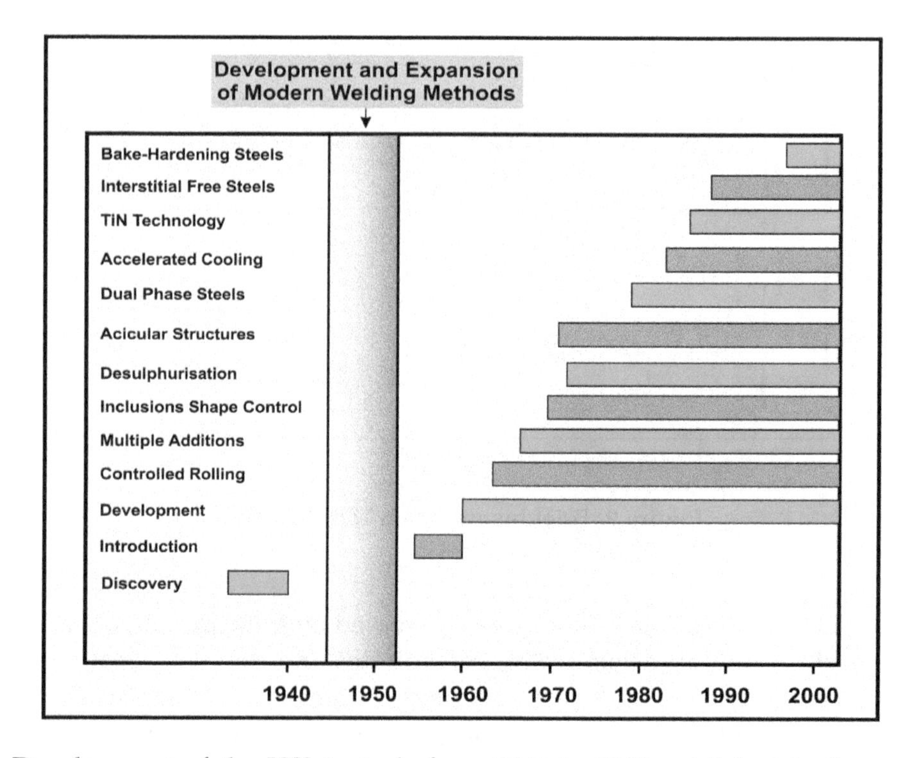

Figure 11. Development of the HSLA steels from 1940 to 2000 published in the proceedings of international conferences [67–71].

Table 6. Relevant topics reported in literature from the 1940s to the present [1–167].

Decade	Relevant Topics
1940s	Patent No. 2,264,355, "Steel" by F.M. Becket and F. Russell.
1950s	Deformation and ageing of mild steel. Cleavage strength. Metallurgy of Microalloyed steel. Columbium(Nb)-treated steels. Effect of small Nb additions to steels.
1960s	Small Nb addition to C-Mg steels. Dislocations and plastic flow. Effects of controlled rolling. Strong Tough Structural Steel.
1970s	Theory of hydrogen embrittlement. New model for hydrogen embrittlement. Assessment of precipitation kinetics. Stages of the controlled-rolling. Control of inclusions. Controlling inclusions by injecting Ca. Materials for hydrogen pipelines. Analysis of hydrogen trapping.
1980s	Recrystallization of austenite during hot deformation. Hydrogen degradation. Effect of accelerated cooling. Niobium carbonitride precipitation. Environmentally assisted cracking.
1990s	Specifications requirements for modern linepipe. Strain induced precipitation of Nb in austenite. Hydrogen interactions with defect. Nitrogen in steels. TiN-MnS Addition for Improvement toughness. Sulfide Stress Cracking. Softening and flow stress behavior. Influence of titanium and carbon contents. Titanium technology.
2000s	Effects of coiling temperature on microstructure. Effects of sulfide-forming elements. Effect of chromium on the microstructure. Hydrogen induced blister cracking. Effects of thermo-mechanical control process. Influence of Ti on the hot ductility. Trap-governed hydrogen diffusivity. Comparison of acicular ferrite and ultrafine ferrite. Influence of Mo content. Ultra-Fast Cooling. Role of Nb, B and Mo hardenability. Ductile crack propagation in pipes. Effect of bainite/martensite mixed microstructure. Effect of tempering and carbide free bainite. Microstructural evolution. Influence of Mn content. Steels processed through CSP thin-slab technology. Effect of Mo on continuous cooling bainite transformation. Correlation of microstructure and Charpy impact properties. Dual phase versus TRIP strip steels. The effect of niobium in Castrip® steel. Spray evaporative cooling.

Table 6. *Cont.*

Decade	Relevant Topics
2010s	Ti-alloyed high strength microalloyed steel. Hot strip steels. The first direct observation of hydrogen trapping sites. Evolution during thermomechanical processing. Ultra-High Strength X120 pipeline steel. Niobium high carbon applications. Modern HSLA steels. Ultra-low Carbon steels. Ultra-Fast Cooling. Mechanical anisotropy in steels. Effect of dissolution and precipitation of Nb. Influence of nanoparticle reinforcements. Reversible hydrogen trapping. Strengthening by multiply nanoscale microstructures. Effects of TMCP schedule on precipitation.

5. Sulfur Content in Microalloyed Steels

Some applications of microalloyed steels, such as X-80 or ultra-high strength X-100/120, require very high Charpy V-notch or DWTT (Drop Weight Tear Test) energies for ductile fracture control and can only be achieved in clean low sulfur steels [72]. Currently, this technology has advanced enough to obtain very low levels of sulfur (less than 10 ppm), allowing the industry to produce large quantities of steel with high strength and excellent toughness.

Over the years, the sulfur content has been reduced in the alloys of microalloyed steels, which has allowed for continuously improving the toughness of steels (Figure 12).

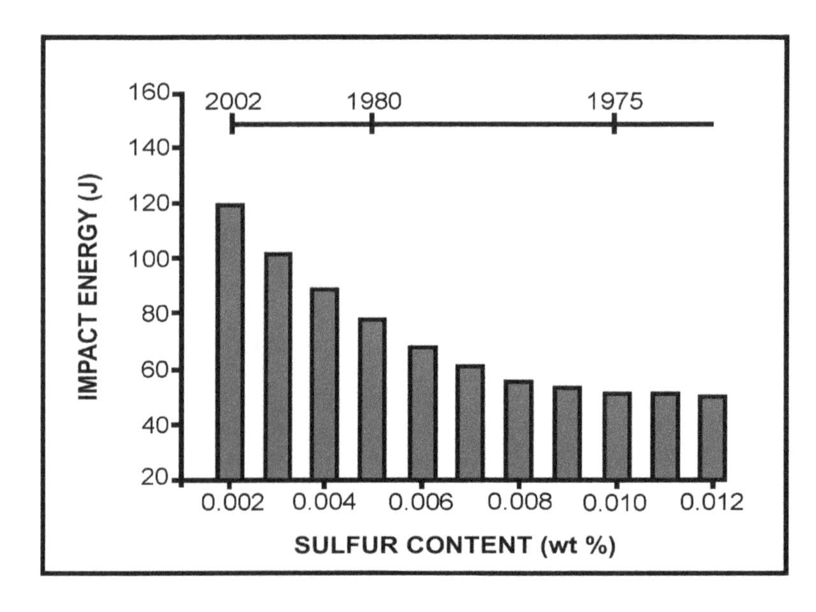

Figure 12. Effect of sulfur content on toughness of linepipe through the years.

Advances in desulfurization have been of great importance because they contribute to the control of the microstructure. The control by sulfides has been considered to improve the toughness of microalloyed steels [73,74], microstructure of heat-affected zone of welded metals [75], silicon steel sheets [76], etc.; this is mainly due to the fact that sulfur can retard the growth of the austenitic grain since it is relatively stable between the precipitates at high temperatures of rolling, besides promoting the formation of intergranular ferrite.

Some researchers have continued to investigate the effects of sulfur content on microalloyed steels. Tsunekage and Tsubakino [77] reported nine ferrite-pearlitic microalloyed steels, in which the sulfur contents were varied from 0.001 to 0.176 mass %, compensating for the increase in the formation of MnS inclusions with manganese in the amount equivalent to atomic % of S content. They determined that the Charpy impact values in the longitudinal direction of the studied steel increased with the addition of sulfur between 0.05–0.1 mass %; however, in the transverse direction, the Charpy impact values were reduced with content of 0.1 mass % of sulfur. On the other hand, the Charpy impact values in the transverse direction of the steel with 0.1 mass % sulfur content increased with the addition of Ca or Mg without affecting the Charpy impact values in the longitudinal direction.

Tomita et al. [75] also performed studies on the influence of MnS precipitates coupled with TiN precipitates on steel for offshore structures and thus clarify which are the main mechanisms responsible for the improvement of the heat affected zone toughness. This study determined that the heat affected zone contained intragranular ferrite formed by TiN-MgS and intergranular ferrite inhibited by ultra-low niobium. These conditions reduced the size of the fracture facet of the heat affected area and improved the steel toughness.

6. Mechanical Properties of Microalloyed Steels

The main purpose for designing microalloyed steels was to develop a good combination of mechanical properties (high strength and good toughness) in the as rolled condition, which allows for reducing the amount of required material, reduces the weight in specific applications, fabricates more effective mechanical designs, and reduces costs. These types of steels can achieve the same strength as cold worked C-steels and produce similar mechanical properties to quenched and tempered air cooled hot worked steels.

In the 1960s, the publications of Hall [60], Petch [61] and Cottrell [78] provided the first real understanding of the variables that control the mechanical properties of crystalline materials. The Hall–Petch equation allows for determining the yield strength as a function of the ferrite grain size d:

$$\sigma_y = \sigma_0 + k_y d^{-1/2}$$

where σ_0 and k_y are experimental constants. An example of the grain size effect on the strength of a plain carbon steel and C-Mn-Nb steel is shown in Figure 13.

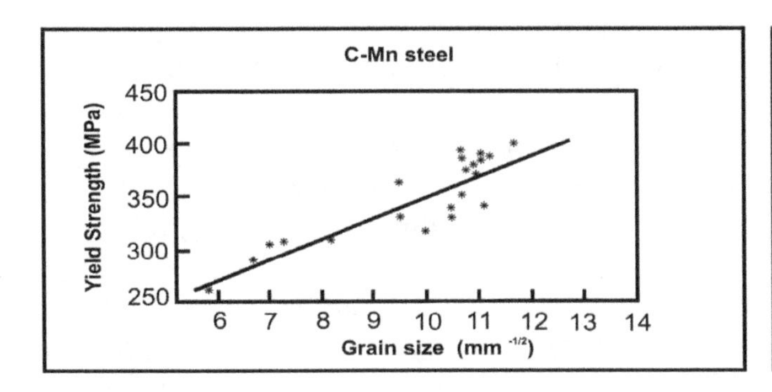

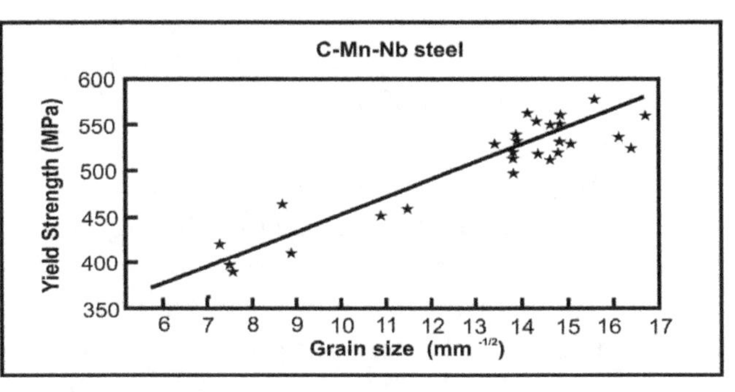

Figure 13. Experimental Hall–Petch relationship of grain size effect on strength of plain carbon and microalloyed steels [79].

The austenite grain size is critical to obtain a specific final microstructure, strength level and toughness of microalloyed steels after being thermomechanically processed and controlled cooling.

The earlier microalloyed steels avoided transformations of acicular microstructures, in the years 1960–1980, microstructures of microalloyed steels were ferrite-pearlite principally, obtained after hot rolling and air cooled [80]; this was triggered due to the low hardenability, reaching yield strength up to 450 MPa, and the principal characteristic was higher C content and relatively high

Mn levels [17]. The principal applications were for pipeline and automotive components, but the constant requirements of high mechanical properties demanded a higher strength microstructure such as non-polygonal ferrite, acicular ferrite, bainites and martensites achieved by higher hardenability and high cooling rates, using water cooling as principal media. The common processes for improving mechanical properties were interrupted accelerated cooling and interrupted direct quenching, controlling the cooling rate independently of microstructural control and composition. With the evolution of controlled rolling and high cooling rate, a low C steel containing 2 wt % of Mn could reach 690 MPa without heat treatments [81].

The first report on the use of microalloyed steels was a normalized API Grade X-52 vanadium steel around 1952 and extended later to X-56 and X-60 API Grade [82].

In the last two decades, high strength steels thermomechanically processed and accelerated cooled were developed with yield strengths in the range of 350–800 MPa and substituted the mild steels with lower yield strength (150–200 MPa) [83] for their use in the transportation, construction and energy industries. Figure 14 shows the evolution of the mechanical properties of pipeline steels.

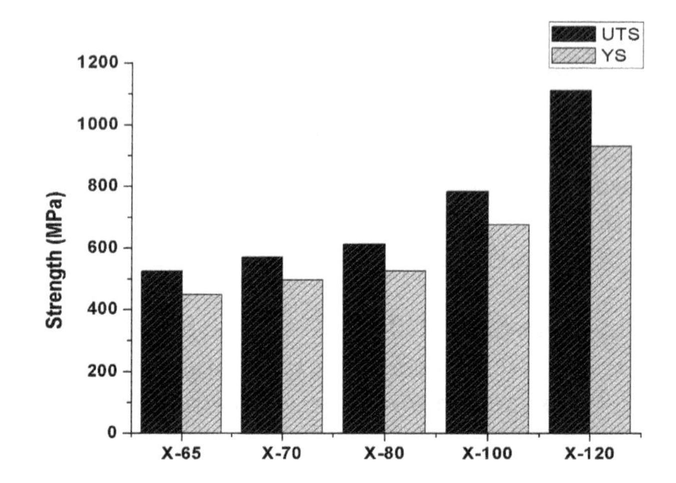

Figure 14. Mechanical properties of pipeline steels [19].

In order to achieve higher mechanical properties, the transformation temperatures should be lowered by addition of microalloying elements as well as fast cooling quenching to promote the formation of martensite–bainite. The addition of V had a small effect on transformations temperature due to the formation of nitrides in higher N content [84]. The evolution of common microstructures used for pipeline steels and its processing is shown in Figure 15 as a function of API grade steels.

Controlled rolling processes have been developed and used in the manufacture of microalloyed steels, with the basic objective of refining and/or deforming the austenite grains during rolling to obtain fine grains of ferrite during cooling.

The mechanical properties, as well as the fine-grained microstructure, can be modified by the strict control of TMCP (heating temperatures, percentage of reduction, time between rolling passes, start and end of rolling) and accelerated cooling (cooling start temperature, cooling rate, final cooling temperature) [85,86]. Depending on these conditions, different microstructural combinations can be obtained [50,87].

In order to optimize the control of TMCP, a clear understanding of the behavior of austenite recrystallization, precipitation kinetics of microalloying elements and the effect of cooling rate on the final microstructure of these steels is necessary [88]. Some studies have reported that deformation in the non-recrystallization region may increase the number of dislocations and induce precipitation during the rolling process [89–91]; this also increases the fraction of acicular ferrite volume after cooling. Figure 16 shows a processing comparison of steels obtained via conventional rolling, controlled rolling and controlled rolling-accelerated controlled cooling (CR-ACC). The main benefits of the CR-ACC process come from the band deformation that occurs in the austenite deformed grain thats act as a nucleus for ferrite grains during accelerated cooling and consequently facilitates the formation of finer ferrite grains.

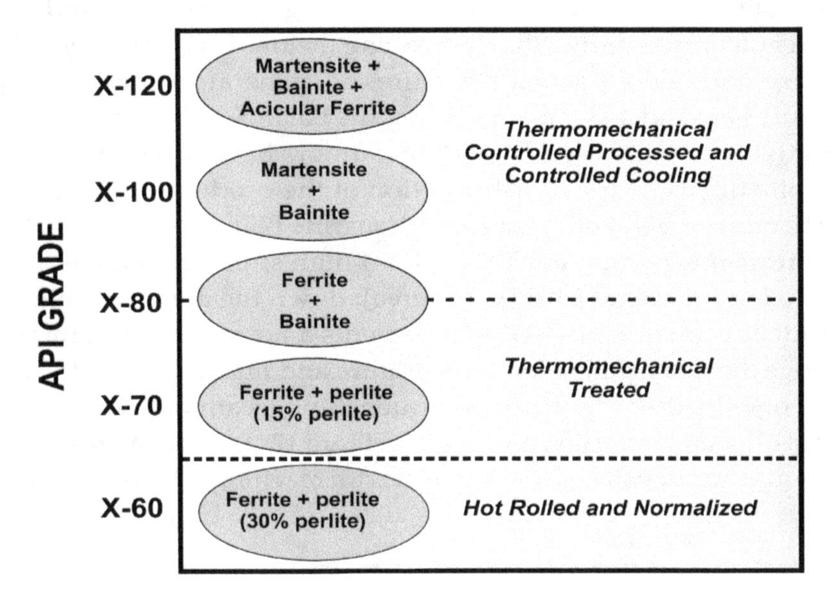

Figure 15. API grade steel evolution as a function of microstructure and thermomechanical processing.

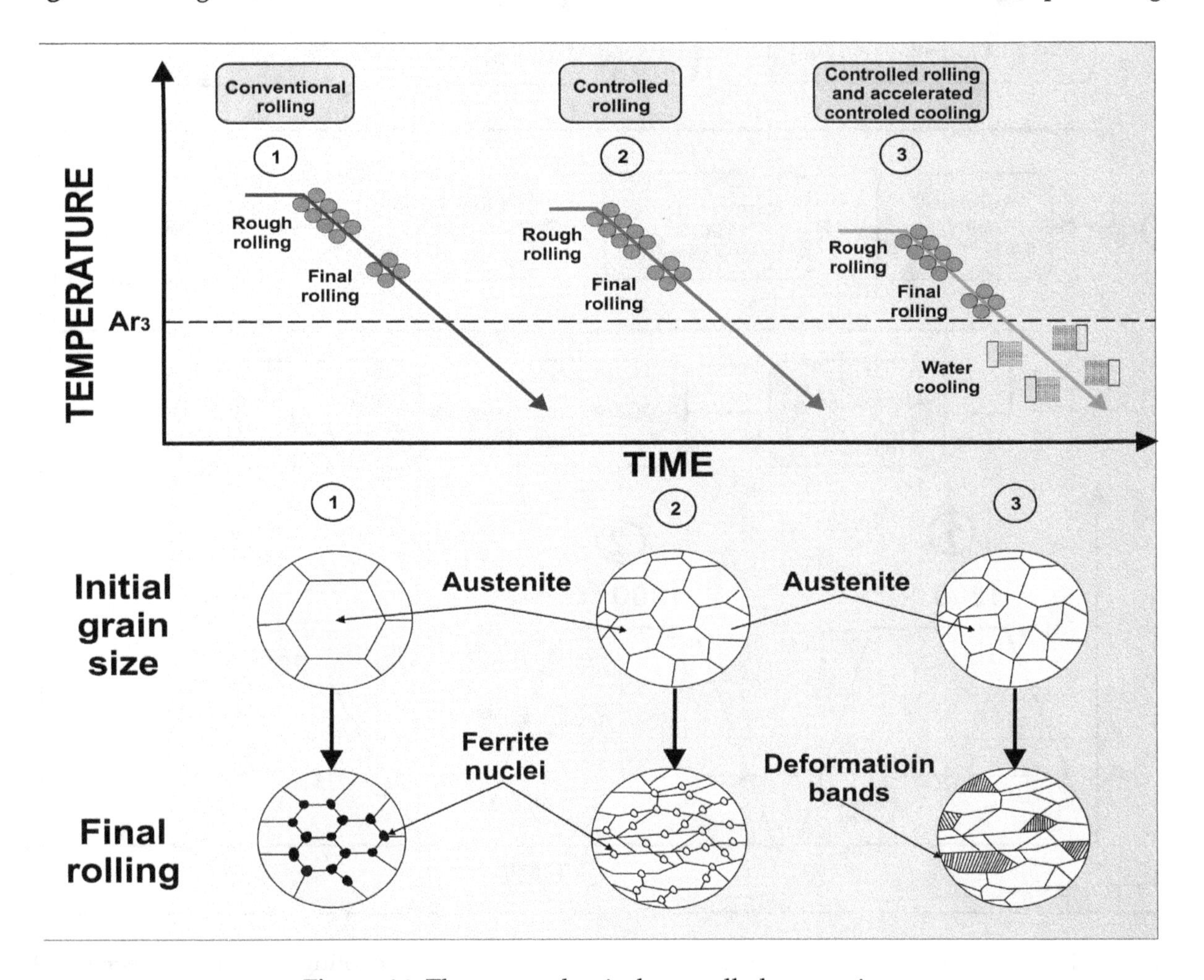

Figure 16. Thermomechanical controlled processing.

In the same way, the total degree of deformation has a great influence on the strength and toughness, Manuel Gomez et al. [92] studied the effect of hardening of the austenite in an API 5L-X80 microalloyed steel, during and at the end of the thermomechanical processing, as well as in the final microstructure after cooling. They observed that the ferrite grains are finer and more equiaxed when the austenite is severely deformed during the final rolling (below Tnr); however, when the deformation percent is lower, a large number of acicular structures are generated due to precipitation of (Ti, Nb) (C, N) with particles sizes around 4–6 nm, improving the balance between the mechanical properties, reaching yield strengths even up to 840 MPa. Many authors have observed that the high density of dislocations in austenite displaces the transformation of the products of bainitic microstructures to acicular ferrite [93,94], or even polygonal ferrite and pearlite [95].

Figure 17 illustrates an experimental TMCP for a high strength microalloyed steels. Figure 17 ①, shows an austenitizing process at 1200 °C for break down the dendritic microstructure obtained from melting slab and air cooled. Figure 17 ② represents a normalized treatment at 1200 °C for 1 h and rough rolling above the recrystallization temperature and final rolling in the non-recrystallization temperature and air cooled. After this process, a quenching treatment was carried out from 900 °C using as cooling media oil-water emulsion and water (Figure 17 ③). The microstructures obtained were formed mainly by bainite–martensite and acicular ferrite reaching a UTS (Ultimate Tensile Strength) around 1100 MPa.

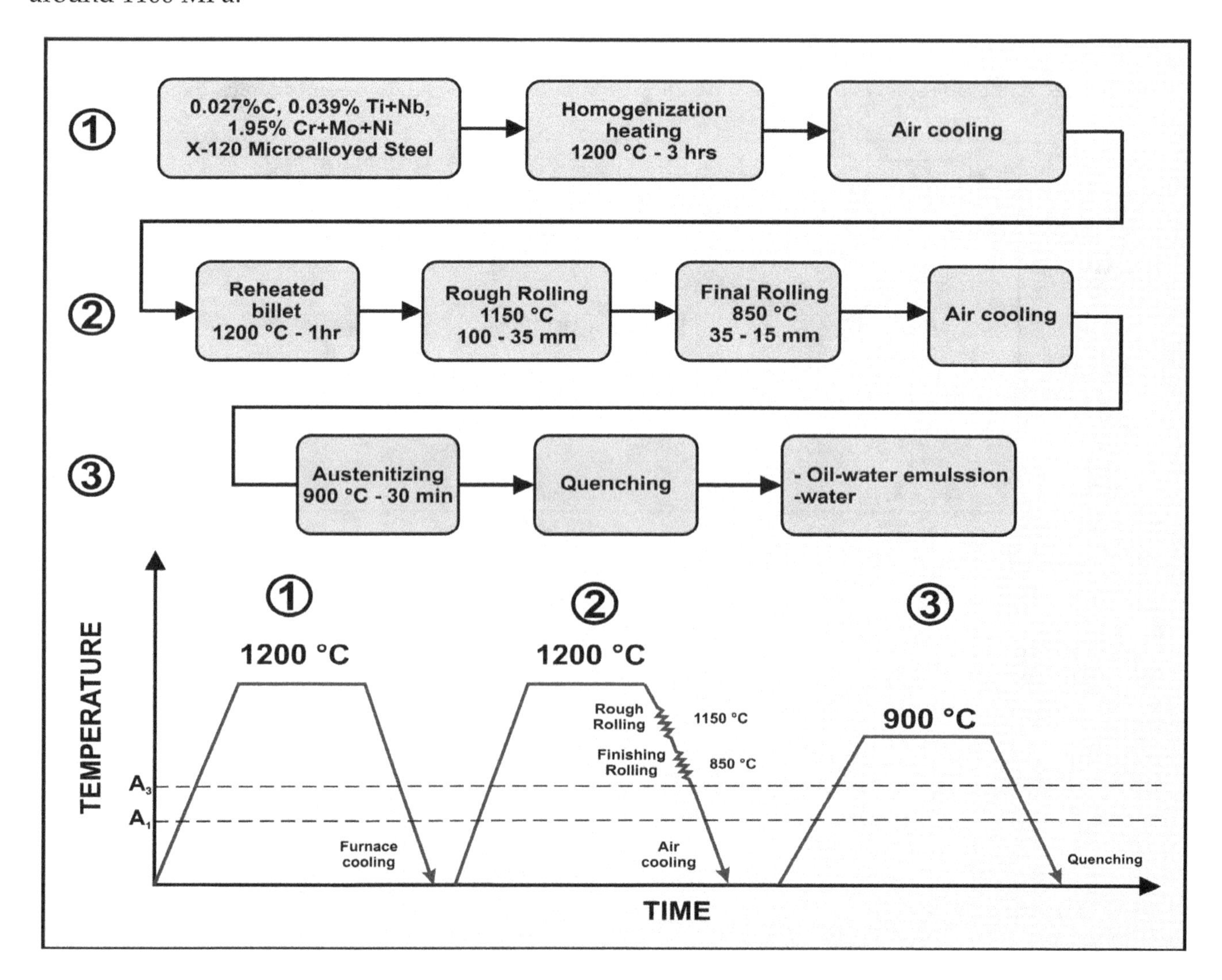

Figure 17. ① Schematic of steel processing, ② heat billing and air cooling, ③ thermomechanical processing and quenching processes.

In the following pictures, micrographs obtained by different API grade steel with different processing and cooling rates [96] of unpublished work are compared. Figure 18a shows an API X60 grade steel with a UTS around 425 MPa and a ferrite-pearlite microstructure; this steel was hot rolled and cooled in air. The use of thermomechanical processing allowed for achieving higher API grade steels; as it is shown in Figure 18b,c, those steels were thermomechanically controlled processed and cooled in oil and in an emulsion of water-oil, reaching a strength around 600–700 MPa and microstructures formed mainly by bainite–martensite. Figure 18d show the microstructure of API X120 steels. It consists of a complex mixture of bainite–martensite and acicular ferrite, allowing it to obtain a good combination of strength, ductility and toughness. This steel was thermomechanically controlled processed and cooled in water, obtaining strength values over 1100 MPa and an absorbed impact energy of 84 J.

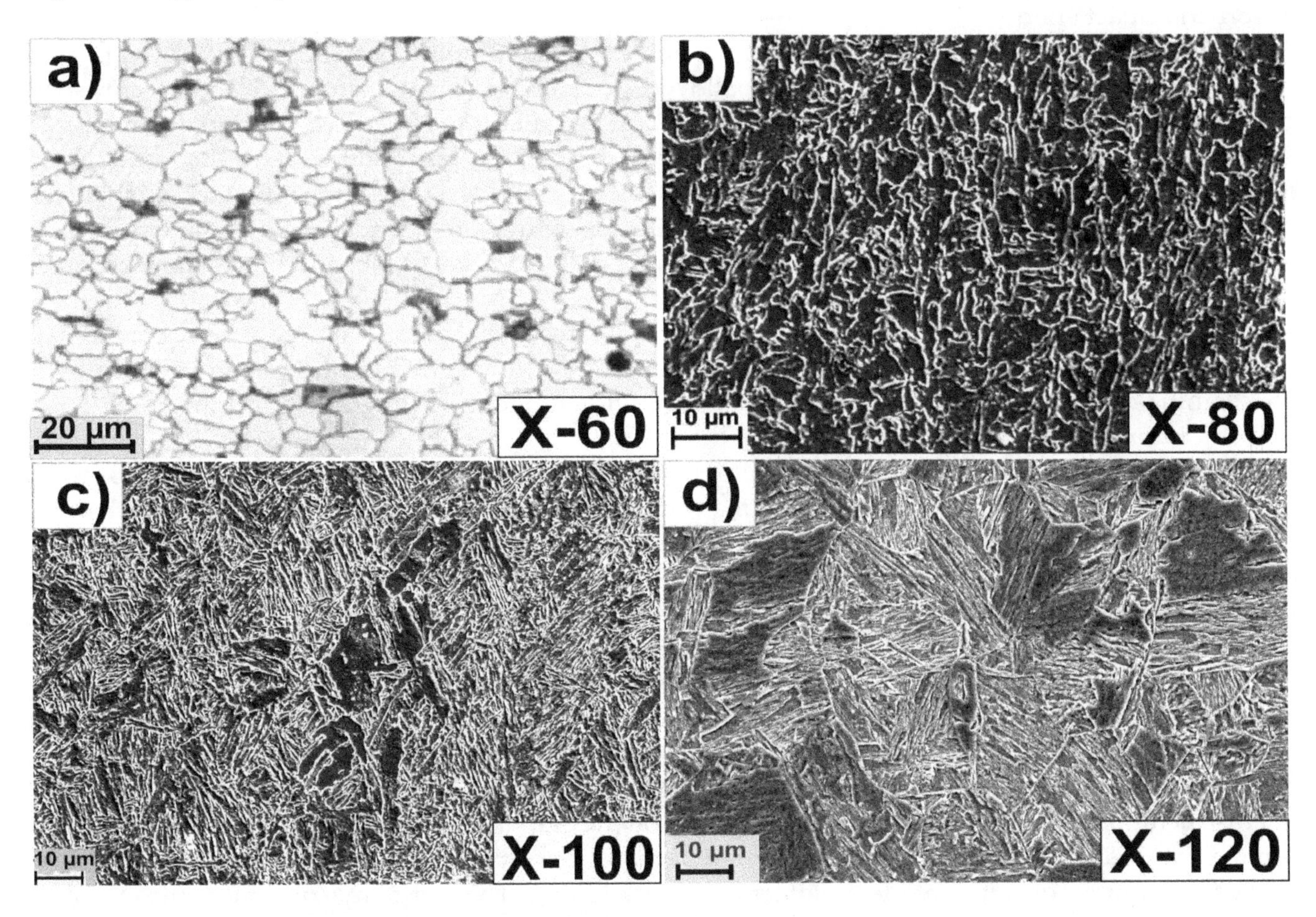

Figure 18. A typical microstructure of API microalloyed steels as a function of thermomechanical processing. (**a**) ferrite-perlite, (**b**) ferrite-bainite, (**c**) acicular ferrite-martensite, (**d**) acicular ferrite-bainite-martensite.

The addition of microalloying elements may intensify the effects generated by the TMCP due to the displacement of the transformation temperatures (A_1–A_3), and as a result the austenite transformation to acicular ferrite occurs at lower temperatures during the cooling. This process can generate a transformation of finer ferrite grain. A lower transformation temperature induces the formation of non-polygonal ferrite, acicular ferrite, bainite and martensite. However, to generate this type of microstructure, high hardenability and high cooling rates are required [17].

In addition to the ferritic grain refinement, the micro-alloying elements promote precipitation hardening and solid solution strengthening [97]. This precipitation hardening effect can be increased by the addition of Ti and Mo, due to the Nano-precipitation of (Ti, Mo) C according to Kim [89], which are thermally more stable compared to other types of carbonitrides, because they are more homogeneous and thinner. However, increasing Ti, Mo and C content conveys a decrease of impact toughness properties due to the improvement in strength of the steel.

It has been observed that the C content decreases toughness due to the high-volume fraction of martensitic-austenitic microconstituents [98,99]. In order to compensate the decrease in toughness, it is necessary to increase Ni and Cr content. Previous studies have shown that Cr promotes a higher toughness and could increase the amount of acicular ferrite. On the other hand, the addition of Mn and Ni can reduce the transition temperatures, therefore obtaining good toughness values at low temperatures [100]. According to other studies [101,102], the energy absorbed in the impact tests is also affected by the type, volume fraction, size and morphology of the individual phases that composed the material. Kang et al. [103] observed that materials with finer ferrite grain sizes provide greater absorbed impact energy.

The addition of Nb and Mo is used to achieve a favorable combination of strength and toughness. It is well known that Nb in microalloyed steels improves microstructure and precipitation hardening as well as grain refinement. Chen et al. [33] studied the effect of Nb on the formation of ferrite-bainite in a low-C HSLA steel and concluded that Nb (C, N) precipitates retard the formation of ferrite-bainite during accelerated cooling processes. Wang et al. [104] studied the influence of Nb on the microstructure and mechanical properties of low-C steels with bainitic and air-cooled microstructure and concluded that the amount of bainite is increased by the addition of Nb, and the agglomeration of precipitates of Nb are finer in this type of steel.

On the other hand, the addition of Mo reduces the initial temperature of bainitic transformation [105]. The addition of Mo in steels can separate the bainitic transformation lines and bainitic microstructures can be obtained over a wide range of cooling rates, and in combination with Nb improve their hardenability [106,107].

Ti is an effective and economical microalloying element, due to the fact that it has a great influence on the microstructural evolution during the manufacturing processes that change the final mechanical properties [103]. Ti as well as B is deliberately added to improve the hot ductility of microalloyed steels and Nb usually is added to prevent cracking on the surface during continuous casting. It also usually induces precipitation at high temperatures in austenite phase [108]. Along with the addition of Nb and Mo, microalloyed steels could reach yield strengths above 700 MPa due to the coprecipitation of TiC.

The accelerated cooling modifies the transformation microstructure [109,110], obtaining microstructures of acicular ferrite and dual phase, with martensitic–bainitic microconstituents dispersed in a polygonal ferrite matrix. Li [111] has studied the impact of TMCP on microalloyed steels, and, with the addition of Nb and Ti, they obtained a wide range of mechanical properties through microstructural changes, depending on the cooling rate. Compared to continuous cooling, interrupted cooling was the most suitable for the generation of dual microstructures (polygonal ferrite-martensite) because they are more susceptible to deformation hardening. The ferrite matrix contributes to good ductility, while the second phase (bainite, martensite and retained austenite) provides the high strength and high strain rate [111]. Other studies have focused on hardening mechanisms in Ti and low C microalloyed steels, using a combination of TMCP and ultra-fast cooling (UFC, 65 °C/s) [112]. This process improves the precipitation of carbides, nitrides with an average size smaller than 35 nm, and promotes the formation of martensite–austenite morphologies with 400 nm thickness, with yield strengths up to 650 MPa, and impact toughness of 93 J at −20 °C.

7. Microalloyed Steels Welding

Throughout history, multiple techniques for joining metals have been developed, all of them with different characteristics, advantages and defects. Among the most used today by the industrial sector are oxy-acetylene welding, electric arc welding, TIG (Tungsten Inert Gas) welding, MIG (Metal Inert Gas) welding, spot welding and the novel friction stir welding process (Figure 19), most of them used in different types of welding, such as circumferential welding joint (Figure 20).

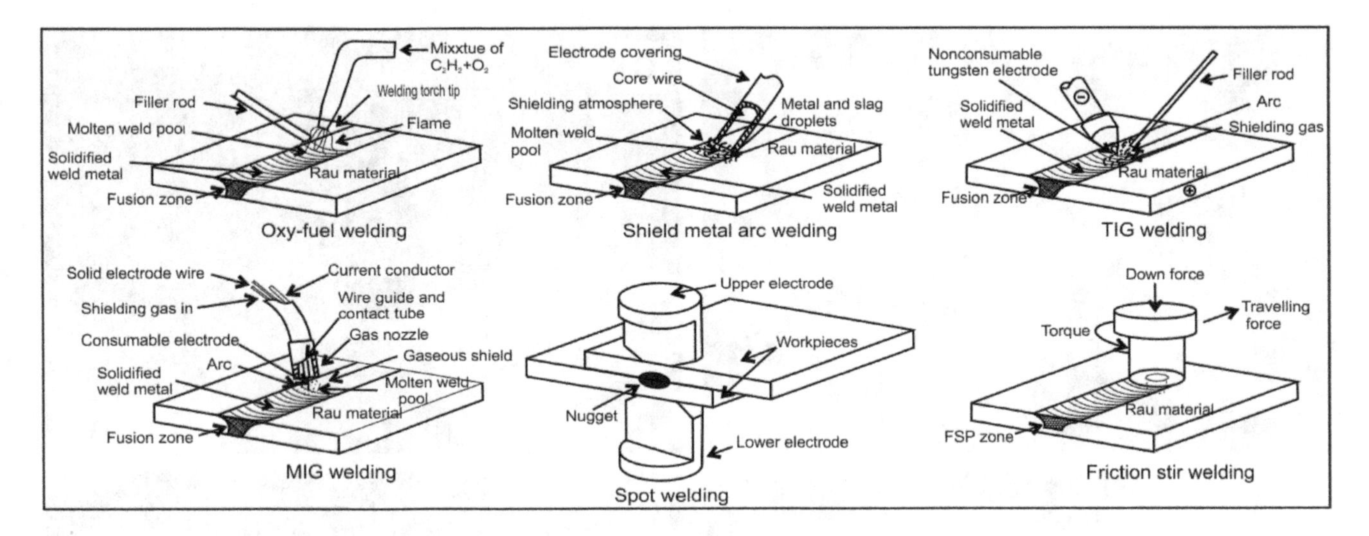

Figure 19. Most used types of welding processes.

All welding techniques modify to a greater or lesser extent the microstructure of the steels. The welding can be divided into zones: base metal (BM), heat affected zone (HAZ), partially melted region, interface, and weld metal (WM). In Figure 21, these zones are represented schematically. Figure 22 shows micrographs in optical microscope (Figure 22a) and scanning electron microscope (SEM) in the areas WM (Figure 22a,b), MB (Figure 22c) and HAZ (Figure 22d).

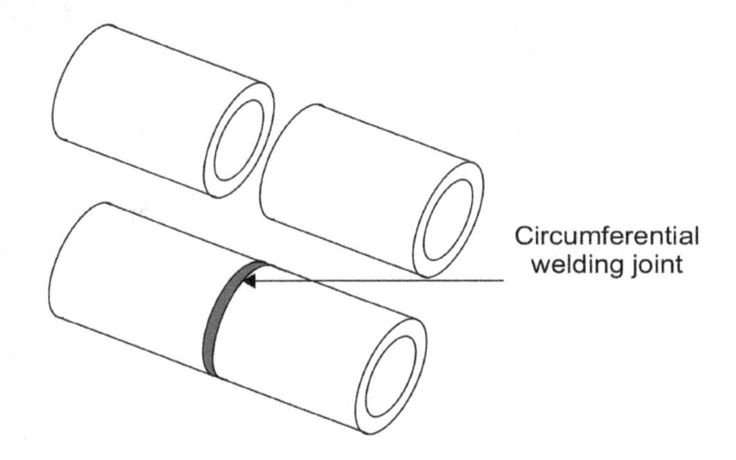

Figure 20. Circumferential welding joint.

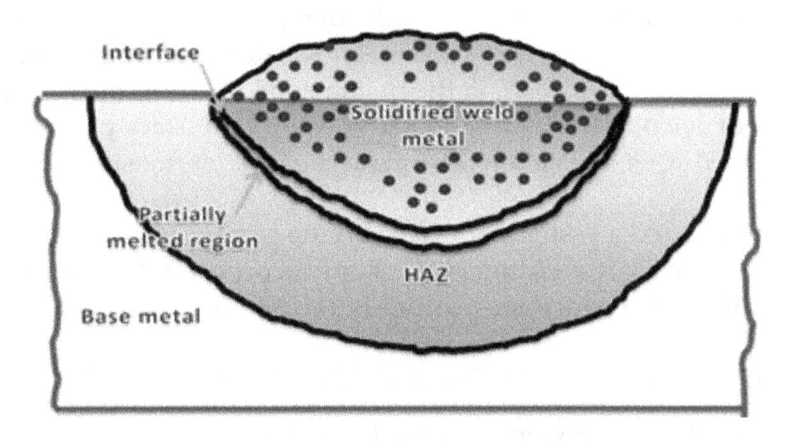

Figure 21. Schematic of welding regions.

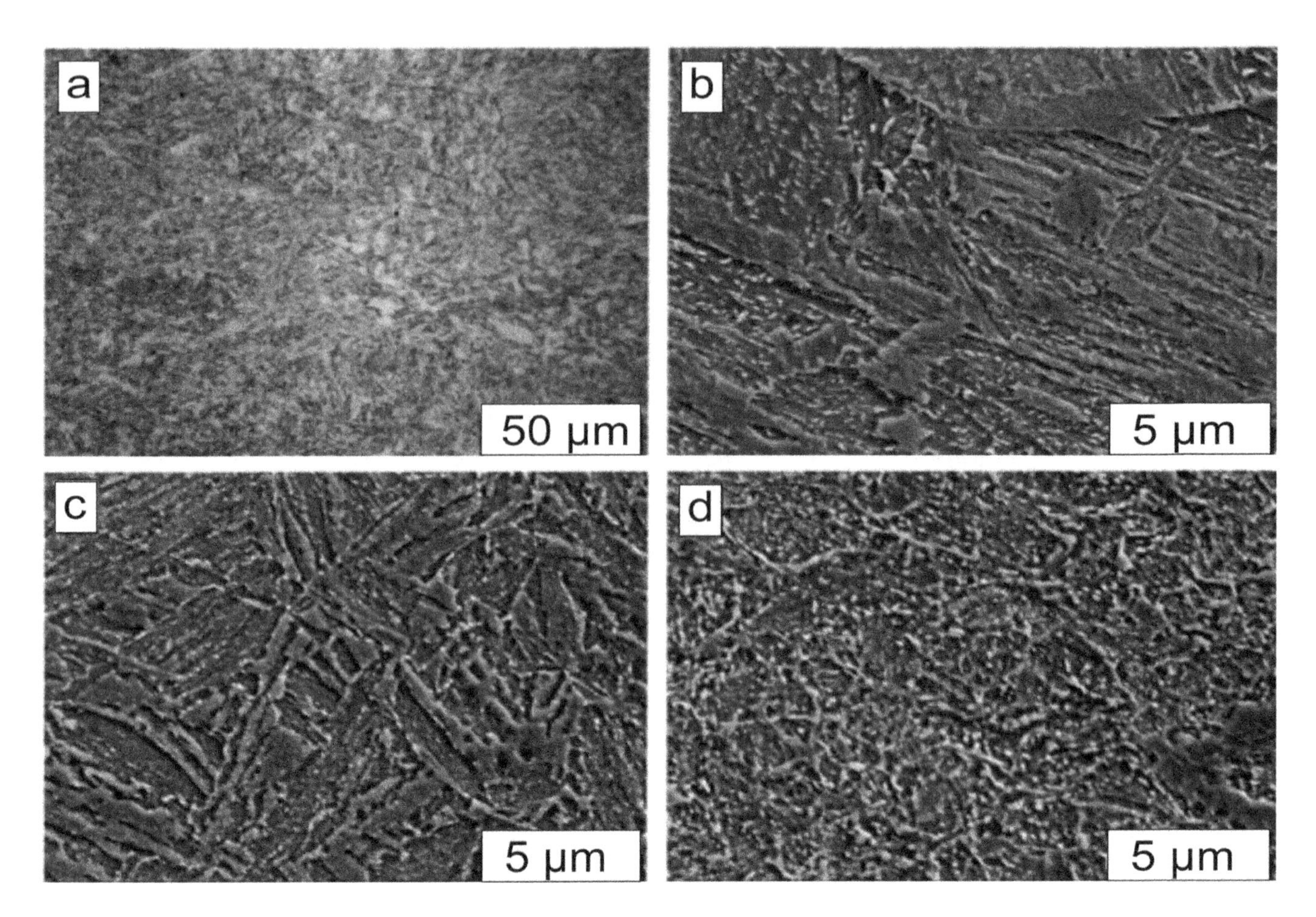

Figure 22. Optical (**a**) and SEM images of the WM (**a**,**b**), BM (**c**) and HAZ (**d**).

HSLA steels have been widely used for their excellent properties obtained through their microstructure, which is strongly altered by the selection of alloying elements and thermomechanical processes. [113]. In the year 2007, Bose-Filho et al. studied the effect of Ti, Ni, Mo and Cr on the microstructural development in the welding of an HSLA steel and reported that increasing the Ti content from 50 to 400 ppm does not contribute considerably to microstructural development; however, by increasing the content of Ni, Mo and Cr to increase the hardenability, they observed a change in the microstructural morphology from a mixture of allotriomorphic ferrite, Widmanstätten ferrite, acicular ferrite and microphases to a mixture of acicular ferrite, bainite, low carbon martensite and microphases [114].

Currently, the industry requires steels with high strength and toughness, and one of the ways to achieve this is through microalloy elements such as Ti. Studies have been carried out on steels with different Ti contents, and it was found that, in the fusion zone, Ti forms nano-sized carbonitrides that facilitate the formation of acicular ferrite due to a low energy interface of inclusion/matrix and the depletion of C in austenite during nucleation. The main cause of martensite formation is due to local C enrichment [115].

Bailey and Jones conducted a systematic study on the compositional variables on solidification cracking for submerged electric arc welding [116]. They incorporated a formula for obtaining crack susceptibility in terms of units of crack susceptibility (UCS), in function of the composition of the welded metal:

$$UCS = 230\ C + 190\ S + 75\ P + 45\ Nb - 12.3\ Si - 5.4\ Mn - 1$$

This formula is valid in the following composition ranges:

0.08–0.23% C	0.010–0.050% S	0.010–0.045% P	0.15–0.65% Si	0.45–1.6 Mn	0–0.07% Nb

They concluded that there is no risk of cracking in compositions above the following compositions:

1% Ni	0.5% Cr	0.4% Mo	0.07% V	0.3% Cu	0.02% Ti	0.03% Al	0.002% B

In another study, it is indicated that adding V in pipes welded by submerged arc with 0.16% C and 1.4% Mn reduces the risk of solidification cracking [117].

The microstructure developed during the solidification of the welding pool highly relates to the obtained mechanical properties of the metals that have been fused using the welding technique. The union of HSLA steels tends to occur due to fracture toughness defects associated with the HAZ, and some of the problems in the HAZ have been reported, such as grain growth, formation of fragile phases close to the melting line, non-metallic inclusions and excessive softening [118,119]. Many studies have been carried out in order to minimize these defects. One of the techniques that has been studied is the application of an electromagnetic field (electromagnetic stirring), which prevents the deflection of the electrode and the electric arc during the welding process. By means of electromagnetic agitation, the formation of acicular ferrite during solidification is propitiated, which reduces epitaxial and columnar growth [120].

The grain growth in the HAZ is highly dependent on the stability of the nitrides during the thermal cycles of the weld. Loberg et al. have found that only TiN remains stable during the high temperatures present in the welding process. The best characteristics of fracture toughness and grain growth were obtained for optimal ratios of Ti/V/N with low Al levels [121].

For the determination of the kinetic parameters in the welding area, techniques such as synchrotron-based spatially resolved X-ray diffraction (SRXRD) can be used. Zhang, Elmer and DebRoy studied the welding of AISI 1005 steel during gas tungsten arc (GTA) welding using the SRXRD technique and found that over time there was a decrease in nucleation rate of austenite from a ferrite matrix [122].

Many studies have been carried out to determine the characteristics and formation conditions of acicular ferrite in welded steel, and some of the techniques used have been the computer-aided three-dimensional reconstruction technique and electron backscattered diffraction analysis. It has been observed that the acicular ferrite laths are formed by multiple nucleations in inclusions mainly formed by Ti, Al, Si, Mn and O and by sympathetic nucleation. The elongation of acicular ferrite takes place mainly in the early stages of grain formation, while the extensive hard impingement and mutual intersection are presented in the last stages of formation [123].

Recently, the techniques to improve fracture toughness in the HAZ have been studied extensively. The proposed methods usually focus on reducing the C content to avoid the formation of martensite–austenite in the HAZ and to add elements such as Ti, Zr and Ca to the alloy, in order to form fine oxide inclusions with high melting points that inhibit the thickening of austenitic grain [124–128]. The change in the microstructure in the HAZ due to thermal cycles derived from the welding process are the cause of a decrease in fracture toughness in localized areas, known as local brittle zones, coarse grained heat affected zone (CGHAZ) and intercritically reheated CGHAZ (ICCGHAZ) [129,130]. Lan et al. investigated the relationship between the fracture toughness and the microstructure of CGHAZ in low carbon bainitic steel. They reported a change from lath martensite to coarse bainite with the increase in welding cooling time, decrease in fracture toughness in cooling ranges between 10 and 50 s and an even a greater decrease when cooling times of 90 s or greater were used [131].

Another way of controlling the microstructure in the HAZ is through the selection of welding parameters such as current. It has been shown that, by changing the current intensity during welding, the hardness and tensile strength are changed [132]. The increase in current intensity causes an increase in the welding temperature, which increases the area of the HAZ, in addition to a change in the microstructure. In general, it is possible that, by increasing the current intensity, an increase of widmanstatten ferrite or martensite structures take place in the center of the weld due to higher cooling rates [133].

One of the methods that has been studied for the determination of the rate of transformation of phases on microalloyed steels has been the quantitative determination of the kinetics of the phase transformation during welding by combination of mapping using X-ray diffraction and numerical modeling of the transport phenomenon [134].

In addition to the control of the microstructure through the selection of welding parameters, mechanical properties can also be controlled by another important parameter, such as protective gas. The variation of the protective gas composition can improve or decrease the final mechanical properties of the weld. Ekici et al. studied the mechanical properties in the welding of microalloyed steel, where they modify the composition of the protective gas. They used three compositions: 100% argon, a mixture of 15% CO_2 and 85% argon, and a mixture of 25% CO_2 and 75% argon. The welding was carried out using two techniques: gas metal arc welding (GMAW) and electric arc welding. They used two types of filler cables: coded as ER 100 SG and SG3 (G4Si1). They concluded in their article that the highest values of strength, hardness and ductility were obtained with the combination of 15% of CO_2 and 85% of argon using the filling cable ER 100 SG by means of the GMAW welding technique. They determined that electric arc welding can cause faults such as gas hole and slag formation. In addition, the mechanical properties obtained by this welding technique are lower [135]. By controlling the cooling time, the mechanical properties can be controlled. Dunder et al. determined that, for weld cooling times of 800–500 °C for 8–10 s in a microalloyed steel with increased resistance TStE 420, a decrease in the hardness of the HAZ is observed. In addition, the reduction of the risk of cold cracks and the values of impact toughness are increased [136].

The selection of the filling material also plays a very important role, since different compositions will generate different properties. Sharma and Shahi conducted a study on quenched and tempered low alloy abrasion resistant steels, which were welded using two types of austenitic stainless steel fillers. They found that by using fillers containing Cr and Mo with Nb, Ti, Al, V, Cu and N as microalloying additions, it resulted in martensitic refinement in the weld metal, which showed a microhardness of more than 400 HV as well as the percentage of elongation being reduced and its UTS increased displaying the highest joint efficiency [137].

One of the most recent techniques of welding is friction stir welding (FSW), which has great relevance in the industry because it has the potential ability to join dissimilar materials. Recent studies have focused on improving the properties of this technique, seeking a better appearance and defect-free joints [138]. The appearance of the weld and the reduction of the defects can be controlled by the tool rotation rate. Liu et al. studied the parameters involved to improve the welding of aluminum alloy with copper and determined that the best appearance was obtained under the condition of tool rotation rate of 600 or 1000 rev min^{-1}, welding speed of 100 mm min^{-1}, tool offset of zero and Cu sheet on the AS [138]. In Figure 23, an example of a weld with defects and poor appearance (Figure 23a) and a defectless weld with good appearance can be seen (Figure 23b).

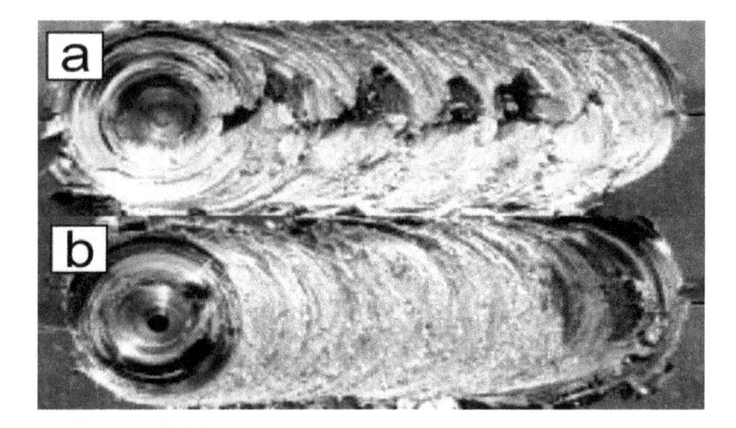

Figure 23. Comparison between a weld with defects (**a**) and one without defects (**b**).

Among the problems associated with welding is HAZ hydrogen cracking, which is highly relevant in oil-gas pipelines failures. One of the factors that has been found to have some adverse influence on the HAZ hydrogen cracking is the presence of V in the alloy, although, on the other hand, a more recent study in steel with 0.18% V showed that it is not necessary to include this element in the alloy [139].

Possible trapping sites in the weld metal are shown in Figure 24. The coexistence of the retained martensite and the carbides causes sites of hydrogen entrapment in the spaces between the acicular ferrites, since they have retained austenite and many interfaces causing the phenomenon of HAZ hydrogen cracking [140].

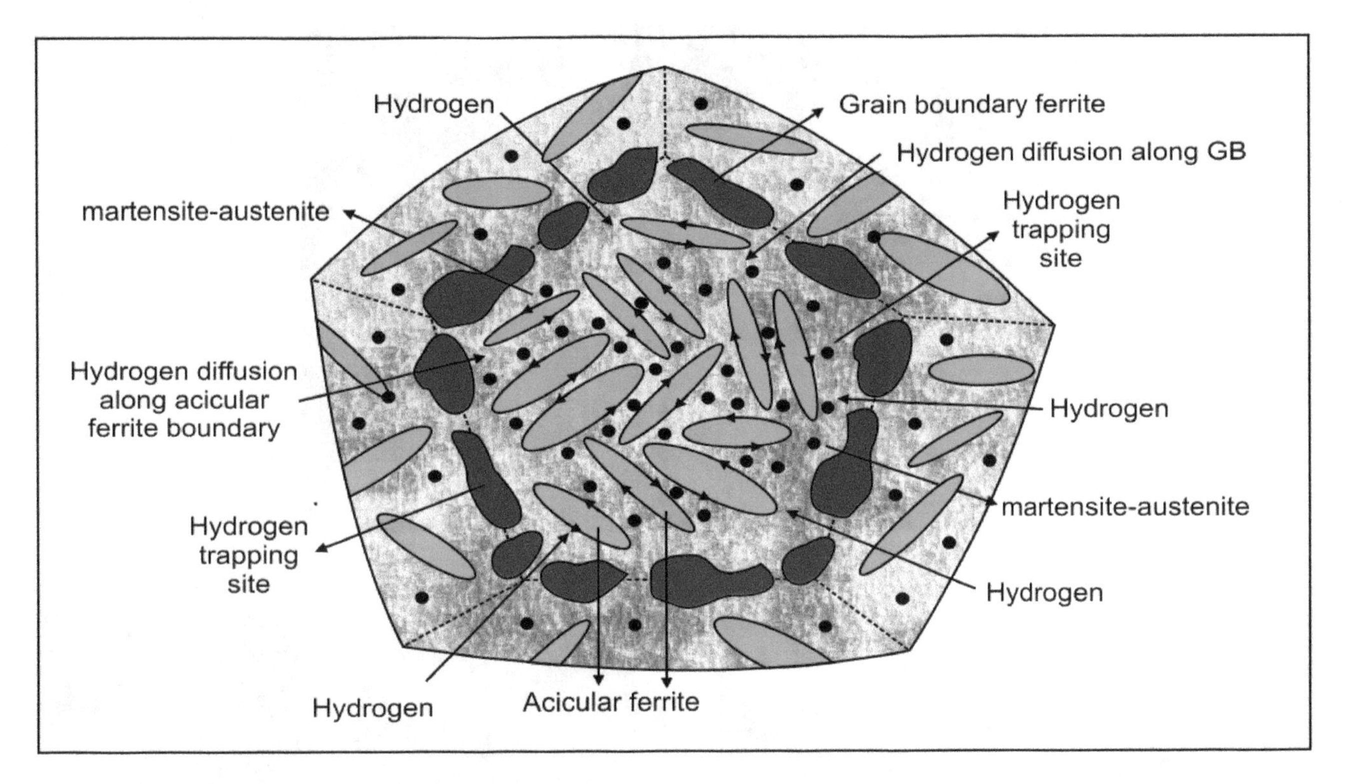

Figure 24. Schematic diagram of the hydrogen trapping sites in the weld metal [140].

In a study carried out by Hart and Harrison to determine the critical cooling times to avoid HAZ cracking, it was determined that these are related to the composition of the steel by the expression [139]:

$$\log \Delta t\ 800 - 500\ ^{\circ}\mathrm{C}\ (\mathrm{crit}) = 3.7\left(\mathrm{C} + \frac{\mathrm{Mn}}{13} + \frac{\mathrm{V}}{6} + \frac{\mathrm{Ni}}{40} + \frac{\mathrm{Mo}}{10}\right) - 0.31$$

Mechanical properties obtained in different welding zones are a function of microstructure. One of the most used techniques to know the mechanical properties in the welding zone is the determination of the microhardness by microindentation, which is under the "Standard Test Method for Microindentation Hardness of Materials" ASTM E384. Figure 25a shows the micrograph of a welded sample in which the microindentation was made. In Figure 25b the weld pool area is seen, in Figure 25c,d, the transition zones 1 and 2 are appreciated, and Figure 25e shows the heat affected zone.

Figure 26 shows the microhardness distribution as a function of distance from the welding center line. The tests were carried out at 2 mm distance from the top surface of the weld joints. The average microhardness of FZ (fusion zone), HAZ and BM was 356 HV, 325 HV and 298 HV, respectively (Table 7). These values correspond to the microhardness obtained in bainite and acicular ferrite morphology. FZ presented a higher microhardness due to its microstructure that consists of lath martensite. Wang et al [141] concluded that microalloying elements such as Cu, Ni and Cr could be dissolved in martensite causing an increment in microhardness [142]. The microstructure in HAZ zone consists of ferrite and martensitic microconstituents, and the content of lath martensite was decreased but granular bainite increased gradually causing a reduction in microhardness. The microstructure and grain size are the most significant parameters that affect microhardness. Generally, the microhardness follows the relationship: martensite > bainite > perlite > ferrite [143]. The microhardness of the lath martensite and granular bainite was more than that of the ferrite, which resulted in a higher microhardness value in HAZ than in the BM.

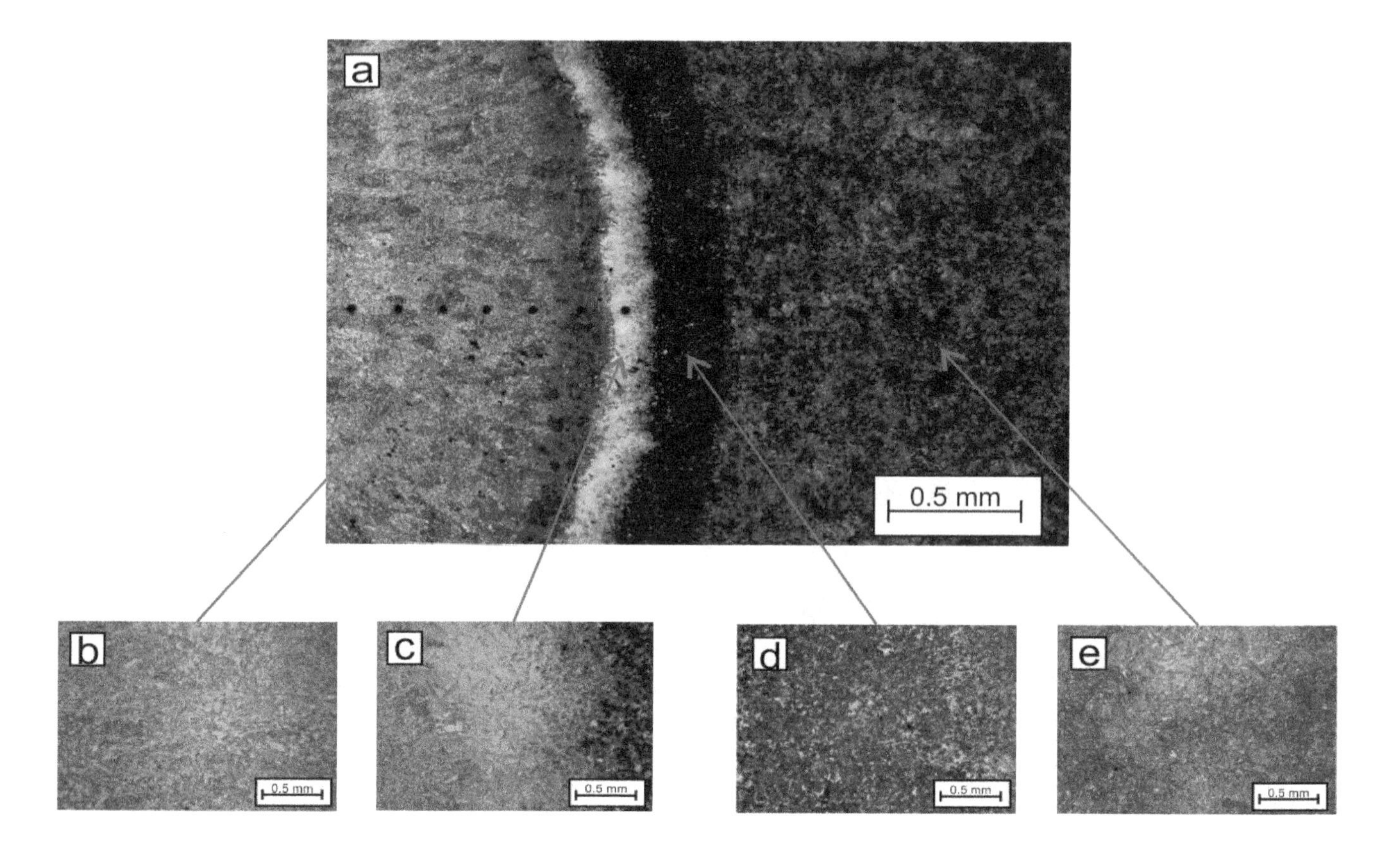

Figure 25. (**a**) Micrograph of a welded zone, (**b**) weld pool area, (**c**) and (**d**) transition zones 1 and 2, and (**e**) heat affected zone.

YS and UTS tend to decrease across the weld metal to the base metal, due to the variation in volume fraction of different ferritic morphologies and grain size. The temperature increment during welding, increases the diffusion rate causing that ferrite nucleates at austenite grain boundaries. As the temperature decreases in the weld pool, a ferrite transformation occurs and the grain boundary ferrite growth is suppressed; meanwhile, the acicular ferrite is produced in the austenite-acicular ferrite interface [144]; this causes acicular ferrite fine grains that reduce the strength and increase the toughness by inhibiting the interlocking mechanisms.

Bainite microstructure presents higher strength and lower ductility due to the high-density dislocations formed by accelerated cooling rate in weld metals that can produce acicular ferrite microphases [145]. The fine austenite grain size in HAZ contributes to the nucleation of bainite in the form of sheaves of small platelets [146]. The transformation of HAZ depends principally on two factors: a low transformation austenite-ferrite temperature during reheating that causes a complete austenite transformation, and high hardenability that allows a complete transformation into a mixture of acicular ferrite and bainite that have a good combination of mechanical properties [146].

Table 7. Average microhardness for different welding zones.

Welding Zone	Microhardness (HV)
Fusion zone	356
Heat affected zone	325
Base metal	298

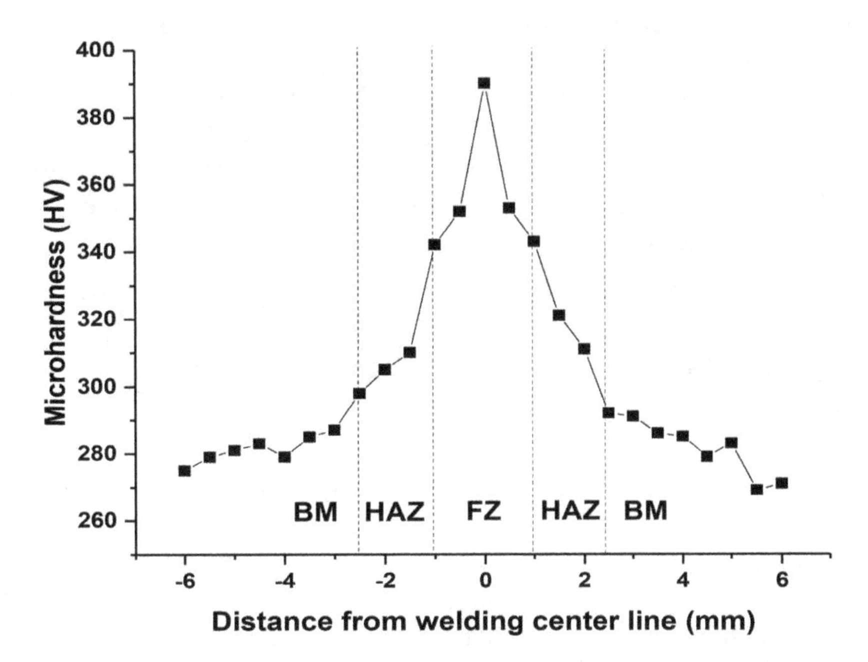

Figure 26. Microhardness profile as a function of distance from welding center line.

8. Hydrogen Embrittlement

Microalloy steels have been widely used in the hydrocarbon delivery industry, through the use of long-distance pipelines since their implementation in 1959 [147], due to their chemical composition and microstructures generated during their processing. These steels provide a good combination of mechanical properties such as high strength, toughness and good weldability, being some of the most economical options for the transport of natural gas. In the last two decades, microalloyed steels standardized by the American Petroleum Institute (API) have been used, the most common being the API 5L X-70, X-80 steels and recently the X-100 and X-120 steels [148]. With the growth of the petroleum industry, the oil-gas pipelines have been rapidly improved in terms of safety, economy and efficiency. With the increase of gas pressure, high strength pipeline steels are being widely used in various projects [80]. There are many applications in which the yielding stress of a material is a limiting factor in the design. If the strength of the material is increased, this would allow a structure to be lighter. Thus, the materials would offer a high resistance-weight ratio [87] and the pipeline wall thicknesses can be reduced, thus decreasing the production costs. Steels can be produced with yield stresses of up to 2000 MPa, but unfortunately as the yield stress is increased, other mechanical properties tend to decrease [149]. In general, materials become more susceptible to brittle fracture, especially when they are affected by ambient phenomena such as: hydrogen embrittlement, stress corrosion cracking (SCC), or corrosion-fatigue.

Hydrogen is a common element in the oil industry that can be generated as a corrosion by-product or by the application of welding with coated electrodes in the fitting of pipelines or valves. Variables such as: microstructure, strength, dislocation-recovery and carbide nano-coprecipitation can be very important for the hydrogen effects over the steel mechanical properties. These phenomena are associated with the amount of hydrogen absorbed by the steel, and its accumulation in the lattice and other defects; it enhances cracking, blistering and further failure.

The hydrogen embrittlement effects are characterized by a slight descent in the ductility up to a brittle fracture with a relatively low applied stress, ($<\sigma_y$) [150]. Even a few ppm of hydrogen dissolved in the steel can bring about cracking and loss of ductility, particularly in high strength steels. Few studies of the combined effects of microalloyed steels thermomechanically processed and hydrogen embrittlement susceptibility modify the mechanical properties—accomplished due to the hydrogen rate diffusion, trap site density and the exposed stress level in the presence of hydrogen, giving rise to local concentration of segregated hydrogen in tri-axial strain fields or on strain-induced defects.

Problems related to the formation of hydrogen-induced blisters tend to be more of a problem in low-strength steels used for pipelines intended for the transport of crude oil.

It is now established that high strength steels are susceptible to embrittlement by hydrogen dissolution and mainly to failure by SCC cracking attributed to hydrogen embrittlement [151]. There are two main classes of hydrogen effects:

- Quasi-brittle fracture in high strength materials that can occur with relatively low concentrations of hydrogen.
- Internal cracking and surface blistering in low strength materials (mainly C steels) due to very high internal hydrogen fugacity, allowing hydrogen pressure induced cracking, commonly referred to as HIC.

X80 pipeline steel, as high strength pipeline steel, is becoming one of the most widely applied pipe materials because of its high strength and toughness, which not only saves lots of steel, but also has a better performance [92]. However, hydrogen induced cracking (HIC) has been acknowledged as one of the predominant failures in pipeline steel in humid environments with H_2S and other sour materials, which causes the heavy leakage of oil as well as serious economic losses and casualties [152].

Up to now, a lot of work has been done to investigate the properties of X80 pipeline steel. Hardie et al. [153] compared the susceptibility to hydrogen embrittlement of three kinds of API pipeline steels (X-60, X-80, and X-100 grades), showing that the increase of strength level tends to decrease the resistance of steels to HIC. Shin et al. [154] studied the relationship between the microstructure and Charpy impact properties of X-70 and X-80 pipeline steels. Shterenlikht et al. [155] reported the capacity of the anti-ductile crack growth of X-80 pipeline steels. Kong et al. [105] reported the effect of Mo on the microstructure and mechanical properties of X-80 pipeline steel. However, the specifics for HIC of X-80 pipeline steel, and additionally the impact of inclusions on HIC and the mechanical properties of the steel are still unknown. In view of this, it is extremely necessary to work out the details of hydrogen-induced damage and how these failures deteriorate the properties of X-80 pipeline. Hydrogen induced cracking in X-80 pipeline steel under various electrochemical hydrogen-charging conditions as well as the influence of inclusions and microstructure on the origin of HIC were investigated in this work.

Reversible hydrogen embrittlement can occur after small concentrations have been absorbed from the environment. However, local concentrations of hydrogen may be larger than the average values because the absorbed hydrogen diffuses between the grains or preferably at the grain boundaries. The hydrogen embrittlement is affected by the rate of deformation, which suggests that diffusion is time dependent as a control factor. On the other hand, embrittlement is more severe at room temperature during sustained stress or in tests conducted at slow deformation rates. Another factor such as elevated temperatures can cause hydrogen to diffuse in areas of concentration. The effect of hydrogen is also strongly influenced by other variables, such as (Figure 27):

- The level of stress (or hardness) of the alloy;
- The microstructure;
- The amount of stress applied;
- The presence of localized tri-axial stress;
- The previous amount of cold work;

- The degree of stress segregation of low melting point elements such as: P, S, N, Ti or Sb at the grain boundaries.

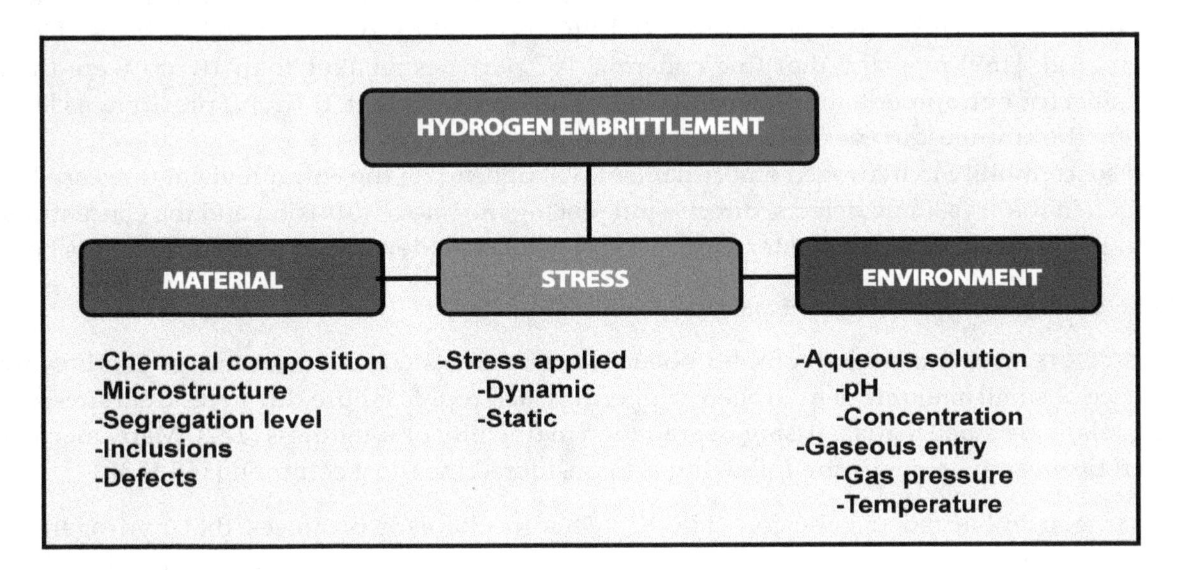

Figure 27. Principal variables involved in the hydrogen embrittlement phenomenon [156].

8.1. Hydrogen Trapping

The hydrogen absorbed in the steel can exist in different forms, mainly monoatomic:

- Interstitial hydrogen, dissolved in solid solution in the steel matrix.
- Hydrogen associated with structural defects, such as dislocations or second phase particles.
- Hydrogen accumulated in voids or blisters in gaseous form.

These different forms present different mobility and solubility in the crystalline structure of the steel, which affects the mechanical properties to different degrees. Hydrogen atoms are strongly attracted to defects in metals; these are referred to as hydrogen traps [157]. These traps include vacancies, dislocations, grain boundaries, second-phase particles and non-metallic particles [152].

Hydrogen traps are commonly characterized as irreversible or reversible [158] and are related to the number of hydrogen atoms that can be retained. The influence of each trapping site will depend on its density and its activation energy. Irreversible traps are those sites with high activation energy, and where the residence time of hydrogen is higher; hydrogen is usually regarded as non-diffusible. On the other hand, hydrogen trapping sites with low activation energy are considered as reversible traps, in which case the hydrogen will have higher diffusivity [159].

Trapping sites may be generated and/or modified during thermal and mechanical processes due to the kinetics of second phase particle precipitation, inclusions distribution, dislocation density, grain size change, dimple formation and microcracks [160,161]. According to Liu et al. [162], reversible traps may have a greater influence on the diffusion of hydrogen and the susceptibility to embrittlement, since the residence time of the hydrogen atoms is smaller, leading to high diffusivity. Hirth [163] suggested that a material with finely distributed irreversible traps is less susceptible to hydrogen embrittlement.

The segregation of C, form and distribution of precipitates (Nb, V, Ti) (C, N), as well as other nonmetallic impurities may increase hydrogen trapping. Some studies have concluded that carbide interfaces and the presence of incoherent precipitates, present higher trapping energy compared to grain boundaries and dislocations [164]. However, the latter play an important role in mechanisms of hydrogen embrittlement [165]. Wei et al. [166] studied the influence of TiC precipitates coherence and observed that coherent and semi-coherent precipitates of TiC showed a considerable difference in their trapping energy (55.8 kJ/mol), as well as in the activation energy to release hydrogen

(95 kJ/mol), concluding that the activation energy in this type of precipitate is smaller and therefore behaves as reversible traps. Escobar et al. [167] observed a similar behavior, and the activation energy of semiconducting TiC precipitates is lower than the incoherent particles. On the other hand, Wallaert et al. [168] has reported that (Nb) (C, N) tend to form irreversible traps. However, Takahashi et al. [169] reported that fine coherent TiC particles smaller than 10 nm were the most effective sites for entrapment and Valentini et al. [170] observed that Ti (C, N) precipitates less than 35 nm were the strongest irreversible traps in microalloyed steels.

This susceptibility to hydrogen embrittlement will depend on the entrapment and release kinetics of hydrogen in the crystalline defects, directly influencing the rate of diffusion and the concentration of hydrogen in the steel. McNabb–Foster models [171] have been developed to determine this behavior.

8.2. Hydrogen Embrittlement Mechanisms

Hydrogen embrittlement is a complex phenomenon manifested in various forms. This is caused by the presence of small amounts of hydrogen, triggering catastrophic failure due to residual stresses or by applying relatively small loads, causing degradation in ductility or toughness [172]. Many mechanisms have been proposed; however, the following are considered the most common [165,173]:

(1) Hydrogen enhanced decohesion (HEDE). This mechanism proposes that hydrogen causes a reduction in the bond strength of metallic atoms, allowing weakness under tensile loads, in addition to promoting a propagation of fragile cracks [174].
(2) Hydrogen enhanced local plasticity (HELP). This mechanism proposes that the presence of hydrogen increases the mobility of dislocations, causing a highly localized plastic deformation [175]. Because this deformation is concentrated in a small volume, the macroscopic ductility is low.
(3) Absorption induced dislocation emission (AIDE). This mechanism is very similar to the HELP mechanism because it also involves localized plasticity. However, the main difference is that the AIDE mechanism proposes that localized plasticity occurs close to the surface in regions of stress concentration, such as cracks [176]. The hydrogen causes the movement of dislocations towards the crack tips, causing the growth of the same, as well as an intense deformation in the vicinity of the crack.

Depending on the interaction of the hydrogen with the dislocations (during the deformation processes), the movement of dislocations can be promoted, inducing a localized plasticity. In addition, the dislocations are accumulated in microcracks, causing their subsequent propagation [177]. On the other hand, during the plastic deformation processes, the movement of dislocations can cause trapped hydrogen to diffuse to zones of higher stress intensity crack points, causing their recombination in molecular hydrogen, causing blistering and increasing their local concentration [153]. This significantly affects the performance of the steels under mechanical load and the type of fracture prevalent during failure.

8.3. Hydrogen Entry

Hydrogen can enter the metals in gaseous form or by an electrochemical reduction of the hydrogen-containing species of aqueous phases. This process depends on many parameters and is the first step for the development of hydrogen embrittlement. For the first case, these factors are mainly environmental, such as applied pressure, gas purity and temperature; meanwhile, for the second case, the predominant factors are the potential of the electrolyte-metal interface, the applied current density, the electrolyte composition and the pH of the solution [178].

8.4. Hydrogen Gaseous Entry

Many models have been proposed for the hydrogen gaseous entry, and the details of this process are still uncertain. In general terms, however, the reactions involved are adsorption of molecular hydrogen, dissociation of the hydrogen molecule and thus hydrogen atoms are adsorbed on the surface, and the subsequent diffusion of atoms adsorbed on the surface within the crystal lattice of the metal [179]:

$$H_2 \rightarrow 2H_{ads},$$

$$H_{ads} \leftrightarrow H_{metal}$$

For most metals exposed to gaseous hydrogen atmosphere, the gas–solid interaction is defined by three steps: physisorption, chemisorption and absorption [180], these processes are depicted in the Figure 28.

- Physisorption: is the result of Van Der Waals forces between the metal surface and an adsorbent. It is completely reversible, and usually occurs instantly (direct adsorption of the hydrogen molecule on the surface).
- Chemisorption: is a chemical reaction that occurs between an atom of the metal surface and the adsorbent molecule. The chemical forces involved are short range and are limited to single layers. Chemisorption is usually slow and may be slowly reversible or irreversible. This process may be related to the formation of covalent bonds between an atom or adsorbent molecule and a surface atom (direct dissociation to atomic hydrogen).
- Absorption: it is a gas–solid interaction, which involves the incorporation of the products of the chemisorption within the crystalline network of the steel and its subsequent diffusion. Depending on the input mechanism, hydrogen absorption may be in atomic or ionic form (H^+) [181].

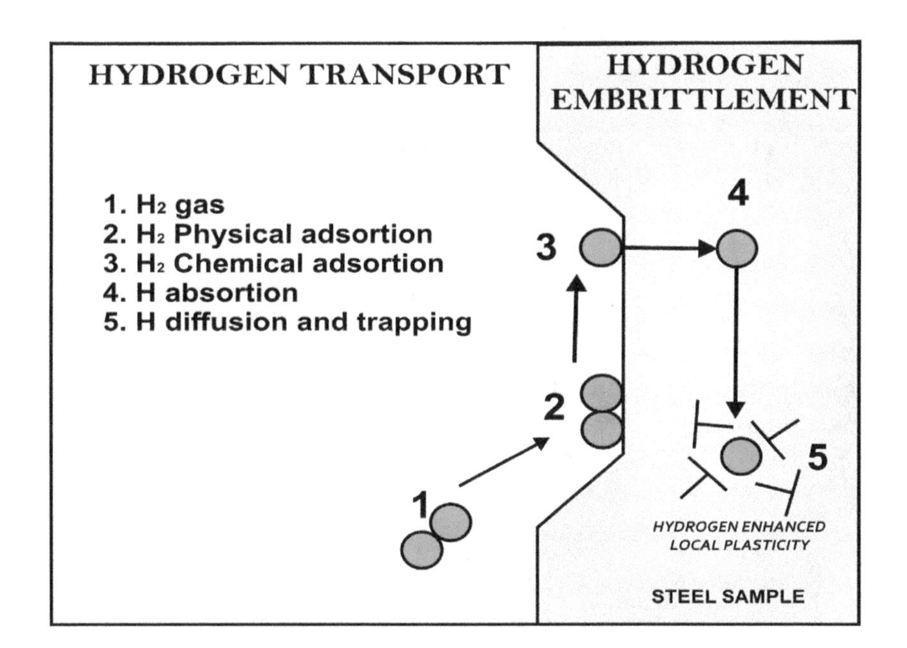

Figure 28. Hydrogen gas entry mechanism in metals.

The hydrogen embrittlement effect is strongly affected by environmental factors such as hydrogen gas pressure, hydrogen purity and temperature. The HE susceptibility at high pressure (70 MPa) is influenced by hydrogen solubility (*S*) and diffusivity (*D*). The *S* are a function of hydrogen surface concentration (C_0) and hydrogen pressure (*P*) according to the Sievert's law [182]:

$$C_0 = SP^{1/2}$$

Due to the direct relation between C_0 and *P* applied, the increment of gas pressure is related to high hydrogen embrittlement susceptibility [183]. That is, ductility tends to reduce proportionally to the square root of hydrogen pressure [184].

8.5. Entry of Hydrogen into Aqueous Phase

The mechanism of electrochemical production of hydrogen in steels in aqueous solution has received much attention. It is accepted that the reaction occurs in several stages. The first of these is an initial charge transfer that produces adsorption of the atomic hydrogen. In acid solutions, this involves the reduction of a hydrogen ion:

$$H_3O^+ + e^- \rightarrow H_{ads} + H_2O$$

In neutral and alkaline solutions, where the hydrogen concentration is very low, the reaction changes for the reduction of the water molecules:

$$H_2O + e^- \rightarrow H_{ads} + OH^-$$

The second stage of the reaction to produce molecular hydrogen can occur through two mechanisms. In the first of these, known as chemical desorption, chemical recombination or Tafel reaction, two adsorbed hydrogen atoms combine to produce molecular hydrogen:

$$H_{ads} + H_{ads} \rightarrow H_2$$

Alternatively, adsorbed hydrogen atoms may participate in a second electrochemical reaction, known as electrochemical desorption, or Heyrovsky reaction [151]:

$$H_{ads} + H_3O^+ + e^- \rightarrow H_2 + H_2O \text{ (acid)}$$

$$H_{ads} + H_2O + e^- \rightarrow H_2 + OH^- \text{ (neutral or alkaline)}$$

A third reaction, which goes in parallel with the desorption reaction, is the entry of the atomic hydrogen into the metal from a surface adsorption state:

$$H_{ads} \rightarrow H_{metal}$$

In many circumstances, the kinetics of these reactions are controlled by the rate at which hydrogen can diffuse into the metal.

If the adsorbed hydrogen is produced from a gas phase or an aqueous solution, it appears that the presence of the hydrogen atoms distorts the crystal structure of the metal surface [185].

In Figure 29, it is shown a hydrogen charging cell using 0.5 M H_2SO_4 + 0.2 gr of As_2O_3.

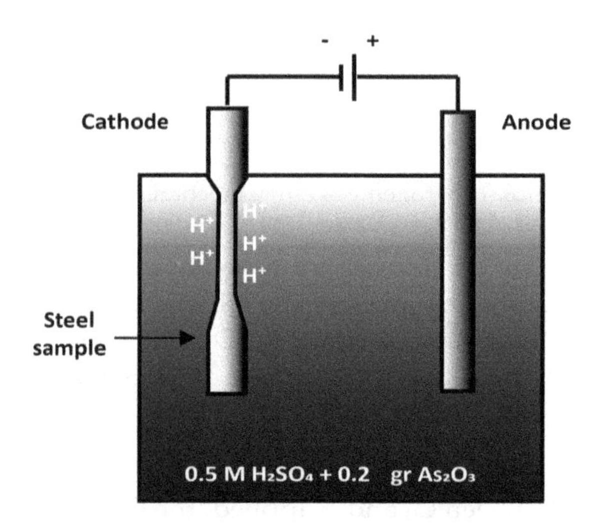

Figure 29. Hydrogen charging in a 0.5 M H_2SO_4 + 0.2 gr As_2O_3.

8.6. Hydrogen Embrittlement Effect over Mechanical Properties of Tempered Treated Microalloyed Steels

The HSLA steels are used for hydrogen storage and transportation, due to their high strength, toughness and good weldability. Theses steels are subjected during its processing to thermomechanical processing (TMCP), as well as heat treatment to adjust the final mechanical properties [186,187]. The TMCP is an important aspect in the production of API grade steels, parameters such as reheated temperature, rolling temperature and cooling rates play a significant role in the obtainment of final microstructure and mechanical properties [188]. The Nb addition as a microalloying element tends to refine the austenite grain size [189], obtaining banded ferrite–perlite microstructures if fast cooling after controlled rolling is not applied.

Zhou [190] studied the mechanical properties on microalloyed steels as a function of its obtained microstructures through different thermo-mechanical processes that consisted of direct quenching and a posterior tempering at 500 °C for 1 h. They studied the mechanical properties of these steels in relation to their microstructure, which was obtained by different thermo-mechanical processes consisting of direct annealing and tempering at 500 °C for 1 h after the process controlled rolling mill (DQT); and direct annealing, annealing and subsequent quenching and tempering (RQT), obtaining microstructures of deformed bands of retained austenite and martensite; in addition, by means of a second thermo-mechanical treatment, a typical martensite microstructure is obtained in the form of needles without deformed bands. When this microstructure is precipitated, the precipitated carbides grow at the borders of the martensite needles; some of the precipitates have been identified as (Nb, Ti) (C, N) by EDS (Energy Dispersive Spectrometry). These studies conclude that, as the tempering temperature increases, the yield stress keeps fluctuating slightly near 1033 MPa, but the tensile stress decreases drastically, as well as an increase in the percentage of elongation of the material. The impact tests carried out indicate that higher tenacity values are achieved at tempering temperatures ranging from 500 to 650 °C in both types of microstructures [190].

Zhong Ping [191] evaluated the effect of heat treatment of tempering on the microstructure and mechanical properties of an ultra-high strength steel with a final tensile stress of 2230 MPa that is subjected to tempering thermal treatments during six hours in a temperature range of 100 °C to 650 °C. The effect of secondary hardening on tempering heat treatments over a range of temperatures resulted in a substantial change in mechanical properties particularly in impact toughness, fracture toughness, as well as an increase in yield stress and tensile strength in ranges from 300 °C to 480 °C; on the other hand, at 650 °C, the strength decreased, as well as also exhibiting lower tenacity properties at a range of 400–470 °C; nonetheless, this increased considerably at 510 °C. The optimum combination between strength and toughness is achieved in a temperature range of 480–510 °C. The results of the studies by TEM indicate that, in the early stages of tempering, the precipitates could be in the form of agglomerates; meanwhile, the increases of strength at 470 °C may be the result of a microstructure consisting of uniform dispersion, irregular shapes, coherent zones or carbides; however, at 510 °C, due to the loss of coherence in the precipitates, the strength decreased.

8.7. Hydrogen Embrittlement Effect over the Mechanical Properties

Microalloyed steels are intended to be used in the transportation of sour gas, since the extracted gas contains dissolved hydrogen sulfide (H_2S), the corrosion process in the pipelines is aggravated. Under these conditions, the cracking of these pipes is favored by the presence of atomic hydrogen produced on the surface as a corrosion byproduct of the exposed steel [192].

Microalloyed steels may fail due to severe H_2S degradation, which is present in crude oil and natural gas, so attention should be paid to the size, morphology and distribution of non-metallic inclusions in microalloyed steels to avoid hydrogen induced cracking (HIC), as well as the type of microstructure present in the steel and the strength level of the material.

The effect of material factors on the diffusivity of hydrogen depends on two main factors, which are: the concentration of atomic hydrogen on the metal surface due to environmental factors such as the pH of the solution and the partial pressure of hydrogen gas; the other factor is microstructural consisting of primary and secondary phases including nonmetallic inclusions and precipitates that can affect the entrapment and diffusivity of the steel.

Some of the factors that can affect the mechanical properties of steels, and their effects are described below:

Elastic constants. There is evidence of small changes in the elastic properties of steels as a result of dissolved hydrogen. These changes are small and therefore impractical. This is perhaps from the point of view of the low hydrogen solubility in the iron crystal lattice and the small effect on the metal–metal bond strength [193].

Bonding effort. The effect on the yield stress of iron and steels is unpredictable. For pure iron, polycrystalline or monocrystalline, the yield stress frequently shows a decrease due to the effect of hydrogen but can be increased or maintained, depending on the dislocation structure, crystalline orientation and iron purity [193]. In the next picture (Figure 30), the hydrogen embrittlement effect on mechanical properties of API X-120 microalloyed steels are shown. Strength (Figure 30a,b) and strain fracture tend to decrease as hydrogen charging time increases, and the stiffness increases as time increases (Figure 30c). These results show the embrittlement effect of hydrogen and the detrimental effect on mechanical properties.

Plastic behavior. The effect of hydrogen on the plastic behavior of iron and steel is very complex, hydrogen can harden or soften the material, according to its structure and the slip mode. Lunarska [193] concluded that hydrogen segregation around dislocations reduced its elastic range. At room temperature (where the diffusion rate of hydrogen is sufficiently high that it can maintain the movement of dislocations), this allows a softening of monocrystals when only a sliding system is operative. Hydrogen can also suppress the slippage of screw dislocations, and this results in hardening when multiple slip systems are activated [151].

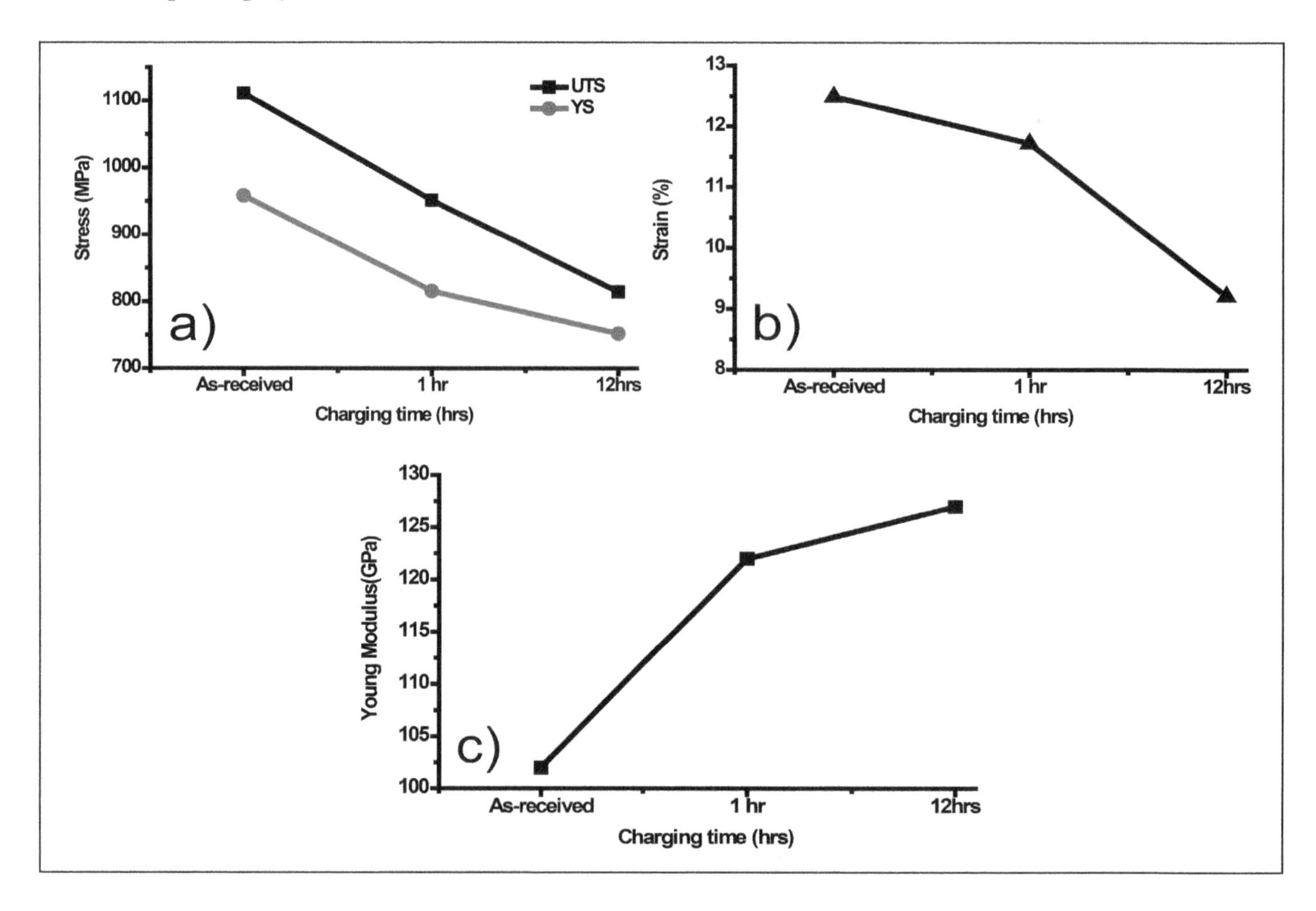

Figure 30. Hydrogen embrittlement effect, (**a**) UTS and YS, (**b**) strain percent and (**c**) young modulus; as a function of charging time.

In 2006, Hardie [153] carried out comparative studies on three different grade API steels of different resistance levels from the point of view of susceptibility to hydrogen embrittlement. The microstructure of these steels consists of bands of pearlite distributed in a ferrite phase in an elongated structure in the longitudinal direction to the rolling. The average grain sizes are 8, 4 and 2 μm, with microhardness of 200, 224 and 252 HV (Vickers Hardness) respectively. Some of their results show that there are different susceptibilities to the loss of ductility after a cathodic hydrogen charge and this susceptibility tends to increase with the resistance level of the steel when a current density of 0.44 mA/mm^2 is applied. Figure 31 shows the current density applied over area reduction in different API grade steels. These presented a significant loss of ductility when subjected to cathodic cracking, and the degree of embrittlement increased with the increase in current density, attributable to a higher uptake of hydrogen during this process.

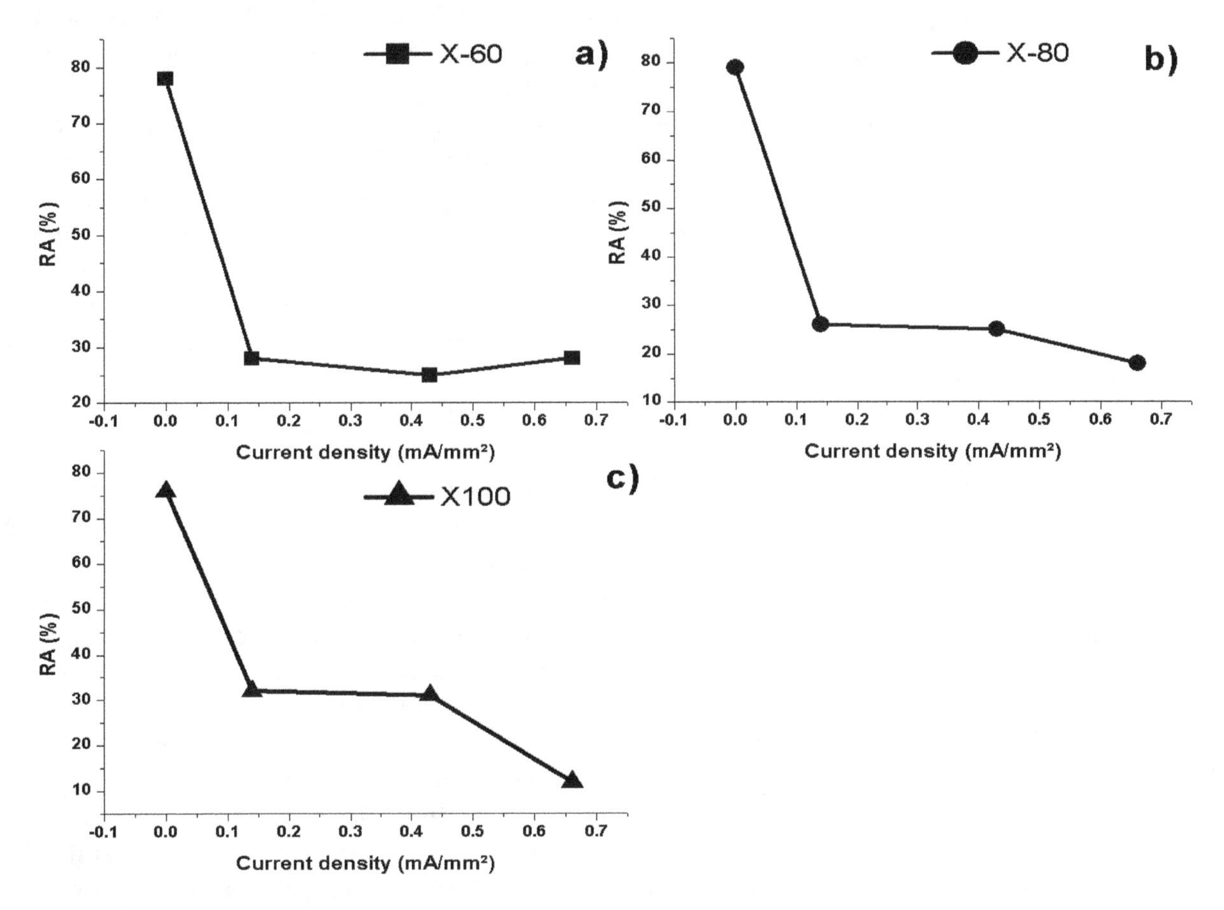

Figure 31. Reduction area change in different API grade steels. (**a**) X-60, (**b**) X-80 and (**c**) X-100, as a function of current density applied [153].

Other studies carried out by Gyu Tae Park [194] tried to evaluate the effect of the microstructure on the susceptibility to hydrogen embrittlement and compare the entrapment efficiency of hydrogen in microstructures consisting mainly of pearlite- ferrite, acicular ferrite, and ferrite-bainite; to evaluate the entrapment of hydrogen they calculated the permeability ($J_{SS}L$) and apparent diffusivity (D_{app}), where the entrapment efficiency of hydrogen increases in the following order of microstructures: perlite–ferrite, ferrite–bainite and acicular ferrite, the latter being the most susceptible to hydrogen embrittlement, as well as certain cracking initiation sites that are caused by nonmetallic inclusions, usually reported as crack initiation sites. As the microstructure changes from acicular to ferrite-bainite ferrite, the susceptibility to embrittlement by hydrogen increases, this being explained by the high tenacity of the acicular ferrite, preventing the propagation of cracks; low values of $J_{SS}L$ and D_{app}, and high values of C_{app} exhibit higher entrapment of hydrogen in the steel.

On the other hand, Liu et al. [195] evaluates the effect of microalloyed steel subjected to heat treatments consisting of air cooling to obtain a microstructure of bainite–martensite, which shows a superior combination of strength and toughness than when a martensitic microstructure is obtained from a quench and then tempered at intermediate temperature. It is concluded that the susceptibility to hydrogen embrittlement of steel with bainite–martensite microstructure decreased as the tempering temperature increased from 280 °C to 370 °C. The results show that the susceptibility to hydrogen embrittlement decreases as the tempering temperature increases, where, after 370 °C, the bainitic-martensitic microstructure is less susceptible to embrittlement. One reason is that area reduction (%RA) has greater stability after tempering at this temperature; another reason is that the internal stresses have been relieved.

Other studies indicate that the increase in the pre-charged time, the increase of the applied current density or the decrease of pH in the test solution allows the increase in the content of hydrogen in an API-X80 grade steel, and this promotes the reduction in the percentage of area reduction in stress tests and consequently a higher index of embrittlement by hydrogen.

Dong [196] studied the relationship between the microstructure and the Charpy impact properties in API grade X-70 and X-80 steels, using current densities of 20, 50 and 100 mA/cm^2 and times of 1, 3, 5 and 8 h exposure to an H_2SO_4 medium, where the mechanical properties of these steels are assessed by stress tests. It was found that there is a reduction in the strength of the material as well as a considerable decrease in elongation and reduction of area after longer exposure times. The fractographic study carried out showed a greater brittleness and cracking at the fracture surface with a longer pre-loading time. The type of fracture present shows evidence of a type of brittle fracture due to the presence of cleavage zones.

9. Future Trends

The growth of petroleum distribution systems has always been closely associated with the development of microalloyed steel grades having high strength, toughness, and weldability. In many ways, the research in microalloyed steel drew impetus from the huge demands of such steels from the petroleum pipeline projects. The first specification for steel for pipelines introduced by the American Petroleum Institute (API) in 1948 was X-42 with a yield strength of 42 ksi or 290 MPa. The strength levels underwent large scale changes over the decades, and, at the present time, the X-80 is a commonplace alloy, while X-100 and X-120 are still in sight. High temperature processes (HTP) are an important factor for consideration in pipelines development. The necessity of incorporating facilities for producing high-strength pipeline steels in up-coming steel plants has been briefly dealt with. With increasing strength levels in pipeline steels and higher operating pressures coming into vogue, pipeline steels emerge as a subject of fresh research, innovation, and area of specialization [197].

Today, heavy plates for longitudinally welded pipes in the X-80 strength grade are produced by Mannesmannroöhren Mülheim (MRM) with the same reliability as lower grades [198].

The manufacturing of products with excellent quality is the ultimate goal of technological development. To achieve this goal, the properties of manufactured products must be predicted, and both the chemistry and the production process must be carefully designed. Recent advances in physical metallurgy, rolling technology and computer control have made great contributions to the field of structure and property prediction and to the development of a control model that allows the prediction of microstructural evolution and mechanical properties. It is expected that the maximum benefits from manufacturing control will be obtained by using predictions derived from metallurgical models [197].

Resource conservation, energy reduction, yield improvement, recycling and reduction in the weight of parts are becoming increasingly important. In the future, microstructure control will therefore be expected to assist in the development of new processing technologies that offer improvements in the above areas. One example of this is the manufacture of parts with high strength and good machinability. This kind of new technology will be applied to many microalloyed steels. It will prove necessary to develop a precise system for the simulation of microstructural evolution during forming in order to optimize process design for these types of functional parts.

The Latin American steel industry continues its effort to modernize its facilities, including in the complicated context of the global situation generated by China's overcapacity and the government's subsidies to steel exports. A part of the modernization effort is reflected in the construction of new plants, which meets different needs.

A good part of these plants has been installed in the center and north of Mexico, primarily focusing on the strong development of the automotive industry, and its consequent exports to the USA. Another important part of the new plants corresponds to the states of Pará, Maranhão and Ceará, in the north and northeast of Brazil, where local governments have offered very favorable conditions for investments.

Innovative technologies are being introduced: the first Jumbo Coking Reactor (JCR) plant in the Americas; the world's first Quantum electric furnace; the first continuous casting of Castrip straps outside the USA and the first ecological pickling plant in Latin America (TYASA). One of the plants uses pig iron produced from biomass (Aço Verde Brasil).

The routes implemented range from electric steel mills that use scrap to integrated mills based on blast furnaces and oxygen mills, also including routes used only in Latin America, such as integrated blast furnace mills and charcoal mills in addition to plants included that are dedicated exclusively to cold rolling or hot dip galvanizing. One particular feature is the installation of flat-bed laminators in a single box, for relatively small production compared to conventional laminators (Gerdau Ouro Branco, AHMSA, TYASA) [199].

At the moment, analytical techniques have been used, such as:

High resolution synchrotron X-ray experiments that can be carried out to accurately measure the NbC volume fraction and corresponding dissolved Nb in solution after reheating conditions, which is otherwise difficult to measure by electron microscopy or conventional X-ray diffraction because of very low volume fraction (about 0.0001–0.0002) of niobium carbide precipitates in the investigated alloys [200].

Hot compression experiments with variable strain and strain rates can be carried out to develop equations for recrystallization kinetics of deformed austenite in the investigated Nb-microalloyed rail steels. In addition, the effect of plane–strain compression and other complex deformation schedules on the recrystallization kinetics can be studied [201].

A more in-depth study of non-propagating secondary cracks in various pearlitic and bainitic microstructures would provide valuable insight and complement existing results [202].

The effect of Nb-microalloying on other important rail properties like rolling contact fatigue performance, weldability, hydrogen embrittlement, etc. needs to be investigated.

Author Contributions: S.S. and B.C. got the idea and directed the project and review of the final version; J.C.V., A.D.-P. and J.M. wrote the paper.

Acknowledgments: This work was supported by PRODEP [grant number 511-6/17-14378] and for the support in the payment of the costs of publication, CONACyT [grant number 434894]; we want to thank to Ben Church; René Guardian Tapia and Ivan Puente Lee for the support provided in SEM images and Alejandro Sedano Aguilar for the support in the revision of the English grammar of the article.

References

1. Stalheim, D.G.; Muralidharan, G. The Role of Continuous Cooling Transformation Diagrams in Material Design for High Strength Oil and Gas Transmission Pipeline Steels. In Proceedings of the IPC 2006, Calgary, AB, Canada, 25–29 September 2006.
2. Takahashi, A.; Lino, M. Thermo-mechanical control process as a tool to grain-refine the low manganese containing steel for sour service linepipe. *ISIJ Int.* **1996**, *36*, 235–240. [CrossRef]

3. Fu, J.; Wang, J.B.; Kang, Y.L. Research and development of HSLC steels produced by EAF-CSP technology. In Proceedings of the International Symposium on Thin Slab Casting and Rolling, Guangzhou, China, 3–5 December 2002.
4. Xu, G.; Gan, X.; Ma, G.; Luo, F.; Zou, H. The development of Ti-alloyed high strength microalloy steel. *Mater. Des.* **2010**, *31*, 2891–2896. [CrossRef]
5. Becket, F.M.; Russell, F. Steel. U.S. Patent 2,264,355, 2 December 1941.
6. Gray, J.M.; Siciliano, F. High Strength Microalloyed Linepipe: Half a Century of Evolution. In Proceedings of the Pipeline Technology Meeting, Houston, TX, USA, 22–23 April 2009.
7. Beiser, C.A. The Effect of Small Columbium Additions to semi-killed Medium-Carbon Steels. ASM Preprint No. 138. In Proceedings of the Regional Technical Meeting, Buffalo, NY, USA, 17–19 August 1959.
8. Morrison, W.B. The Influence of Small Niobium Additions on the Properties of Carbon-Manganese Steels. *ISIJ Int.* **1963**, *201*, 317.
9. Irani, J.J.; Burton, D.; Jones, J.D.; Rothwell, A.B. Beneficial Effects of Controlled Rolling in the Process of Structural Steels. In *Strong Tough Structural Steels*; Iron and Steel Institute Special Publication: Scarborough, UK, 1967.
10. Iron and Steel Institute. Strong Tough Structural Steel. In Proceedings of the Joint Conference, Scarborough, UK, 4–6 April 1967.
11. Cuddy, J.L. The Effect of Microalloy Concentration on the Recrystallisation of Austenite During Hot Deformation. In *Thermomechanical Processing of Microalloyed Austenite*; AIME: Pittsburgh, PA, USA, 1981; pp. 129–140.
12. Tata Steel. Available online: http://www.tatasteeleurope.com (accessed on 16 January 2018).
13. Starratt, F.W. Columbium-treated steels. *J. Met.* **1958**, *10*, 799.
14. Oakes, G.; Barraclough, K.C. *Steels*; The Development of Gas Turbine Materials; Applied Science Publishers: London, UK, 1981; pp. 31–61.
15. Llewellyn, D.T. Nitrogen in steels. *Ironmak. Steelmak.* **1993**, *20*, 35–41.
16. Pressouyre, G.M.; Bernstein, I.M. A quantitative analysis of hydrogen trapping. *Metall. Mater. Trans. A* **1978**, *9*, 1571–1580. [CrossRef]
17. DeArdo, A.J.; Hua, M.J.; Cho, K.G.; Garcia, C.I. On strength of microalloyed steels: An interpretive review. *Mater. Sci. Technol.* **2009**, *25*, 1074–1082. [CrossRef]
18. Vervynckt, S.; Verbeken, K.; Lopez, B.; Jonas, J.J. Modern HSLA steels and role of non-recrystallisation temperature. *Int. Mater. Rev.* **2012**, *57*, 187–207. [CrossRef]
19. Baker, T.N. Microalloyed Steels. *Ironmak. Steelmak.* **2016**, *4*, 264–307. [CrossRef]
20. DeArdo, A.J. *Microalloying '95*; ISS-AIME: Warrendale, PA, USA, 1995; Volume 15.
21. DeArdo, A.J. Niobium in Modern Steels. *Int. Mater. Rev.* **2003**, *48*, 371–402. [CrossRef]
22. Gilman, T. *The Physical Metallurgy of Microalloyed Steel*; Cambridge University Press: Cambridge, UK, 1957.
23. Baker, T.N. *Microalloyed Steel*; Future Developments of Metals and Ceramics; Institute of Materials: London, UK, 1992; pp. 75–119.
24. Gladman, T. *Microalloyed Steels*; Institute of Materials: London, UK, 1997.
25. Xie, K.Y.; Zheng, T.; Cairney, J.M.; Kaul, H.; Williams, J.G.; Barbaro, F.J.; Killmore, C.R.; Ringer, S.P. Strengthening from Nb-rich clusters in a Nb-microalloyed steel. *Scr. Mater.* **2012**, *66*, 710–713. [CrossRef]
26. Soto, R.; Saikaly, W.; Bano, X.; Issartel, C.; Rigaut, G.; Charai, A. Statistical and theoretical analysis of precipitates in dual-phase steels microalloyed with titanium and their effect on mechanical properties. *Acta Mater.* **1999**, *47*, 3475–3481. [CrossRef]
27. Mangonon, P.L., Jr.; Heitmann, W.E. *Proceeding Conference Microalloying*; Union Carbide Corporation: Houston, TX, USA, 18–21 November 1977.
28. Zhang, L.; Kannengiesser, T. Austenite grain growth and microstructure control in simulated heat affected zones of microalloyed HSLA steel. *Mater. Sci. Eng. A* **2014**, *613*, 326–335. [CrossRef]
29. Gu, Y.; Tian, P.; Wang, X.; Han, X.; Liao, B.; Xiao, F. Non-isothermal prior austenite grain growth of a high-Nb X100 pipeline steel during a simulated welding heat cycle process. *Mater. Des.* **2016**, *89*, 589–596. [CrossRef]
30. Karmakar, A.; Biswas, S.; Mukherjee, S.; Chakrabarti, D.; Kumar, V. Effect of composition and thermo-mechanical processing schedule on the microstructure, precipitation and strengthening of Nb-microalloyed steels. *Mater. Sci. Eng. A* **2017**, *690*, 158–169. [CrossRef]

31. Chen, G.; Yang, W.; Guo, S.; Sun, Z. Effect of Nb on the transformation kinetics of low carbon (manganese) steel during deformation of undercooled austenite. *J. Univ. Sci. Technol. Beijing Miner. Metall. Mater.* **2006**, *13*, 411–415. [CrossRef]
32. Baker, T.N. *Titanium Technology in Microalloyed Steels*, 2nd ed.; The Institute of Materials, Minerals and Mining: London, UK, 1997.
33. Kojima, A.; Yoshii, K.; Hada, T. Development of high HAZ toughness steel plates for box columns with high heat input welding. *Nippon Steel Tech. Rep.* **2004**, *90*, 39–44.
34. Chen, Y.; Zhang, D.T.; Liu, Y.C.; Li, H.J.; Xu, D.K. Effect of dissolution and precipitation of Nb on the formation of acicular ferrite/bainite ferrite in low-carbon HSLA steels. *Mater. Charact.* **2013**, *84*, 232–239. [CrossRef]
35. Karjalainen, L.P.; Maccagno, T.M.; Jonas, J.J. Softening and Flow Stress Behaviour of Nb Microalloyed Steels during Hot Rolling Simulation. *ISIJ Int.* **1995**, *35*, 1523–1531. [CrossRef]
36. Hansen, S.S.; Sande, J.B.V.; Cohen, M. Niobium Carbonitride Precipitation and Austenite Recrystallization in Hot-Rolled Microalloyed Steels. *Metall. Trans. A* **1987**, *11*, 387–402. [CrossRef]
37. He, X.L.; Shang, C.; Yang, S. *High Performance Low Carbon Bainitic Steel*; Metallurgical Industry Press: Berjing, China, 2008.
38. Hoogendoorn, T.M.; Spanraft, M.J. Quantifying the Effect of Microalloying Elements on Structures during Processing. In Proceedings of the International Symposium on High-Strength, Low-Alloy Steels, Washington, DC, USA, 1–3 October 1975.
39. LeBon, A.B.; de Saint-Martin, L.N. Using Laboratory Simulations to Improve Rolling Schedules and Equipment. In Proceedings of the International Symposium on High-Strength, Low-Alloy Steels, Washington, DC, USA, 1–3 October 1975.
40. Gauthier, G.; LeBon, A.B. Discussion: On the Recrystallization of Austenite. In Proceedings of the International Symposium on High-Strength, Low-Alloy Steels, Washington, DC, USA, 1–3 October 1975.
41. Gauthier, G.; LeBon, A.B. Discussion: Techniques of Assessment of Precipitation Kinetics. In Proceedings of the International Symposium on High-Strength, Low-Alloy Steels, Washington, DC, USA, 1–3 October 1975.
42. Tanaka, T.; Tabata, N.; Hatomura, T.; Shiga, C. Three Stages of the Controlled-Rolling Process. In Proceedings of the International Symposium on High-Strength, Low-Alloy Steels, Washington, DC, USA, 1–3 October 1975.
43. Kozasu, I.; Ouchi, C.; Sampei, T.; Okita, T. Hot Rolling as a High-TemperatureThermo-Mechanical Process. In Proceedings of the International Symposium on High-Strength, Low-Alloy Steels, Washington, DC, USA, 1–3 October 1975.
44. Gray, J.M.; Barbaro, F. Evolution of microalloyed steels since microalloying '75 with specific emphasis on linepipe and plate. In *HSLA Steels 2015, Microalloying 2015 & Offshore Engineering Steels 2015*; The Minerals, Metals & Materials Society: Pittsburgh, PA, USA, 2016; pp. 53–70.
45. Stalheim, D.G. The use of high temperature processing (HTP) steel for high strength oil and gas transmission pipeline application. *Iron Steel* **2005**, *40*, 699–704.
46. Bracke, L.; de Wispelaere, N.; Ahmed, H.; Gungor, O.E. S700MC/Grade 100 in heavy gauges: Industrialisation at ArcelorMittal europe. In Proceedings of the International Symposium on Recent Developments in Plate Steels, Warrendale, PA, USA, 19–22 June 2011; pp. 131–138.
47. Kim, S.J.; Lee, C.G.; Lee, T.H.; Lee, S. Effects of coiling temperature on microstructure and mechanical properties of high-strength hot-rolled steel plates containing Cu, Cr and Ni. *ISIJ Int.* **2000**, *40*, 692–698. [CrossRef]
48. Misra, R.D.K.; Nathani, H.; Hartmann, J.E.; Siciliano, F. Microstructural evolution in a new 770 MPa hot rolled Nb–Ti microalloyed steel. *Mater. Sci. Eng. A* **2005**, *394*, 339–352. [CrossRef]
49. Reip, C.P.; Shanmugam, S.; Misra, R.D.K. High strength microalloyed CMn (V–Nb–Ti) and CMn (V–Nb) pipeline steels processed through CSP thin-slab technology: Microstructure, precipitation and mechanical properties. *Mater. Sci. Eng. A* **2006**, *424*, 307–3017. [CrossRef]
50. Shanmugam, S.; Misra, R.D.K.; Hartmann, J.E.; Jansto, S.G. Microstructure of high strength niobium-containing pipeline steel. *Mater. Sci. Eng. A* **2006**, *441*, 215–229. [CrossRef]
51. Misra, R.D.K.; Jia, Z.; O'Malley, R.; Jansto, S.G. Precipitation behavior during thin slab thermomechanical processing and isothermal aging of copper-bearing niobium-microalloyed high strength structural steels: The effect on mechanical properties. *Mater. Sci. Eng. A* **2011**, *528*, 8772–8780. [CrossRef]

52. Riva, R.; Mapelli, C.; Venturini, R. Effect of Coiling Temperature on Formability and Mechanical Properties of Mild Low Carbon and HSLA Steels Processed by Thin Slab Casting and Direct Rolling. *ISIJ Int.* **2007**, *47*, 1204–1213. [CrossRef]
53. Misra, D.; Jansto, S.G. Niobium-based alloy design for structural applications. In *HSLA Steels 2015, Microalloying 2015 & Offshore Engineering Steels 2015*; The Minerals, Metals & Materials Society: Pittsburgh, PA, USA, 2016; pp. 261–266.
54. Challa, V.S.A.; Zhou, W.H.; Misra, R.D.K.; O'Malley, R.; Jansto, S.G. The effect of coiling temperature on the microstructure and mechanical properties of a niobium–titanium microalloyed steel processed via thin slab casting. *Mater. Sci. Eng. A* **2014**, *595*, 143–153. [CrossRef]
55. Maubane, R.; Banks, K.M.; Tuling, A.S. Hot strength during coiling of low C and Nb microalloyed steels. In *HSLA Steels 2015, Microalloying 2015 & Offshore Engineering Steels 2015*; The Minerals, Metals & Materials Society: Pittsburgh, PA, USA, 2016; pp. 323–328.
56. Burgmann, F.A.; Xie, Y.; Cairney, J.M.; Ringer, S.P.; Killmore, C.R.; Barbaro, F.J.; Williams, J.G. The effect of niobium additions on ferrite formation in Castrip® steel. *Mater. Forum* **2008**, *32*, 9–12.
57. Thelning, K.E. *Steel and Its Heat*; Butterworth & Co.: London, UK, 1984.
58. Song, R.; Ponge, D.; Raabe, D. Influence of Mn content on microstructure and mechanical properties of ultrafine grained C-Mn Steels. *ISIJ Int.* **2005**, *45*, 1721–1726. [CrossRef]
59. Shen, Y.F.; Zuo, L. High-strength low-alloy steel strengthened by multiply nanoscale microstructures. In *HSLA Steels 2015, Microalloying 2015 & Offshore Engineering Steels 2015*; The Minerals, Metals & Materials Society: Pittsburgh, PA, USA, 2016; pp. 187–193.
60. Hall, E.O. The deformation and ageing of mild steel: 3 discussion of results. *Proc. Phys. Soc.* **1951**, *64*, 747–753. [CrossRef]
61. Petch, N.J. The cleavage strength of polycrystals. *J. Iron Steel Inst.* **1953**, *174*, 25–28.
62. Sanz, L.; Pereda, B.; López, B. Effect of thermomechanical treatment and coiling temperature on the strengthening mechanisms of low carbon steels microalloyed with Nb. *Mater. Sci. Eng. A* **2017**, *685*, 377–390. [CrossRef]
63. Lucas, A.; Simon, P.; Bourdon, G.; Herman, J.C.; Riche, P.; Neutjens, J.; Harlet, P. Metallurgical Aspects of Ultra Fast Cooling in front of the Down-Coiler. *Steel Res. Int.* **2004**, *75*, 139–146. [CrossRef]
64. Mohapatra, S.S.; Ravikumar, S.V.; Pal, S.K.; Chakraborty, S. Ultra Fast Cooling of a Hot Steel Plate by Using High Mass Flux Air Atomized Spray. *Steel Res. Int.* **2013**, *84*, 229–236. [CrossRef]
65. Bhattacharya, P.; Samanta, A.N.; Chakraborty, S. Spray evaporative cooling to achieve Ultra Fast cooling in runout table. *Int. J. Therm. Sci.* **2009**, *48*, 1741–1747. [CrossRef]
66. Siciliano, F. High-Strength Linepipe Steels and Physical Simulation of Production Processes. In *Advanced High Strength Steel*; Springer: Singapore, 2018; pp. 71–78.
67. Gilman, T. *The Physical Metallurgy of Microalloyed Steel*; Cambridge University Press: Cambridge, UK, 1977.
68. Korchynsky, M. *Microalloying 75*; Union Carbide: New York, NY, USA, 1977.
69. DeArdo, A.J.; Ratz, G.A.; Wray, J.P. *Thermomechanical Processing of Microalloyed Austenite*; Metallurgical Society of AIME: Warrendale, PA, USA, 1982.
70. Korchynsky, M. *(Ed.) Proceedings of the HSLA—Technology and Applications: Conference Proceedings of International Conference on Technology and Applications of HSLA Steels*; American Society for Metals: Metals Park, OH, USA, 1983.
71. Taylor, K.A.; Thompson, S.W.; Fletcher, F.B. *Physical Metallurgy of Direct-Quenched Steels*; Minerals, Metals and Materials Society: Chicago, IL, USA, 1993.
72. Fazeli, F.; Amirkhiz, B.S.; Scott, C.; Arafin, M.; Collins, L. Kinetics and microstructural change of low-carbon bainite due to vanadium microalloying. *Mater. Sci. Eng. A* **2018**, *720*, 248–256. [CrossRef]
73. Gray, J.M. *An Independent View of Linepipe and Linepipe Steel for High Strength Pipelines: How to Get Pipe That's Right for the Job at the Right Price*; Microalloyed Steel Institute L.P.: Houston, TX, USA, 2002.
74. Tsunekage, N.; Kobayashi, K.; Tsubakino, H. Influence of S and V on toughness of ferrite-pearlite or bainite steels. *Curr. Adv. Mater. Processes* **2000**, *13*, 534.
75. Nomura, I. Metallurgical Technology in Microalloyed Steels for Automotive Parts. *Mater. Jpn.* **1995**, *34*, 705. [CrossRef]
76. Tomita, Y.; Saito, N.; Tsuzuki, T.; Tokunaga, Y.; Okamoto, K. Improvement in HAZ Toughness of Steel by TiN-MnS Addition. *ISIJ Int.* **1994**, *34*, 829–835. [CrossRef]

77. Suzuki, S.; Kuroki, K.; Kobayashi, H.; Takahashi, N. Sn Segregation at Grain Boundary and Interface between MnS and Matrix in Fe-3 mass%Si Alloys Doped with Tin. *Mater. Trans. JIM* **1992**, *33*, 1068–1076. [CrossRef]
78. Tsunekage, N.; Tsubakino, H. Effects of Sulfur Content and Sulfide-forming Elements Addition on Impact Properties of Ferrite–Pearlitic Microalloyed Steels. *ISIJ Int.* **2001**, *41*, 498–505. [CrossRef]
79. DeArdo, A.J. Microalloyed Steels: Past, Present and Future. In *HSLA Steels 2015, Microalloying 2015 and Offshore Engineering Steels, 2015*; The Minerals, Metals & Materials Society: Pittsburgh, PA, USA, 2015.
80. Cottrell, A.H. *Dislocations and Plastic Flow in Crystals*; Clarendon Press: Oxford, UK, 1964.
81. Korchynsky, M. Microalloyed Steels. In *Proc. Conf. Microalloying '75*; Union Carbide Corp.: New York, NY, USA, 1977.
82. Bai, D.; Collins, L.; Hamad, F.; Chen, X.; Klein, R. Microstructure and mechanical properties of high strength linepipes steels. In *Proceedings of the Materials. Science & Technology Congress*; AIST/ASM: Warrendale, PA, USA, 2007; pp. 355–366.
83. Peters, P.A.; Gray, J.M. *Genesis and Development of Specifications and Performance Requirements for Modern Linepipe—Strength, Toughness, Corrosion Resistance and Weldability*; International Convention; Australian Pipeline Industry Association, Inc.: Hobart, Tasmania, Australia, 24–29 October 1992.
84. Siwecki, T.; Eliasson, J.; Lagneborg, R.; Hutchinson, B. Vanadium microalloyed bainitic hot strip steels. *ISIJ Int.* **2010**, *50*, 760–767. [CrossRef]
85. Kim, Y.W.; Song, S.W.; Seo, S.J.; Hong, S.G.; Lee, C.S. Development of Ti and Mo micro-alloyed hot-rolled high strength sheet steel by controlling thermomechanical controlled processing schedule. *Mater. Sci. Eng. A* **2013**, *565*, 430–438. [CrossRef]
86. Juanhua, K.; Lin, Z.; Bin, G.; Pinghe, L.; Aihua, W.; Changsheng, X. Influence of Mo content on microstructures and mechanical properties of high strength pipeline steel. *Mater. Des.* **2004**, *25*, 723–728. [CrossRef]
87. Xiao, F.R.; Liao, B.; Shan, Y.Y.; Qiao, G.Y.; Zhong, Y.; Zhang, C.; Yang, K. Challenge of mechanical properties of an acicular ferrite pipeline steel. *Mater. Sci. Eng. A* **2006**, *431*, 41–52. [CrossRef]
88. Wang, W.; Shan, Y.; Yang, K. Study of high strength pipeline steels with different microstructures. *Mater. Sci. Eng. A* **2009**, *502*, 38–44. [CrossRef]
89. Katsumata, M.; Machida, M.; Kaji, H. Recrystallization of Austenite in High-Temperature Hot-Rolling of Niobium Bearing Steels. In Proceedings of the International Conference on the Thermomechanical Austenite, Pittsburgh, PA, USA, 17–19 August 1981.
90. Suikkanen, P.P.; Komi, J.I.; Karjalainen, L.P. Effect of austenite deformation and chemical composition on the microstructure and hardness of low-carbon and ultra low-carbon bainitic steels. *Met. Sci. Heat Treat.* **2005**, *47*, 507–511. [CrossRef]
91. Gong, P.; Palmiere, E.J.; Rainforth, W.M. Thermomechanical processing route to achieve ultrafine grains in low carbon microalloyed steels. *Acta Mater.* **2016**, *119*, 43–54. [CrossRef]
92. Dutta, B.; Valdes, E.; Sellars, C.M. Mechanism and kinetics of strain induced precipitation of Nb (C, N) in austenite. *Acta Mater.* **1992**, *40*, 653–662. [CrossRef]
93. Gomez, M.; Valles, P.; Medina, S.F. Evolution of microstructure and precipitation state during thermomechanical processing of a X80 microalloyed steel. *Mater. Sci. Eng. A* **2011**, *528*, 4761–4773. [CrossRef]
94. Zhao, M.C.; Yang, K.; Shan, Y. The effects of thermo-mechanical control process on microstructures and mechanical properties of a commercial pipeline steel. *Mater. Sci. Eng. A* **2002**, *335*, 14–20. [CrossRef]
95. Zhao, M.C.; Yang, K.; Shan, Y.Y. Comparison on strength and toughness behaviors of microalloyed pipeline steels with acicular ferrite and ultrafine ferrite. *Mater. Lett.* **2003**, *57*, 1496–1500. [CrossRef]
96. Xu, Y.B.; Yu, Y.M.; Xiao, B.L.; Liu, Z.Y.; Wang, G.D. Microstructural evolution in an ultralow-C and high-Nb bearing steel during continuous cooling. *J. Mater. Sci.* **2009**, *44*, 3928–3935. [CrossRef]
97. Graf, M.K.; Hillenbrand, H.; Peters, P. *Accelerated Cooling of Steel*; TMS-AIME: Warrendale, PA, USA, 1986; pp. 165–179.
98. Xiang-dong, H.; Lie-jun, L.; Zheng-wu, P.; Song-jun, C. Effects of TMCP schedule on precipitation, microstructure and properties of Ti-microalloyed high strength steel. *J. Iron Steel Res. Int.* **2016**, *23*, 593–601.
99. Jorge, J.C.F.; Souza, L.F.G.; Rebello, J.M.A. The effect of chromium on the microstructure/toughness relationship of C- Mn weld metal deposits. *Mater. Charact.* **2001**, *47*, 195–205. [CrossRef]
100. Beidokhti, B.; Koukabi, A.H.; Dolati, A. Influences of titanium and manganese on high strength low alloy SAW weld metal properties. *Mater. Charact.* **2009**, *60*, 225–233. [CrossRef]

101. Dongsheng, L.; Binggui, C.; Yuanyan, C. Strengthening and Toughening of a Heavy Plate Steel for Shipbuilding with Yield Strength of Approximately 690 MPa. *Metall. Mater. Trans. A* **2013**, *44*, 440–445.
102. Shin, S.Y.; Han, S.Y.; Hwang, B.C.; Lee, C.G.; Lee, S.H. Effects of Cu and B addition on microstructure and mechanical properties of high-strength bainitic steels. *Mater. Sci. Eng. A* **2009**, *517*, 212–218. [CrossRef]
103. Han, S.Y.; Shin, S.Y.; Lee, S.H.; Kim, N.J.; Bae, J.H.; Kim, K. Effects of Cooling Conditions on Tensile and Charpy Impact Properties of API X80 Linepipe Steels. *Metall. Mater. Trans. A* **2010**, *41*, 329–340. [CrossRef]
104. Kang, Y.L.; Han, Q.H.; Zhao, X.M.; Cai, M.H. Influence of nanoparticle reinforcements on the strengthening mechanism of an ultrafine-grained dual phase steel containing titanium. *Mater. Des.* **2013**, *44*, 331–339. [CrossRef]
105. Wang, Y.W.; Feng, C.; Xu, F.Y.; Bai, B.Z.; Fang, H.Z. Influence of Nb on Microstructure and Property of Low-Carbon Mn-Series Air-Cooled Bainitic Steel. *J. Iron Steel Res. Int.* **2010**, *17*, 49–53. [CrossRef]
106. Kong, J.H.; Xie, C.S. Effect of molybdenum on continuous cooling bainite transformation of low-carbon microalloyed Steel. *Mater. Des.* **2006**, *27*, 1169–1173. [CrossRef]
107. Zackay, V.F.; Justusson, W.M. *Special Report 76*; Iron and Steel Institute: London, UK, 1962.
108. Hara, T.; Asahi, H.; Uemori, R.; Tamehiro, H. Role of combined addition of niobium and boron and of molybdenum and boron on hardenability in low carbón steels. *ISIJ Int.* **2004**, *44*, 1431–1440. [CrossRef]
109. Luo, H.; Karjalainen, L.P.; Porter, D.A.; Liimatainen, H.M.; Zhang, Y. The influence of Ti on the hot ductility of Nb-bearing steels in simulated continuous casting process. *ISIJ Int.* **2002**, *42*, 273–282. [CrossRef]
110. Shukla, R.; Das, S.K.; Kumar, B.R.; Ghosh, S.K.; Kundu, S.; Chatterjee, S. An Ulta-low Carbon, Thermomechanically Controlled Processed Microalloyed Steel: Microstructure and Mechanical Properties. *Metall. Mater. Trans. A* **2012**, *43*, 4835–4845. [CrossRef]
111. Li, C.N.; Ji, F.Q.; Yua, G.; Kang, J.; Misra, R.D.K.; Wang, G.D. The impact of thermos-mechanical controlled processing on structure-property relationship and strain hardening behavior in dual-phase steels. *Mater. Sci. Eng. A* **2016**, *662*, 100–110. [CrossRef]
112. Oliver, S.; Jones, T.B.; Fourlaris, G. Dual phase versus TRIP strip steels: Microstructural changes as a consequence of quasi-static and dynamic tensile testing. *Mater. Charact.* **2007**, *58*, 390–400. [CrossRef]
113. Li, X.L.; Lei, C.S.; Deng, X.T.; Wang, Z.D.; Yu, Y.G.; Wang, G.D.; Misra, R.D.K. Precipitation strengthening in titanium microalloyed high-strength steel plates with new generation-thermomechanical controlled processing (NG-TMCP). *J. Alloys Compd.* **2016**, *689*, 542–553. [CrossRef]
114. Zhao, M.C.; Yang, K.; Xiao, F.R.; Shan, Y.Y. Continuous cooling transformation of undeformed and deformed low carbon pipeline steels. *Mater. Sci. Eng. A* **2003**, *355*, 126–136. [CrossRef]
115. Bose-Filho, W.W.; Carvalho, A.L.M.; Strangw, M. Effects of alloying elements on the microstructure and inclusion formation in HSLA multipass welds. *Mater. Charact.* **2007**, *58*, 29–39. [CrossRef]
116. Beidokhti, B.; Kokabi, A.H.; Dolati, A. A comprehensive study on the microstructure of high strength low alloy pipeline welds. *J. Alloys Compd.* **2014**, *597*, 142–147. [CrossRef]
117. Bailey, N.; Jones, S.B. *Solidification Cracking of Ferritic Steel during Submerged Arc-Welding*; Welding Institute: Cambridge, UK, 1978; pp. 217–231.
118. Mandelberg, S.L.; Rybacov, A.A.; Sidorenko, B.G. Resistance of pipe steel welded joints to solidification cracks. *Avtom. Svarka* **1972**, *3*, 1–5.
119. Shiga, C. Effects of steelmaking, alloying and rolling variables on the HAZ structure and properties in microalloyed plate and line pipe. In Proceedings of the International Conference on The Metallurgy, Welding, and Qualification of Microalloyed (HSLA) Steel Weldments, Houston, TX, USA, 6–8 November 1990; pp. 327–350.
120. Aihara, S.; Okamoto, K. Influence of Local Brittle Zone on HAZ Toughness of TMCP Steels. In Proceedings of the International Conference on The Metallurgy, Welding, and Qualification of Microalloyed (HSLA) Steel Weldments, Houston, TX, USA, 6–8 November 1990; pp. 402–426.
121. García, R.; López, V.H.; Lazaro, Y.; Aguilera, J. *Grain Refi nement in Electrogas Welding of Microalloyed Steels by Inducing a Centred Magnetic Field*; Soldagem & Inspecao: Sao Paulo, Brasil, 2007; pp. 300–304.
122. Loberg, B.; Nordgren, A.; Strid, J.; Easterling, K.E. The Role of Alloy Composition on the Stability of Nitrides in Ti-Microalloyed Steels during Weld Thermal Cycles. *Metall. Trans. A* **1984**, *15*, 33–41. [CrossRef]
123. Zhang, W.; Elmer, J.W.; DebRoy, T. Modeling and real time mapping of phases during GTA welding of 1005 steel. *Mater. Sci. Eng. A* **2002**, *333*, 320–335. [CrossRef]

124. Wan, X.L.; Wang, H.H.; Cheng, L.; Wu, K.M. The formation mechanisms of interlocked microstructures in low-carbon high-strength steel weld metals. *Mater. Charact.* **2012**, *67*, 41–51. [CrossRef]
125. Terada, Y.; Tamehiro, H.; Morimoto, H.; Hara, T.; Tsuru, E.; Asahi, H. X100 Linepipe With Excellent HAZ Toughness and Deformability. In Proceedings of the 22nd International Conference on Offshore Mechanics and Arctic Engineering ASME, Cancun, Mexico, 8–13 June 2003; pp. 1–8.
126. Matsuda, F.; Ikeuchi, K.; Liao, J.S.; Tanabe, H. Weld HAZ Toughness and Its Improvement of Low Alloy Steel SQV-2A for Pressure Vessels (Report 2): Microstructure and Charpy Impact Behavior of Intercritically Reheated Coarse Grained Heat Affected Zone (ICCGHAZ) (Materials, Metallurgy & Weldability). *Trans. JWRI* **1994**, *23*, 49–57.
127. Shim, J.H.; Cho, Y.W.; Chung, S.H.; Shim, J.D.; Lee, D.N. Nucleation of intragranular ferrite at Ti_2O_3 particle in low carbon steel. *Acta Mater.* **1999**, *47*, 2751–2760. [CrossRef]
128. Guo, A.M.; Li, S.R.; Guo, J.; Li, P.H.; Ding, Q.F.; Wu, K.M.; He, X.L. Effect of zirconium addition on the impact toughness of the heat affected zone in a high strength low alloy pipeline steel. *Mater. Charact.* **2008**, *59*, 134–139. [CrossRef]
129. Lee, J.L.; Pan, Y.T. The Formation of Intragranular Acicular Ferrite in Simulated Heat-affected Zone. *ISIJ Int.* **1995**, *35*, 1027–1033. [CrossRef]
130. Gianetto, J.A.; Braid, J.E.M.; Bowker, J.T.; Tyson, W.R. Heat-Affected Zone Toughness of a TMCP Steel Designed for Low-Temperature Applications. *Trans. ASME* **1997**, *119*, 134–144. [CrossRef]
131. Matsuda, F.; Ikeuchi, K.; Fukada, Y.; Horii, Y.; Okada, H.; Shiwaku, T.; Shiga, C.; Suzuki, S. Review of Mechanical and Metallurgical Investigations of M-A Constituent in Welded Joint in Japan. *Trans. JWRI* **1995**, *24*, 1–24.
132. Lan, L.; Qiu, C.; Zhao, D.; Gao, X.; Du, L. Microstructural characteristics and toughness of the simulated coarse grained heat affected zone of high strength low carbon bainitic steel. *Mater. Sci. Eng. A* **2011**, *529*, 192–200. [CrossRef]
133. Hunt, A.C.; Kluken, A.O.; Edwards, G.R. Heat input and dilution effects in microalloyed steel weld metals. *Weld. J.* **1994**, *73*, 9–15.
134. Karabulut, H.; Türkmen, M.; Erden, M.A.; Gündüz, S. Effect of Different Current Values on Microstructure and Mechanical Properties of Microalloyed Steels Joined by the Submerged Arc Welding Method. *Metals* **2016**, *6*, 281. [CrossRef]
135. Zhang, W.; Elmer, J.W.; DebRoy, T. Kinetics of ferrite to austenite transformation during welding of 1005 steel. *Scr. Mater.* **2002**, *46*, 753–757. [CrossRef]
136. Ekicia, M.; Ozsarac, U. Investigation of Mechanical Properties of Microalloyed Steels Joined by GMAW and Electrical Arc Welding. *Acta Phys. Pol. A* **2013**, *123*, 289–290. [CrossRef]
137. Dunđer, M.; Vuherer, T.; Samardžić, I. Weldability of microalloyed high strength steels TStE 420 and S960QL. *Metalurgija* **2014**, *53*, 335–338.
138. Sharma, V.; Shahi, A.S. Quenched and tempered steel welded with micro-alloyed based ferritic fillers. *J. Mater. Process. Technol.* **2018**, *253*, 2–16. [CrossRef]
139. Liu, H.J.; Shen, J.J.; Zhou, L.; Zhao, Y.Q.; Li, C. Microstructural characterisation and mechanical properties of friction stir welded joints of aluminium alloy to copper. *Sci. Technol. Weld. Join.* **2011**, *16*, 92–99. [CrossRef]
140. Hart, P.H.M.; Harrison, P.L. Compositional Parameters for HAZ Cracking and Hardening in C-Mn Steels. In Proceedings of the 67th Annual AWS Meeting, Atlanta, GA, USA, 14–16 April 1986; pp. 13–18.
141. Wang, S.H.; Luu, W.C.; Ho, K.F.; Wu, J.K. Hydrogen permeation in a submerged arc weldment of TMCP steel. *Mater. Chem. Phys.* **2002**, *77*, 447–454. [CrossRef]
142. Sun, Q.; Di, H.; Li, J.; Wu, B.; Misra, R. A comparative study of the microstructure and properties of 800 MPa microalloyed C-Mn steel weld joints by laser and gas metal arc welding. *Mater Sci Eng A* **2016**, *669*, 150–158. [CrossRef]
143. Wang, X.; Zhang, S.; Zhou, J.; Zhang, M.; Chen, C.; Misra, R. Effect of heat input on microstructure and properties of hybrid fiber laser-arc weld joints of the 800 MPa hot-rolled Nb-Ti-Mo microalloyed steels. *Opt. Lasers Eng.* **2017**, *91*, 86–96. [CrossRef]
144. Zhang, M.; Wang, X.; Zhu, G.; Chen, C.; Hou, J.; Zhang, S.; Jing, H. Effect of laser welding process parameters on microstructure and mechanical properties on butt joint of new hot-rolled nano-scale precipitate strenthen steel. *Acta Metall. Sin.* **2014**, *27*, 521–529. [CrossRef]
145. Bhadesia, H. *Bainite in Steels*; The Institute of Materials: London, UK, 1992.

146. Pamnani, R.; Karthik, V.; Jayamukar, T.; Vasudevan, M.; Sakthivel, T. Evaluation of Mechanical Properties across Micro Alloyed HSLA Steel Weld Joints using Atomated Ball Indentation. *Mater. Sci. Eng. A* **2015**, *651*, 214–223. [CrossRef]
147. Chatzidouros, E.V.; Papazoglou, V.J.; Tsiourva, T.E.; Pantelis, D.I. Hydrogen effect on fracture toughness of pipeline steel welds, with insitu hydrogen charging. *Int. J. Hydrog. Energy* **2011**, *36*, 12626–12643. [CrossRef]
148. Demofonti, G.; Mannucci, G.; Di Biagio, M.; Hillenbrand, H.G.; Harris, D. Fracture propagation resistance evaluation of X100 TMCP steel pipes for high pressure gas transportation pipelines by full scale burst tests. *Pipeline Technol. Proc.* **2004**, *1*, 467–482.
149. Yakubtsov, I.A.; Poruks, P.; Boyd, J.D. Microstructure and mechanical properties of bainitic low carbon high strength plate steels. *Mater. Sci. Eng. A.* **2008**, *480*, 109–116. [CrossRef]
150. Woodtli, J.; Kieselbach, R. Damage due to hydrogen embrittlement and stress corrosion cracking. *Eng. Fail. Anal.* **2000**, *7*, 427–450. [CrossRef]
151. Hui, W.J.; Weng, Y.Q.; Dong, H. *Steels for High Strength Fastener*; Metallurgical Industry Press: Beijing, China, 2009.
152. Cottis, R.A. *Hydrogen Embrittlement*; School of Materials: Manchester, UK, 2010.
153. Domizzi, G.; Anteri, G.; Ovejero-García, J. Influence of sulphur content and inclusion distribution on the hydrogen induced blister cracking in pressure vessel and pipeline steels. *Corros. Sci.* **2001**, *43*, 325–339. [CrossRef]
154. Hardie, D.; Charles, E.A.; Lopez, A.H. Hydrogen embrittlement of high strength pipeline steels. *Corros. Sci.* **2006**, *48*, 4378–4385. [CrossRef]
155. Shin, S.Y.; Hwang, B.; Lee, S.; Kim, N.J.; Ahn, S.S. Correlation of microstructure and charpy impact properties in API X70 and X80 line-pipe steels. *Mater. Sci. Eng. A* **2007**, *458*, 281–289. [CrossRef]
156. Shterenlikht, A.; Hashemi, S.H.; Howard, I.C.; Yates, J.R.; Andrews, R.M. A specimen for studying the resistance to ductile crack propagation in pipes. *Eng. Fract. Mech.* **2004**, *71*, 1997–2013. [CrossRef]
157. Asoaka, T.; Lapasset, G.; Aucouturier, M.; Lacombe, P. Observation of hydrogen trapping in Fe-0.15 wt % Ti alloy by high resolution autoradiography. *Corrosion* **1978**, *34*, 39–47. [CrossRef]
158. Qian, L.; Andrej, A. Reversible hydrogen trapping in a 3.5 NiCrMoV medium strength steel. *Corros. Sci.* **2015**, *96*, 112–120.
159. Mohsen, D.; Martin, M.L.; Nagao, A.; Sofronis, P.; Robertson, I.M. Modeling hydrogen transport by dislocations. *J. Mech. Phys. Solids* **2015**, *78*, 511–525.
160. Marchetti, L.; Herms, E.; Lagoutaris, P.J. Hydrogen embrittlement susceptibility of tempered 9% Cr, 1% Mo steel. *Int. J. Hydrog. Energy* **2011**, *34*, 15880–15887. [CrossRef]
161. Druce, P.D.; Daniel, H.; Ahmed, A.S.; Ayesha, J.H.; Andrzej, C.; Pereloma, V.E. Investigation of the effect of electrolytic hydrogen charging of x70 steel: I. The effect of microstructure on hydrogen-induced cold cracking and blistering. *Int. J. Hydrog. Energy* **2016**, *41*, 12411–12423.
162. Liu, Q.; Venezuela, J.; Zhang, M.; Zhou, Q.; Atrens, A. Hydrogen trapping in some advanced high strength Steels. *Corros. Sci.* **2016**, *111*, 770–785. [CrossRef]
163. Hirth, J. Effects of hydrogen on the properties of iron and steel. *Metall. Trans. A* **1980**, *11*, 861–890. [CrossRef]
164. Thomas, R.L.S.; Li, D.; Gangloff, R.P.; Scully, J.R. Trap-governed hydrogen diffusivity and uptake capacity in ultrahigh-strength AERMET 100 steel. *Metall. Mater. Trans. A* **2002**, *33*, 1991–2004. [CrossRef]
165. Venezuela, J.; Liu, Q.; Zhang, M.; Zhou, Q.; Atrens, A. The influence of hydrogen on the mechanical and fracture properties of some martensitic advanced high strength steels studied using the linearly increasing stress test. *Corros. Sci.* **2015**, *99*, 98–117. [CrossRef]
166. Wei, F.G.; Tsuzaki, K. Quantitative analysis on hydrogen trapping of TiC particles in steel. *Metall. Mater. Trans. A* **2006**, *37*, 331–353. [CrossRef]
167. Depover, T.; Escobar, D.P.; Wallaert, E.; Zermout, Z.; Verbeken, K. Effect of hydrogen charging on the mechanical properties of advanced high strength steels. *Int. J. Hydrog Energy* **2014**, *39*, 4647–4656. [CrossRef]
168. Wallaert, E.; Depover, T.; Arafin, M.; Verbeken, K. Thermal desorption spectroscopy evaluation of the hydrogen-trapping capacity of NbC and NbN precipitatesMetall. *Mater. Trans. A* **2014**, *45*, 2412–2420. [CrossRef]
169. Takahashi, J.; Kawakami, K.; Kobayashi, Y.; Tarui, T. The first direct observation of hydrogen trapping sites in TiC precipitation-hardening steel through atom probe tomography. *Scr. Mater.* **2010**, *63*, 261–264. [CrossRef]

170. Valentini, R.; Solina, A.; Matera, S.; de Gregorio, P. Influence of titanium and carbon contents on the hydrogen trapping of microalloyed steels. *Metall. Mater.* **1996**, *27*, 3773–3780. [CrossRef]
171. McNabb, A.; Foster, P.K. A new analysis of the diffusion of hydrogen in iron and ferritic steels. *Trans. Metall. Soc. AIME* **1963**, *227*, 618–627.
172. Gangloff, R.P. *Hydrogen Assisted Cracking of High Strength Alloys*; In Comprehensive Structural Integrity; Elsevier: New York, NY, USA, 2003; pp. 31–101.
173. Lynch, S. Hydrogen embrittlement phenomena and mechanisms. *Corros. Rev.* **2012**, *30*, 105–123. [CrossRef]
174. Oriani, R.A. A mechanistic theory of hydrogen embrittlement of steels. *Berich. Bunsengesellsch. Phys. Chem.* **1972**, *76*, 848–857.
175. Beachem, C.D. A new model for hydrogen assisted cracking (Hydrogen embrittlement). *Metall. Trans. A* **1972**, *3*, 437–451. [CrossRef]
176. Lynch, S.P. Environmentally assisted cracking: Overview of evidence for an adsorption-induced localized-slip process. *Acta Metall.* **1988**, *20*, 2639–2661. [CrossRef]
177. Tehemiro, H.; Takeda, T.; Matsuda, S.; Yamamoto, K.; Komura, H. Effect of accelerated cooling after controlled rolling on the hydrogen-induced cracking resistance of pipeline steels. *Trans. Iron Steel Inst. Jpn.* **1985**, *25*, 982–988.
178. Ejim, F. Hydrogen Diffusion in Pipeline Steels. Ph.D. Thesis, Dipartimento di Chimica, Materiali e Ingegneria Chimica "Giulio Natta", Milan, Italy, 2011.
179. Pasco, R.W.; Ficalora, P.J. *Hydrogen Degradation of Ferrous Alloys*; Noyes Publications: Park Ridge, NJ, USA, 1984; pp. 199–214.
180. Afrooz, B. *Hydrogen Embrittlement*; Saarland University: Saarbrücken, Germany, 2011.
181. Myers, S.M.; Baskes, M.L.; Brinbaum, H.K.; Corbett, J.W.; Deleo, G.G.; Estreicher, S.K.; Haller, E.E.; Jena, P.; Johnson, N.M.; Kirchheim, R.; et al. Hydrogen interactions with defects in crystalline solids. *Rev. Mod. Phys.* **1992**, *64*, 559. [CrossRef]
182. Xu, K. *Hydrogen Embrittlement of Carbon Steels and Their Welds*; Gaseous Hydrogen Embrittlement of Materials in Energy Technology; Woodhead: Philadelphia, IL, USA, 2012; pp. 526–561.
183. Walter, R.J.; Chandler, W.T. *Effects of High Pressure Hydrogen on Metals at Ambient Temperature*; No. N-70-18637; NASA-CR-102425; Rocketdyne: Canoga Park, CA, USA, 1968.
184. Thompson, A.W.; Bernstein, I.M. Selection of structural materials for hydrogen pipelines and storage vessels. *Int. J. Hydrog. Energy*, **1977**, *2*, 163–173. [CrossRef]
185. Imbihl, R.; Behm, R.J.; Christmann, K.; Ertl, G.; Matsushima, T. Phase transitions of a two-dimensional chemisorbed system: H on Fe(110). *Surface Sci.* **1982**, *117*, 257–266. [CrossRef]
186. Shanmugan, S.; Ramisetti, N.K.; Mirsa, R.D.K. Microstructure and High Strenght-Toughness Combination of a New 700 MPa Nb Microalloyed Pipeline Steel. *Mater. Sci. Eng.* **2008**, *478*, 26–37. [CrossRef]
187. Liu, D.; Xu, H.; Yang, K.; Fang, H. Effect of bainite/Martensite Mixed Microstructure on the strenghth and toughness of low carbon alloy steels. *Acta Metall. Sin.* **2004**, *40*, 882–886.
188. Nayak, S.S.; Misra, R.D.K.; Hartmann, J.; Siciliano, F.; Gray, J.M. Microestructure and properties of low manganese and niobium containing HIC pipeline Steel. *Mater. Sci. Eng. A* **2008**, *494*, 456–463. [CrossRef]
189. Uranga, P.; Fernandez, A.I.; Lopez, B.; Rodriguez-Ibabe, J.M. Transition between static and metadynamic recrystallization kinetics in coarse Nb microalloyed austenite. *Mater. Sci. Eng. A* **2003**, *345*, 319–327. [CrossRef]
190. Zhou, M.; Du, L.X.; Liu, X.H. Relationship among Microstructure and properties and heat treatment of Ultra-High Strenght X120 pipeline steel. *J. Iron Steel Res. Int.* **2011**, *18*, 59–64. [CrossRef]
191. Ping, Z. Microstructure and Mechanical Properties in Isothermal Tempering of High Co-Ni Secondary Hardening Ultrahigh Strength Steel. *Sch. Mater. Sci. Eng. Beijing* **2007**, *14*, 292–295.
192. Asahi, H.; Ueneo, M.; Yonezawa, T. Prediction of Sulfide Stress Cracking in High Strenght Tubulars. *Corrosion* **1994**, *50*, 537. [CrossRef]
193. Lunarska, E.; Zielinski, A. *Hydrogen Degradation of Ferrous Alloys*; Noyes Publications: Park Ridge, NJ, USA, 1985.
194. Park, G.T.; Koh, S.U.; Jung, H.G.; Kim, K.Y. Effect of microstructure on the hydrogen trapping Efficiency induced Cracking of linepipe Steel. *Corros. Sci.* **2008**, *50*, 1865–1871. [CrossRef]
195. Liu, D.; Bai, B.; Fang, H.; Zhang, W.; Gu, J.; Chang, K. Effect of tempering and Carbide Free Bainite on the Mechanical Characteristics of a High Strenght Low Alloy Steel. *Mater. Sci. Eng.* **2004**, *371*, 40–44. [CrossRef]

196. Dong, C.F.; Xiao, K.; Liu, Z.Y.; Yang, W.J.; Li, X.G. Hydrogen Induced Cracking of X80 pipeline Steel. *Int. J. Miner. Metall. Mater.* **2010**, *17*, 579–586. [CrossRef]
197. Das, A.K. The Present and the Future of Line Pipe Steels for Petroleum Industry. *Mater. Manuf. Processes* **2010**, *25*, 1–3. [CrossRef]
198. Meimeth, S.; Grimpe, F.; Meuser, H.; Siegel, H.; Stallybrass, C.; Heckmann, C.J. Development, state of the art and future trends in design and production of heavy plates in X80 steel-grades, steel rolling 2006. In Proceedings of the 9th International & 4th European Conferences, Paris, France, 19–21 June 2006.
199. Madías, J. *Nuevas Plantas Latinoamericanas*; Industria del Acero: San Nicolás, Argentina, 2017; pp. 28–43.
200. Jansto, S. Current development in niobium high carbon applications. In Proceedings of the MS&T 2011, Materials Science & Technology 2011 Conference and Exhibition (MS&T Partner Societies), Columbus, OH, USA, 16–20 October 2011.
201. Jansto, S. Applied metallurgy of the microniobium R alloy approach in long and plate products. In Proceedings of the METAL, Brno, Czech Republic, 23–25 May 2012.
202. Jansto, S. Metallurgical mechanism and niobium efects on improved mechanical properties in high carbon steels. In Proceedings of the Microalloying 2015 & O shore Engineering Steels 2015, Hangzhou, China, 11–13 November 2015; pp. 981–986.

6

The Influence of La and Ce Addition on Inclusion Modification in Cast Niobium Microalloyed Steels

Hadi Torkamani [1,*], Shahram Raygan [1,*], Carlos Garcia Mateo [2,*], Jafar Rassizadehghani [1], Javier Vivas [2], Yahya Palizdar [3] and David San-Martin [2]

1 School of Metallurgy and Materials Engineering, College of Engineering, University of Tehran, 111554563 Tehran, Iran; jghani@ut.ac.ir
2 Materalia Research Group, National Center for Metallurgical Research (CENIM), Consejo Superior de Investigaciones Científicas (CSIC), E-28040 Madrid, Spain; jvm@cenim.csic.es (J.V.); dsm@cenim.csic.es (D.S.-M.)
3 Research Department of Nano-Technology and Advanced Materials, Materials and Energy Research Center, 3177983634 Karaj, Iran; y.palizdar@merc.ac.ir
* Correspondence: h.torkamani@ut.ac.ir (H.T.); shraygan@ut.ac.ir (S.R.); cgm@cenim.csic.es (C.G.M.),

Abstract: The main role of Rare Earth (RE) elements in the steelmaking industry is to affect the nature of inclusions (composition, geometry, size and volume fraction), which can potentially lead to the improvement of some mechanical properties such as the toughness in steels. In this study, different amounts of RE were added to a niobium microalloyed steel in as-cast condition to investigate its influence on: (i) type of inclusions and (ii) precipitation of niobium carbides. The characterization of the microstructure by optical, scanning and transmission electron microscopy shows that: (1) the addition of RE elements change the inclusion formation route during solidification; RE > 200 ppm promote formation of complex inclusions with a (La,Ce)(S,O) matrix instead of Al_2O_3-MnS inclusions; (2) the roundness of inclusions increases with RE, whereas more than 200 ppm addition would increase the area fraction and size of the inclusions; (3) it was found that the presence of MnS in the base and low RE-added steel provide nucleation sites for the precipitation of coarse niobium carbides and/or carbonitrides at the matrix–MnS interface. Thermodynamic calculations show that temperatures of the order of 1200 °C would be necessary to dissolve these coarse Nb-rich carbides so as to reprecipitate them as nanoparticles in the matrix.

Keywords: niobium microalloyed steel; as-cast condition; inclusion; rare earth elements; precipitation

1. Introduction

The chemical composition, population density, and morphology of non-metallic inclusions in metals are among the key factors determining the steels' quality [1–4]. These issues have become the leading subjects in the field of steelmaking processes in the last few decades.

Rare Earth (RE) elements are known as non-metallic inclusion modifiers that can be added into the molten steel in the form of misch metal, a master alloy consisting of rare earth elements such as Ce and La. In contrast to MnS, RE-based inclusions do not deform during hot metal working i.e., they keep their spherical shape, which seems to be more beneficial for the toughness. In fact, despite various roles of RE in steels, the main use of RE in steels concerns the shape control of inclusions, especially MnS particles during the hot deformation processes [5–9]. It has been suggested that the addition of these elements results in a considerable change in inclusion composition and generally leads to the formation of several constituents such as oxysulphides (Ce_2O_2S, La_2O_2S), oxides (Ce_2O_3, La_2O_3) and sulfides (Ce_2S_3, La_2S_3) [10,11].

The standard Gibbs free energy (ΔG°) for the formation of characteristic La- and Ce-based oxides/sulfides is given in Table 1. The values contained in this table have been obtained from different references [12–14]. From this table, it can be discovered that at high temperatures, this energy is so negative that it causes the formation of these components right after their addition into the liquid steel; however, due to their densities, the removal of the RE inclusions from the molten steel is relatively difficult [7,10,11,15–17]. Table 2 illustrates the melting point and density of some typical La and Ce oxides and sulfides. The values shown in this table have been taken from Ref. [11].

Table 1. Standard Gibbs energy (ΔG° = A + BT) of the formation of oxide and sulfide of La and Ce and its value at 1600 °C (1873 K) [12–14].

Compound	A, J mol^{-1}	B, J (mol K)$^{-1}$	ΔG°$_{1873\ K}$ kJ mol^{-1}
Ce_2O_3	-1.30×10^6	374	−600
La_2O_3	-1.44×10^6	337	−810
Ce_2S_3	-1.02×10^6	340	−383
La_2S_3	-1.27×10^6	417	−490

Table 2. Physical properties of oxide and sulfide of La and Ce [11].

Compound	Melting Point °C	Density kg/m^3
Ce_2O_3	~2177	6200
La_2O_3	~2249	6500
Ce_2S_3	~2150	5020
La_2S_3	~2099	5000

Although the effects of RE on the shape, fraction and distribution of inclusions have been widely studied, it seems that there is no unanimity in this regard. For instance, Grajcar et al. [9] suggested that the area fraction of non-metallic inclusions in the steels modified by misch metal was in the range of 0.0012 to 0.0018, which was twice as low as that of untreated steels, while the average area of the particles was the same for both conditions. On the other hand, Handerhan et al. [18] reported that the volume fraction of inclusions has been similar for the base and RE-added steels, but inclusions in the samples with RE were larger, which led to a larger interspacing of the inclusions. The same result was also reported elsewhere [15]. In contrast, Belyakova et al. [19] suggested that the number of inclusions increased by RE addition. In another work, it has been shown that when 0.35 kg/ton of misch metal is added to the molten steel, it results in obtaining a higher volume fraction of inclusions compared to the untreated steel [10]. However, adding higher amounts of RE could change the size distribution of the inclusions. In addition, it has been observed that the size of the inclusions decreases with low level of RE addition while it increases with higher level of RE addition; this somehow implies that the optimum amount of RE should be added to the steel [5].

The reasons for obtaining these different results (sometimes contradictory) could be attributed to the different steel making processes used in the various studies. For example, longer holding time during the ladle treatment after RE addition would cause more deposition or floatation of the inclusions. The location in the casting where the investigated samples have been taken can be another reason for such disagreements. Regarding the latter case, Paul et al. [10] studied the distribution and composition of the different inclusions in RE treated steels and showed that there is a considerable difference between the bottom and top of the ingots.

On the other hand, having reviewed the literature, there has been a large number of investigations pertaining to the use of RE elements through the hot deformation processes to derive benefits from the size and shape control of the inclusions at that stage. However, the contribution of as-cast condition to the inclusion characterization and consequently to the obtained properties has not been completely disclosed. In fact, there is little recent information in the literature concerned with the inclusion modification effects due to RE addition in cast steels while both steelmaking practice and steel compositions have considerably changed over the past decades.

Moreover, few studies have reported that the RE addition could improve the solubility of Nb and consequently nanoprecipitation behavior in steels, but there is no certain consensus about its mechanism [20,21].

In this work, an Nb-containing microalloyed steel has been selected to investigate the effects of RE addition on the modification of non-metallic inclusions as well as on the nanoprecipitation behavior in the as-cast condition. As it is discussed in the next sections, the results allow us to clarify the effects of RE addition (amount) on inclusion characteristics (size and composition) and also on the nanoprecipitation of Nb- and V-carbonitrides, which is directly related to the nature of inclusions formed in each alloy.

2. Experimental Procedure

2.1. Casting and History of Samples

Clean scrap steel was melted in a 100 kg capacity induction furnace under the air atmosphere. Once the melt reached to 1650 °C, the amounts of alloying elements were adjusted and the chemical composition was measured by using the Optical Emission Spectrometry (OES) technique on site. Since rare earth elements are strong oxide forming elements, it is desirable to add misch metal to the molten steel when the oxygen level is as low as possible. In order to meet this requirement, aluminum was added as deoxidizer to the melt prior to pouring the melt into the carrying ladle and adding the RE.

Three different amounts (2.5, 7 and 9 gr) of misch metal, containing 37.8 wt. % La and 62.1 wt. % Ce, were placed at the bottom of the 25 kg capacity carrying ladle as the last addition. In fact, the same melt with a base composition (Table 3) was used for all the castings while different amounts of RE were added to the melts to ensure obtaining the same compositions as the base steel but with different amounts of RE. The loss of RE elements in steelmaking is remarkably practice-dependent while many conducted studies only reported the amount of added rare earth elements per kg of molten steel. Hereupon, the amounts of RE in the ingots were measured by the Inductively Coupled Plasma (ICP) technique, the results of which are given in Table 4. In addition, the amount of O and N in the ingots was measured by using a gas analyzing equipment (model: LECO TC-436 AR (LECO Corporation, Saint Joseph, MI, USA) for the studied steels; the results are shown in Table 4. Regarding the amount of sulfur in these steels, it did not change compared to that of the base steel (Table 3). This result seems reasonable because the misch metal was added to the ladle after deoxidation and removal of floated impurities (steel slag); thus, RE would not have a considerable effect on the S content and this element would place in solid solution as well as forming the sulfide particles distributed through the as-cast ingot.

The ingots experienced homogenization treatment at 1100 °C for 5 h, and then smaller test samples were normalized at 950 °C for 30 min prior to their inspection under the microscopes.

Table 3. Chemical composition of the base microalloyed steel (Fe to balance).

Elements	C	Si	Mn	S	P	V	Nb	Mo	Cu	Al	Cr
wt. %	0.16	0.30	1.00	0.01	0.02	0.11	0.05	0.01	0.09	0.04	0.06

Table 4. Amounts of rare earth (RE) elements (La and Ce), O and N in different samples.

Steels	Elements, ppm				
	Ce	*La*	*Ce + La*	O	N
RE1	<10	<10	—	96 ± 10	113 ± 4
RE2	37.5	17.5	55.0	116 ± 35	114 ± 3
RE3	127.0	72.5	199.5	93 ± 6	112 ± 3
RE4	192.0	100.0	292.0	110 ± 14	114 ± 1

2.2. Sample Preparation and Metallographic Observations

Prior to inspection by Optical Microscopy (OM), the steel samples were ground and polished using standard metallographic procedures. Inclusion characterizations are usually carried out on optical micrographs taken from the polished surface to achieve better contrasts between the inclusions and the matrix. Therefore, the microstructure was characterized in the as-polished condition. Since the surface of the samples might react with water, the samples were dryly ground. In addition, special care was taken during grinding and polishing; controlled force was exerted on the samples during grinding to prevent removal of inclusions from the surface. In addition, a lubricant, which was a mix of ethanol and DP-Lubricant Blue (DP stands for Diamond Polishing; Struers Aps, Ballerup, Denmark), was used for polishing to ensure prevention of oxidation or any possible errors committed through the assessment of inclusions. In the last step of this preparation, ethanol was also used to remove any products coming from the polishing steps. The characterization of the area fraction, average area and roundness factor of the inclusions was carried out with the aid of an image analyzing program (Image J 1.47v, free software developed at the National Institute of Mental Health (NIMH), Bethesda, MD, USA) on at least five random micrographs using the same magnification for all the steel samples. For this characterization, the samples were extracted from the middle and 3 cm from the bottom of each Y block ingots (Figure 1). Considering the inclusions as circular in 2D, the average size (d) was calculated from their average area (A) according to $d = 2 \sqrt{(A/\pi)}$.

Scanning Electron Microscopy (SEM) in both Secondary and Back Scattered-Electron (SE and BSE) imaging modes plus microanalyses of inclusions were carried out by using a scanning electron microscope, model Hitachi S 4800 J (Hitachi Ltd., Chiyoda, Tokyo, Japan) with an Energy Dispersive X-ray Spectroscopy (EDS, Oxford INCA (Oxford Instruments plc., Abington, Oxfordshire, UK) capability. Similar to OM inspection, inclusion characterization by SEM was done on the polished samples.

Transmission Electron Microscopy (TEM) observations were carried out using a microscope model JEOL JEM 3000F (JEOL Ltd., Tokyo, Japan) equipped with an EDS unit (Oxford INCA) for elemental analyses. The samples were prepared from 3 mm diameter discs ground to ~80 μm thickness and then electropolished by Tenupol 5 (Struers Aps, Ballerup, Denmark) using 95/5: acetic/perchloric acid electrolyte at room temperature and the voltage of 40 V.

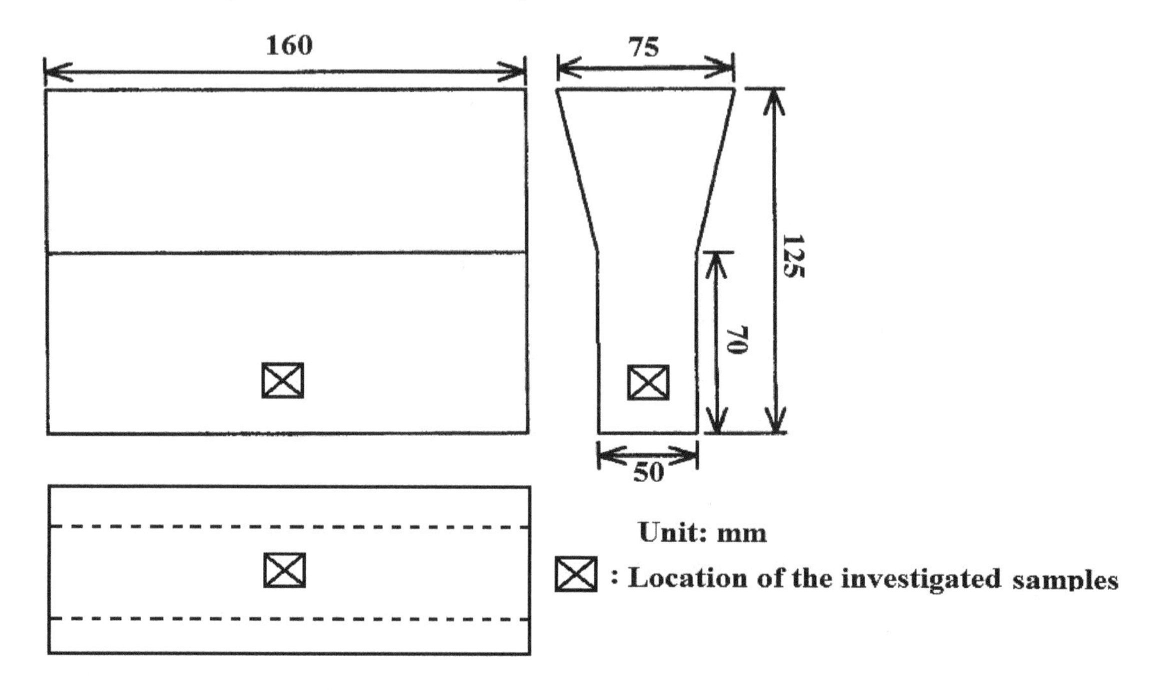

Figure 1. Scheme of the Y-block ingot and the location of investigated sample in it.

3. Results

3.1. OM Observations and Image Analyses of the Inclusions

Figure 2 shows characteristic optical images of distribution of the inclusions in different polished samples for steels R1, R2, R3 and R4, respectively. The average area fraction, average surface area and average roundness factor of the inclusions for the different steels are given in Table 5. These parameters were calculated from the images similar to those shown in Figure 3.

Considering the errors reported in Table 5, these values comprise of standard error (E) and also the errors imposed by the holes/gaps appeared on the polished surface. The value of E was calculated from the standard deviation (σ) and the number of measured inclusions (n) according to American Society for Testing and Materials (ASTM E2586): $E = \sigma/\sqrt{n}$. It should be noted that the average area of the inclusions have been estimated with respect to the total area covered by the inclusions and also the gaps appeared around some of the inclusions in the microstructure. It is very difficult to conclude whether the black particles, appeared in the OM images, are inclusions or pores. From the OM images, the gaps are not easily distinguishable as they appear with a similar color/contrast as the inclusions. However, they can be better differentiated in the SEM images, especially when using both SE and BSE imaging modes. The area covered by these gaps has been estimated and considered in calculation of the error of the reported results. The gaps themselves can be divided into two groups: (i) those caused during cooling by different thermal expansion coefficient between the inclusion and matrix; and (ii) those in which a broken part of an inclusion has been removed (a broken MnS particle as an example). This latter case has been avoided in this work by undertaking a careful metallographic preparation of the samples (which has been described in Section 2.2) and its contribution can be regarded as negligible. By considering the area covered by the gaps and by undertaking this correction, the intention of the authors has been to give the most accurate value for the area fraction of the inclusions. It should be noted that the gaps caused by thermal contraction have been more often observed in RE1 and RE2 rather than RE3 and RE4 steels. However, because of the reasons mentioned, there might be a minor error in the results of RE1 and RE2. The sum of the error given in Table 5 represents $\leq$5% of the average value measured for each sample.

It is known that area fraction/volume fraction (V) and mean diameter (dm) of the particles would affect the magnitude of mean free path (λ) between those particles according to Equations (1) and (2) [22–24], both of which result in a larger mean free path between the inclusions for the data obtained for steel RE3. According to Equation 1, this value was calculated to be around 12.4, 11.6, 14.1 and 11.4 μm for the data obtained for steels RE1, RE2, RE3 and RE4, respectively:

$$\lambda = \frac{4(1-V)}{3V}dm, \tag{1}$$

$$\lambda = \frac{(1-V)}{V}dm. \tag{2}$$

In addition, the results in this table also show that inclusions in samples RE3 and RE4 have an average roundness factor closer to 1 compared to the steels RE1 and RE2. This is clearly depicted in the images shown at higher magnification in Figure 3; these optical micrographs show that in RE1 (sample without RE) and RE2 (sample with 55 ppm RE), the roundness of inclusions is low (Figure 3a,b) while the roundness of the particles in RE3 and RE4 samples is closer to 1 (Figure 3c,d). It should be noted that, although the difference between average roundness factors is about 10–15 percent, the micrographs show a remarkable difference in roundness factor for the coarser inclusions. This is due to the fact that small inclusions in all samples look spherical (with roundness close to 1), which would affect the magnitude of the average roundness factor of inclusions for the different samples. Furthermore, in the micrographs shown in Figure 3, dark areas surrounded by gray envelope can be distinguished almost in all cases.

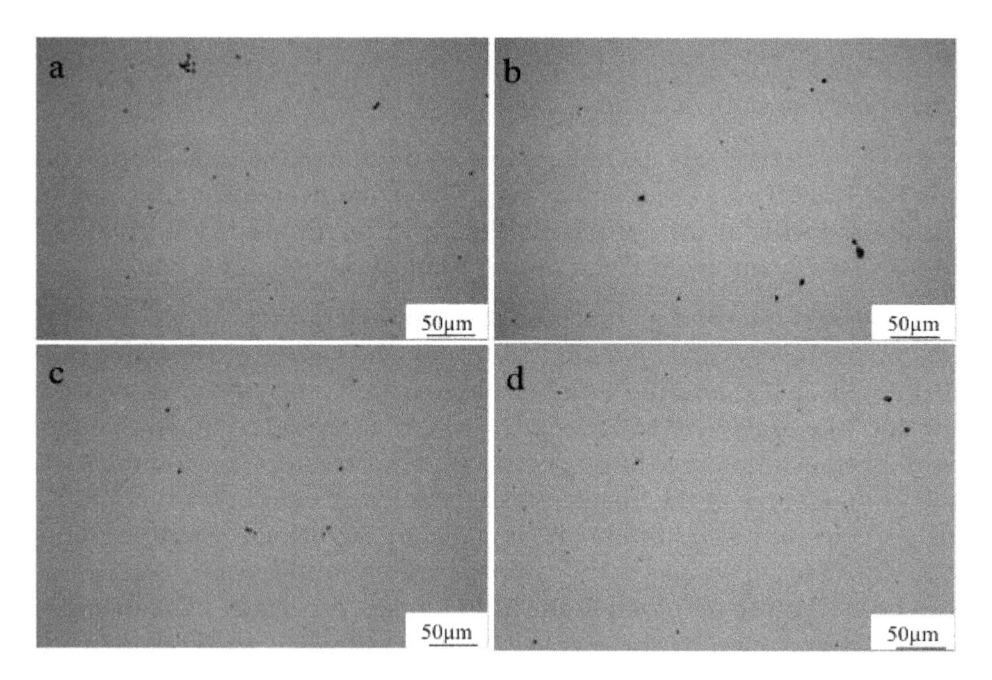

Figure 2. Optical micrographs showing the characteristic distribution of inclusions in steels: (**a**) RE1, (**b**) RE2, (**c**) RE3 and (**d**) RE4.

Table 5. Inclusion characteristics in different samples.

Steels	Area Fraction, %	Average Area, (μm^2)	Average Size, (μm)	Average Roundness Factor
RE1	0.123 ± 0.004	1.35 ± 0.06	1.31	0.71 ± 0.02
RE2	0.138 ± 0.005	1.54 ± 0.07	1.40	0.74 ± 0.03
RE3	0.105 ± 0.004	1.21 ± 0.04	1.24	0.83 ± 0.02
RE4	0.134 ± 0.005	1.38 ± 0.05	1.32	0.82 ± 0.03

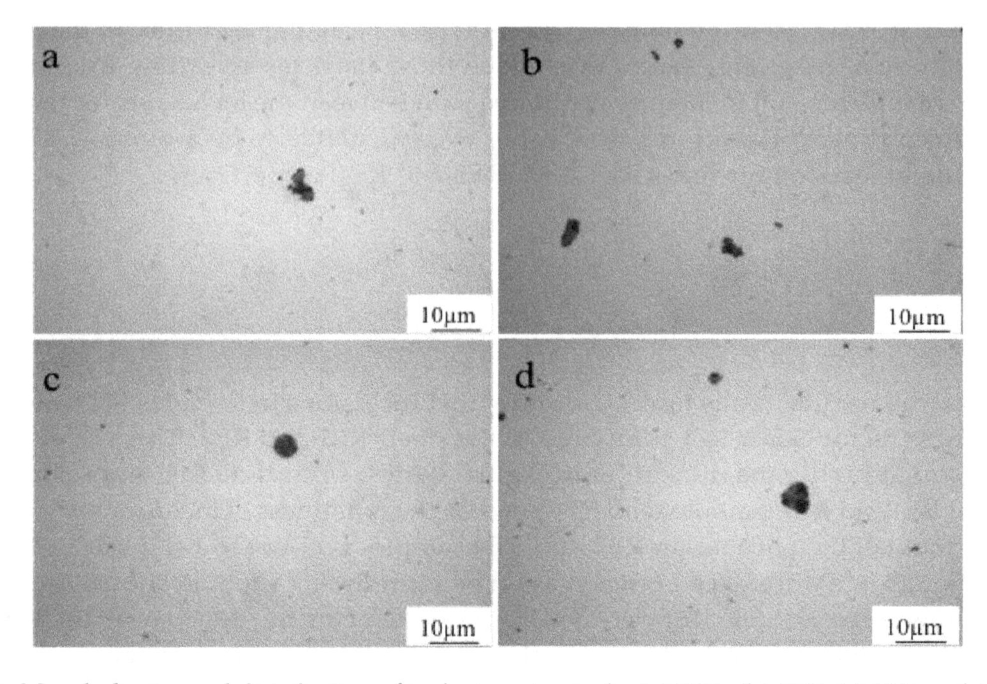

Figure 3. Morphologies and distribution of inclusions in steels: (**a**) RE1, (**b**) RE2, (**c**) RE3 and (**d**) RE4 at higher magnification.

3.2. *Inclusion Characterization by SEM*

3.2.1. Sample RE1

Figure 4 shows SEM images of characteristic inclusions observed in sample RE1 along with their microanalyses. It can be seen in Figure 4a that the roundness of the inclusion particles is low. In addition, in this figure, black areas could be considered as a gap/hole between the inclusion and the matrix. These gaps have been reported as one of the reasons for steels susceptibility to brittleness [25]. Figure 4b (spectrum No. 4) shows a considerable accumulation of Nb in the vicinity of these inclusions. A detailed evaluation of the larger inclusions in this sample using SE and BSE imaging modes (Figure 5a,b) revealed that there are white areas around the MnS particles. Microanalyses of these areas illustrate that there exist aggregations of Nb-rich phases (likely NbC) on the surface of MnS particle (Figure 5c). It is important to be mentioned that the use of Wavelength Dispersive Spectroscopy (WDS) should be considered if better limits of detection or accurate and precision performance is searched for light elements (C, O, N); thus, the results obtained for the light elements should be taken with caution.

The elemental distribution map of an inclusion in RE1 (Figure 6) confirms the accumulation of considerable amounts of Nb around the MnS. In addition, the micrographs show an Al_2O_3 inclusion surrounded by a MnS particle, which suggests the possibility of MnS nucleation on these oxides. This type of synergy between Al_2O_3 and MnS particles has been often observed in the microstructures. It is noteworthy that, for the steels deoxidized with aluminum, Al_2O_3 particles exist as non-metallic inclusions having unique faceted shapes, clusters of which tend to remain in solidified steels [26]. Apart from the Al_2O_3, another dark area can be seen in the bottom left part of this inclusion, which, according to the microanalyses, is suggested to be an Si-oxide particle probably originated from casting in the sand mold. As it can be seen, some parts around this particle have been probably removed during the preparation process.

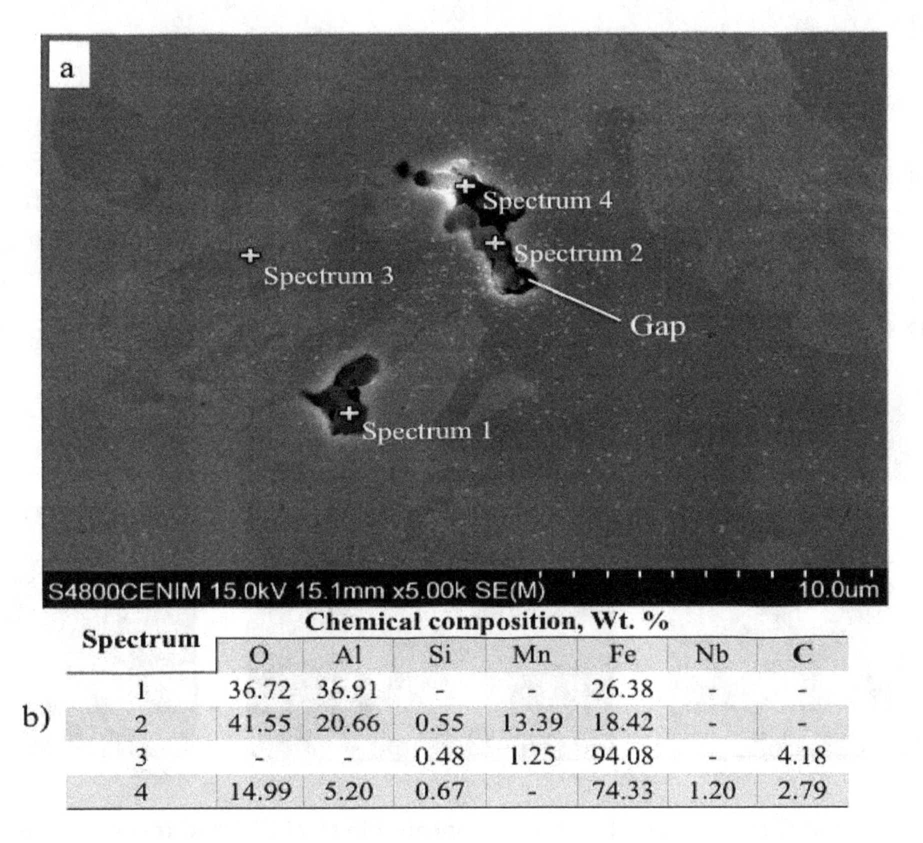

Spectrum	Chemical composition, Wt. %						
	O	Al	Si	Mn	Fe	Nb	C
1	36.72	36.91	-	-	26.38	-	-
2	41.55	20.66	0.55	13.39	18.42	-	-
3	-	-	0.48	1.25	94.08	-	4.18
4	14.99	5.20	0.67	-	74.33	1.20	2.79

Figure 4. (**a**) SEM (Scanning Electron Microscopy) micrograph and (**b**) EDS (Energy Dispersive X-ray Spectroscopy) results of the inclusions in steel RE1.

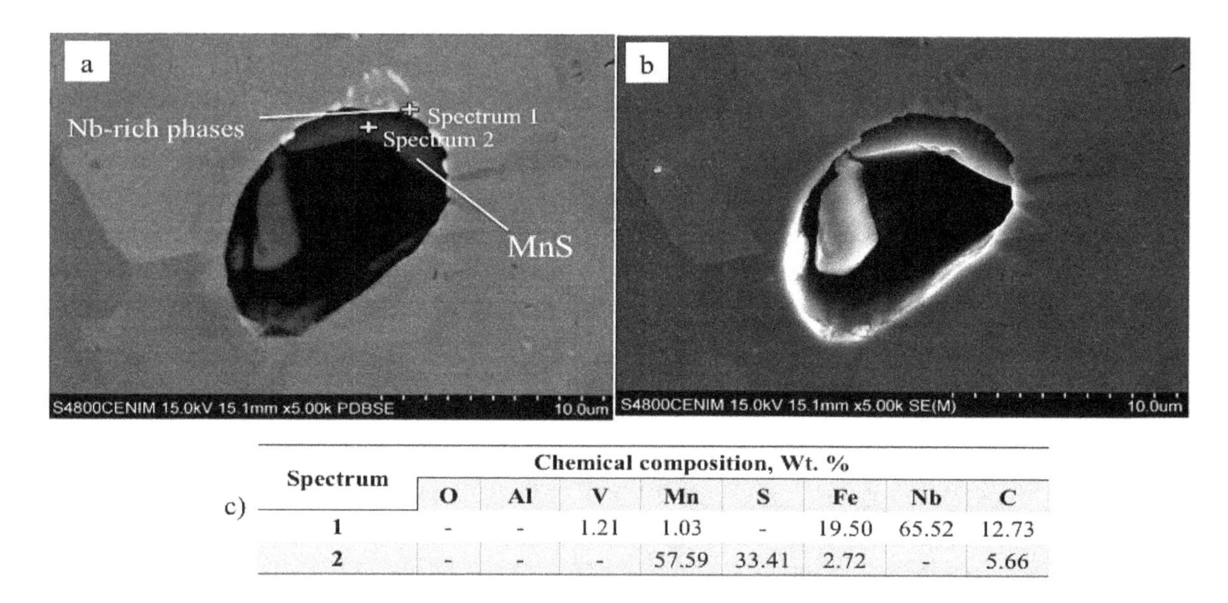

Spectrum	Chemical composition, Wt. %							
	O	Al	V	Mn	S	Fe	Nb	C
1	-	-	1.21	1.03	-	19.50	65.52	12.73
2	-	-	-	57.59	33.41	2.72	-	5.66

Figure 5. SEM micrographs in (**a**) BSE (Back Scattered Electron) and (**b**) SE (Secondary Electron) modes and (**c**) the results of EDS analysis of the inclusion observed in steel RE1.

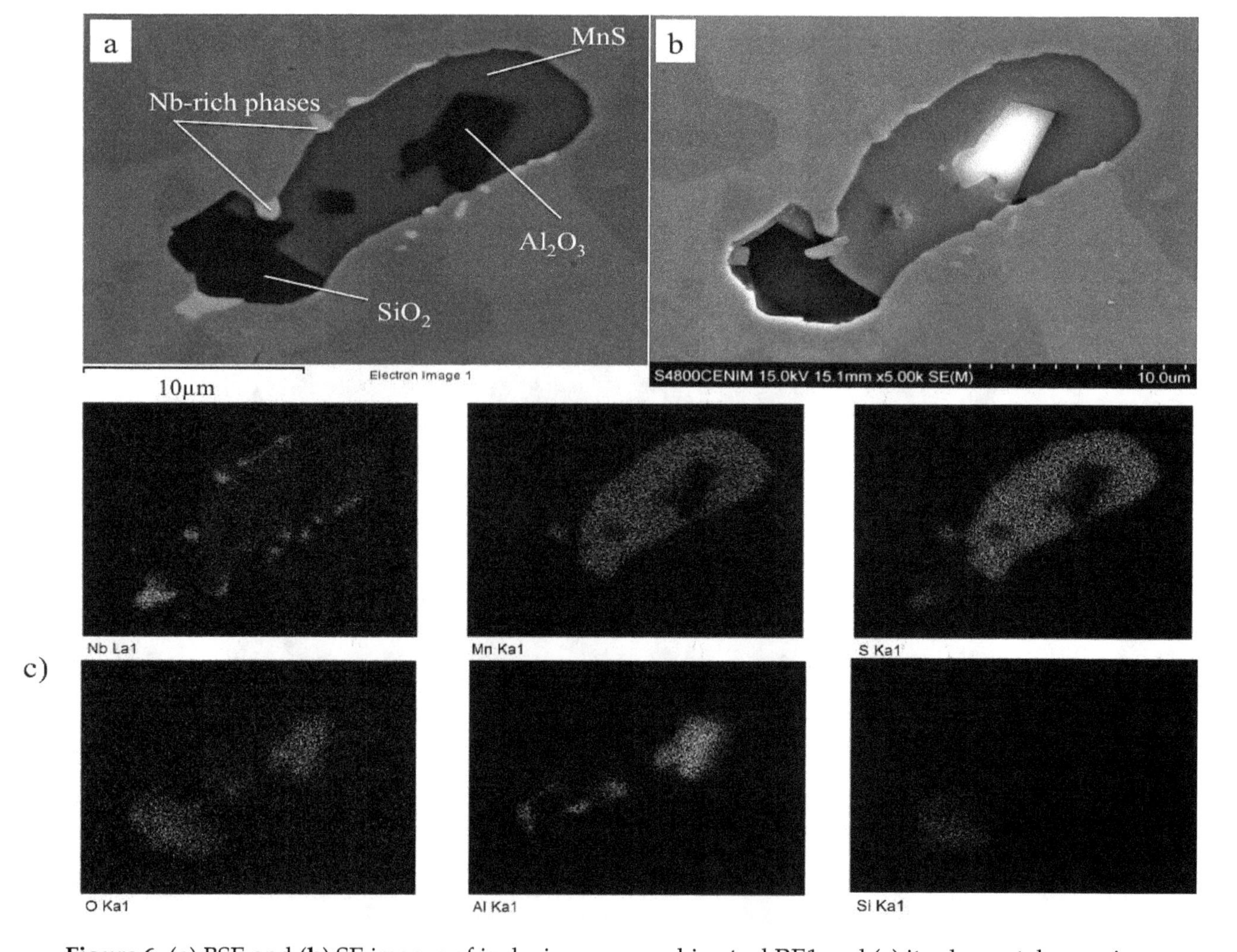

Figure 6. (**a**) BSE and (**b**) SE images of inclusion appeared in steel RE1 and (**c**) its elemental mapping.

3.2.2. Sample RE2

Figure 7 shows the SEM images and elemental map of an inclusion in sample RE2. Despite the addition of RE to this sample, there is not a significant change in the nature of inclusions; MnS could be considered as a dominant inclusion surrounding Al_2O_3. According to the elemental mappings, the existence of (La,Ce)-rich phases in the vicinity of Al-oxide would unveil the possibility of formation of these components on the preexisted Al_2O_3. Similar to steel RE1, precipitation of considerable amount of Nb-rich phases (white area) can be clearly seen around MnS in this steel (Figure 7a). Finding these NbC precipitates at the surface of MnS inclusions was not surprising, as the elements like Al, Mn, La or Ce form inclusions (oxy-sulphides) in the melt or in the pasty region first, while NbC particles would nucleate and grow/coarsen after the formation of these inclusions has taken place.

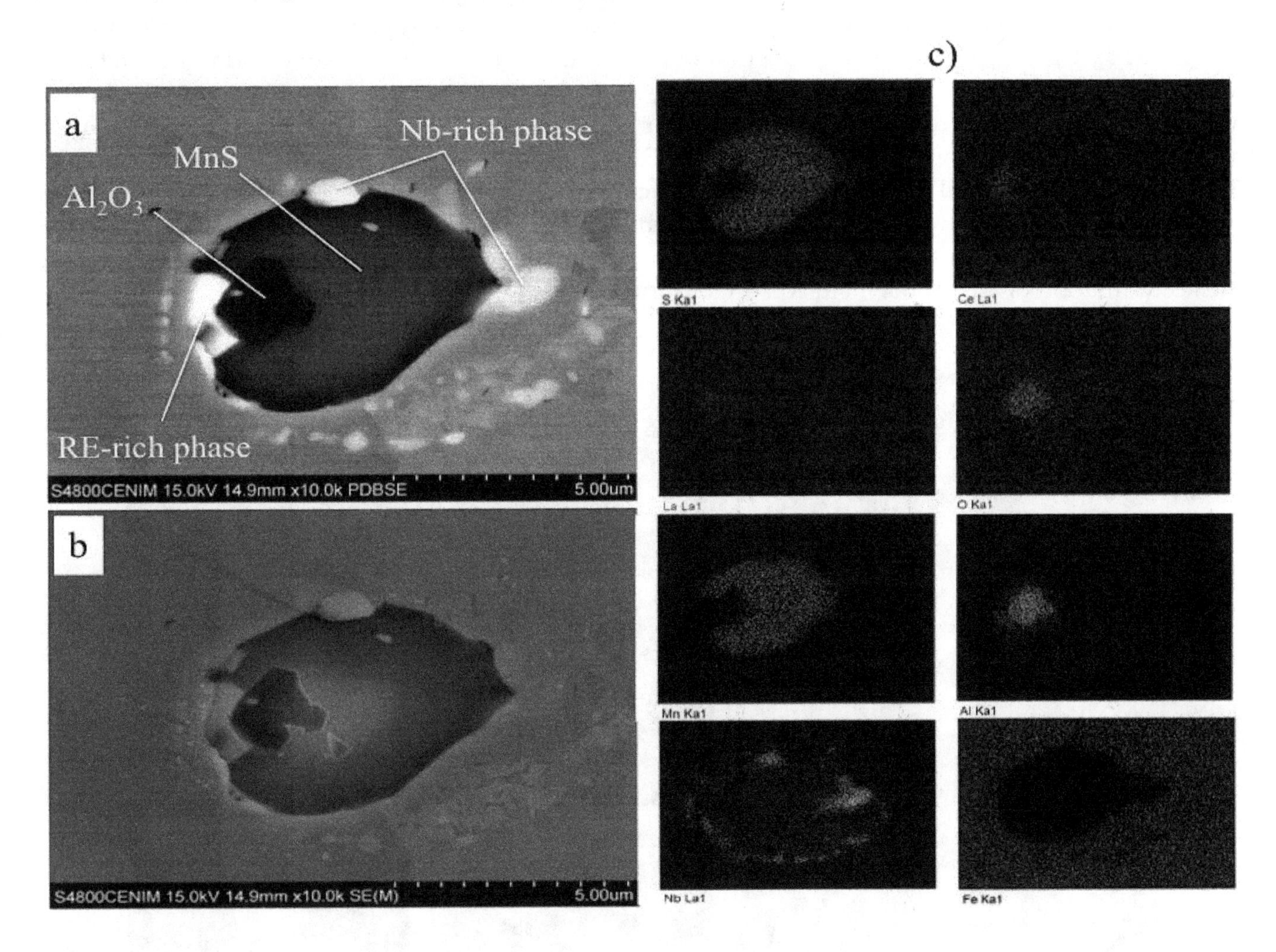

Figure 7. SEM micrographs in (**a**) BSE and (**b**) SE imaging modes of an inclusion appeared in steel RE2 and (**c**) its elemental mappings.

3.2.3. Samples RE3 and RE4

Figure 8 illustrates a complex inclusion observed in the microstructure of sample RE3. According to the microanalyses (elemental map), this complex inclusion is mainly composed of a cluster of cubic light particles all over this inclusion. Due to the high content in La, Ce, Al and O of these cubic particles, they seem to be (RE,Al)-based oxides (likely $(RE,Al)_2O_3$). The results of the EDS microanalysis performed on one of these cubic (Figure 8d) approve that due to the high oxygen content, the cubic light particles are oxides as labeled in Figure 8a. Previous reports indicated and discussed the agglomeration tendency of these cubic inclusions to lower the contact area with molten steel [26]. In addition, some small gray particles can be seen in SEM images, which are based on the microanalyses, are believed to be (RE,Mn)S. In addition to these particles, a darker phase similar to those observed in Figures 6 and 7 can be distinguished in this complex inclusion, which, according to the elemental mappings, is proposed to be Al_2O_3 type. These particles are distributed in the matrix of this inclusion, which seems to consist of RE-sulfides. Despite the addition of RE to sample RE2, such a complex inclusion has not been observed in that sample. Regarding the complex inclusion illustrated in Figure 8, there is no sign of Nb-rich areas in the outer surface of the inclusions, which have been noticed in RE1 and RE2 (Figures 5–7). It is worth mentioning that, although the Mn-containing particles co-exist with Al_2O_3 in RE1, RE2 and RE3, in the latter steel, the presence of Mn is much scarcer.

A characteristic inclusion in sample RE4 and its microanalyses are illustrated in Figure 9. As observed in steel RE3, the EDS analysis shows the co-existence of Al_2O_3 particles with RE inclusions in sample RE4. It should be also mentioned that in this sample the presence of Mn could not be detected as part of the inclusion composition, suggesting that MnS has not been formed (Figure 9c) in this sample. This is possibly due to the fact that sulfur has been linked to La/Ce and there is little sulfur available for the formation of MnS. In fact, when RE consumes S to form RE(S,O)/RES, the content of S in solid solution as well as its activity will be decreased, lowering the possibility of MnS formation in the presence of RE. Thus, it can be proposed that the rest of the sulfur exists in the form of solid solution in the matrix. In addition, in a similar way as for RE3, inclusions in sample RE4 do not show the accumulation of Nb on the inclusion–matrix interfaces.

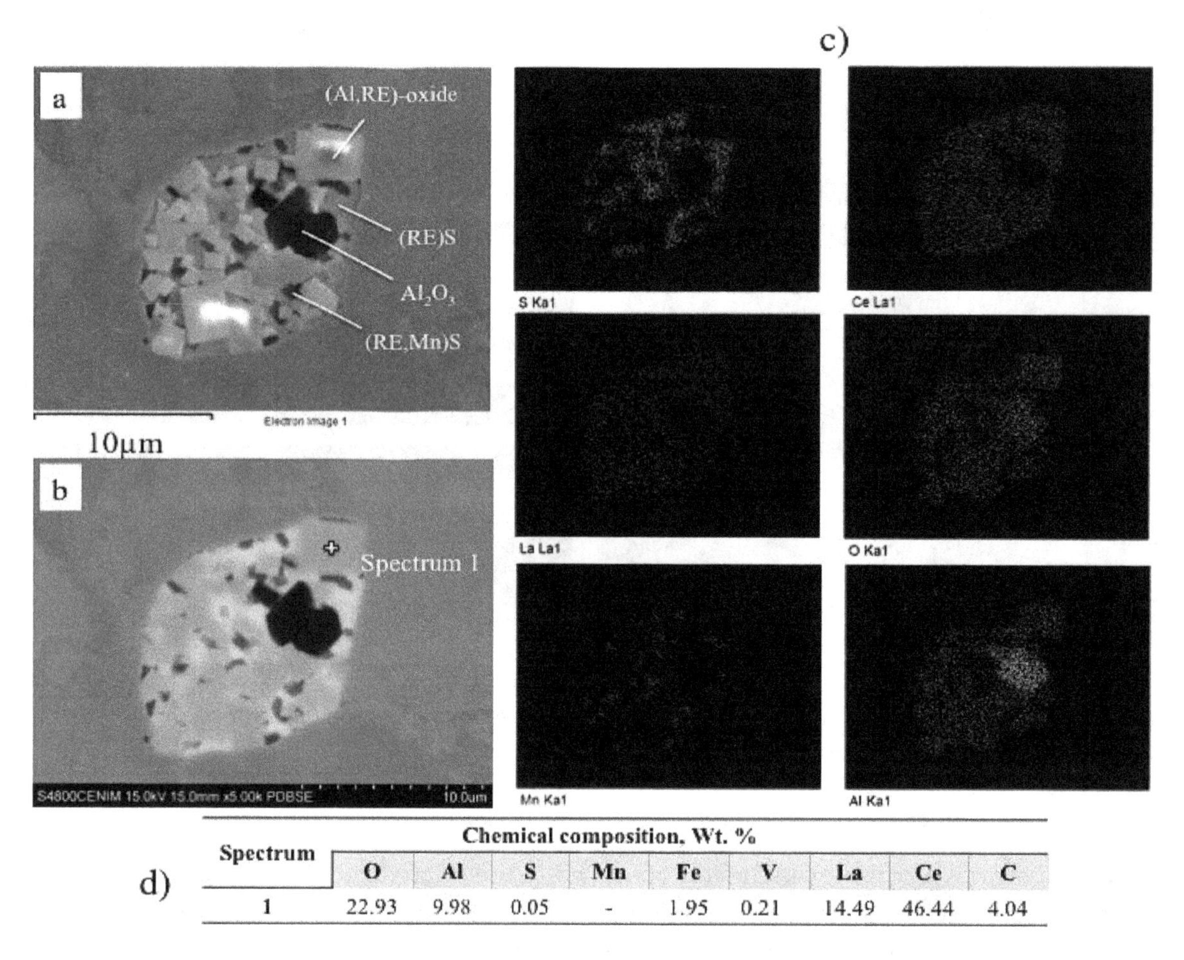

Spectrum	Chemical composition, Wt. %								
	O	Al	S	Mn	Fe	V	La	Ce	C
1	22.93	9.98	0.05	-	1.95	0.21	14.49	46.44	4.04

Figure 8. SEM micrographs in (**a**) SE and (**b**) BSE modes of a complex inclusion modified by 199.5 ppm RE, (**c**) elemental map of the corresponding inclusion in RE3 steel, and (**d**) the EDS results of spectrum 1.

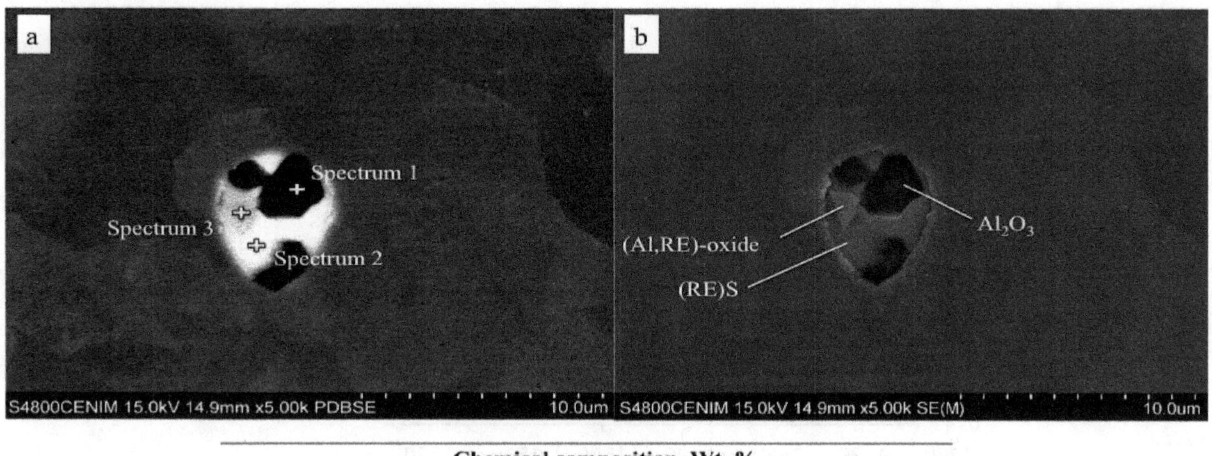

c)

Spectrum	Chemical composition, Wt. %						
	O	Al	S	Fe	Ce	La	C
1	39.56	28.77	1.30	1.39	11.52	8.16	7.16
2	1.83	0.97	19.17	2.65	41.40	21.89	12.07
3	12.07	5.04	7.65	5.32	38.73	20.23	11.28

Figure 9. SEM micrograph in (**a**) BSE and (**b**) SE modes, (**c**) EDS analysis of a complex inclusion observed in steel RE4.

4. Discussion

4.1. Nature of Inclusions: Volume Fraction, Size and Roundness

In order to complement and discuss the results obtained in this investigation, equilibrium phases and their transformation temperature ranges were calculated by means of Thermo-Calc® (Solna, Sweden), which is a thermodynamic software based on the CALPHAD (Computer Coupling of Phase Diagrams and Thermo-chemistry), using TCFE8 database and using the chemical composition of the base steel (Table 3). It should be mentioned that the database does not contain information regarding the influence of RE elements on equilibrium phase formation. Although these simulations do not take into account the influence of RE-alloying elements, they still give very useful information to understand some of the experimental observations presented in this investigation.

Figure 10 reveals that Al_2O_3 exists in the molten steel at temperatures even above 1500 °C, but MnS is present at the lower temperature range of the pasty region (<1464 °C). These data predict that, during solidification, alumina (Al_2O_3) would be formed first, followed by MnS. This sequence of formation would explain observations like that provided in Figure 6; alumina inclusions already present in the molten steel at high temperatures would be used as nucleation sites by MnS particles, which do form at lower temperatures. As a result, complex inclusions with an inner alumina core and MnS crust would be formed.

As it has been mentioned above, the Gibbs free energy of the formation of RE sulfides is so negative (Table 1); thus, these components form right after the RE addition into the molten steel. It seems that 55 ppm addition of RE into the steel (RE2) was not sufficient to consume a considerable amount of the sulfur in molten steel and, thus, some free sulfur also combined with Mn to promote the formation of MnS in RE2 (same as in RE1). Higher level of RE additions (RE3 and RE4) would lead to almost the complete consumption of the sulfur to form sulfides or oxysulphides, reducing the concentration of free sulfur in the molten steel considerably. In this case, the amount of sulfur available to form MnS, would be negligible, which could explain why this inclusion can hardly be found in steel RE3 (Figure 8) or has not been observed in the analysis presented in Figure 9. In other words, when the amount of RE addition is high enough to consume the entire or considerable amount of sulfur, the formation of MnS would be avoided and all the Mn would remain in solid solution.

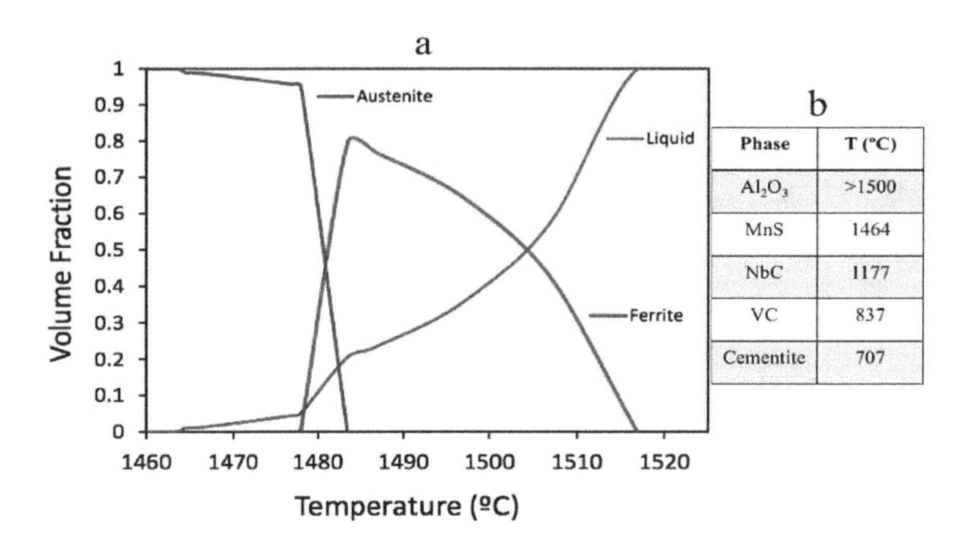

Phase	T (°C)
Al_2O_3	>1500
MnS	1464
NbC	1177
VC	837
Cementite	707

Figure 10. (**a**) equilibrium phases and solidification temperature range and (**b**) maximum temperature at which different equilibrium phases are present in the base steel according to Thermo-Calc predictions.

It was shown in Table 5 that, in comparison with the base steel, a low level of RE addition (55 ppm, RE2) results in having a higher area fraction of inclusions with a larger size while a higher level of rare earth addition (199.5 ppm, RE3) could decrease both their area fraction and average size. The same results (higher fraction of inclusions with low level of RE addition) has been reported earlier [10], where the authors claimed that this outcome would be contributed to the floatability of the inclusions (oxides, sulfides and oxy-sulphides) in the presence of RE and also to the location in the ingot from which the studied samples have been taken. In fact, the modification of the oxide inclusions to oxysulphides improves their floatability because of the lower density of oxysulphides/sulfides compared to oxides. Hence, with the formation of RE sulfides/oxysulphides, especially regarding the modification of Al_2O_3 clusters, RE3 steel achieved the lowest area fraction of inclusions by promoting the floatability of inclusions towards upper part of the ingot (this kind of complex inclusion that has been trapped during solidification is shown in Figure 8). In steel RE4, it seems that an excessive amount of RE has promoted a higher area fraction of inclusions with a larger average size, which could be caused by the higher activity of RE elements. Considering the cleanliness and associated mechanical properties of steels, it has been pointed out that the excessive amount of RE in steel should be avoided [5,25].

In this case, for the investigated type of microalloyed steel, it seems that 200 ppm of RE would be enough in order to avoid the formation of MnS as preferential sites for Nb accumulation and reach a high roundness factor. A higher level of RE addition beyond this amount would result, as discussed before, in having a greater volume fraction of inclusions with larger size, which is detrimental for steel properties [3].

In addition, it was found that, in comparison with the inclusions observed in samples RE1 and RE2, the roundness factor of inclusions in RE3 and RE4 is closer to 1, which could be attributed to the formation of RE(S,O)-Al_2O_3 inclusions in the molten steel and reaching the minimum surface energy with the melt [23].

4.2. Influence of RE Addition on the Accumulation of Nb-Rich Phases Around MnS and Nanoprecipitation

It is well documented that, among the microalloying elements, Nb plays the most important role as a solid solution strengthener and it also forms very fine precipitates in the matrix that can contribute to grain refinement and precipitation hardening in the microalloyed steels [27–35]. Therefore, the formation of coarse Nb(C,N) precipitates would reduce the amount of Nb available in solid solution to strengthen through nanoprecipitation. As it has been shown previously, it is evident that Nb accumulates around MnS in RE1 and RE2 samples (Figures 5–7), likely forming large NbC and/or

Nb(C,N) precipitates. According to the thermodynamic calculations, NbC precipitates form in the solid state (austenite) below ~1177 °C (Figure 10).

In addition, it is known that the presence of heterogeneous nucleation sites in the matrix, like MnS inclusions for Nb-rich phases, can even alter the formation range of Nb(C,N) precipitates, shifting it to higher temperatures [23]. To dissolve these primary large Nb-rich carbonitrides, the steel would have to be heated to very high temperatures [36], which is not always easy to reach. In contrast to MnS particles observed in steels RE1 and RE2, (La,Ce)(S,O)-Al_2O_3 inclusions do not seem to be preferential sites for Nb(C,N) nucleation in steels RE3 and RE4 (Figures 8 and 9).

The presence of the nanoprecipitates in the matrix has been characterized by TEM in steels RE2 and RE3. Figure 11 reveals the presence of few V-rich precipitates in the microstructure of sample RE2; these precipitates are also rich in Nb, which suggests that complex (Nb,V)(C,N) precipitates have been formed. According to the thermodynamic calculations, V-rich precipitates would only form at temperatures much lower than that of Nb(C,N) (<837 °C). As it has been shown in Figure 7, large Nb(C,N) particles have precipitated at the surface of MnS inclusions in steel RE2, reducing the amount of Nb in solid solution available to promote nanoprecipitates in the steel matrix, which is the reason why they have not been detected so easily in this steel. In contrast to RE2, large Nb(C,N) precipitates have not been observed at inclusion surfaces and the microstructure of RE3 sample shows the presence of several Nb-rich precipitates (Figure 12). These precipitates are also rich in V, although its presence is much lower than Nb. As mentioned above, previous studies [20,21] have suggested that RE addition increases the amount of Nb dissolved in solid solution in the austenite, which would allow forming Nb(C,N) nanoprecipitates during cooling in the matrix.

There is limited information in the literature concerned with the mechanism by which RE addition could affect the formation of NbC precipitates in steels. However, the present results suggest that its formation is associated with the presence of MnS inclusions in the microstructure. The addition of significant concentrations of RE elements in RE3 and RE4 samples would promote the formation of (La,Ce)(O,S) inclusions in the melt and the removal of the sulfur from solid solution. As a consequence, the formation of MnS is inhibited (no sulfur available in solution) and the formation of coarse Nb-rich phases is avoided, as these do not seem to form at the surface of (La,Ce)(O,S) and only at MnS inclusions.

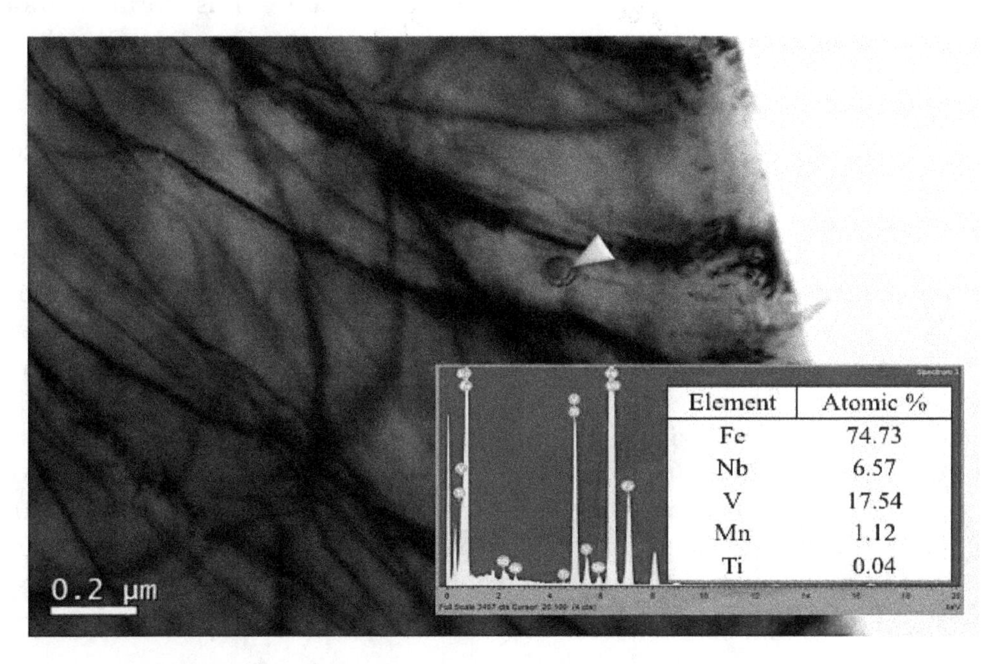

Element	Atomic %
Fe	74.73
Nb	6.57
V	17.54
Mn	1.12
Ti	0.04

Figure 11. TEM (Transmission Electron Microscopy) micrograph of steel RE2 presenting V-rich precipitate along with its EDS microanalysis.

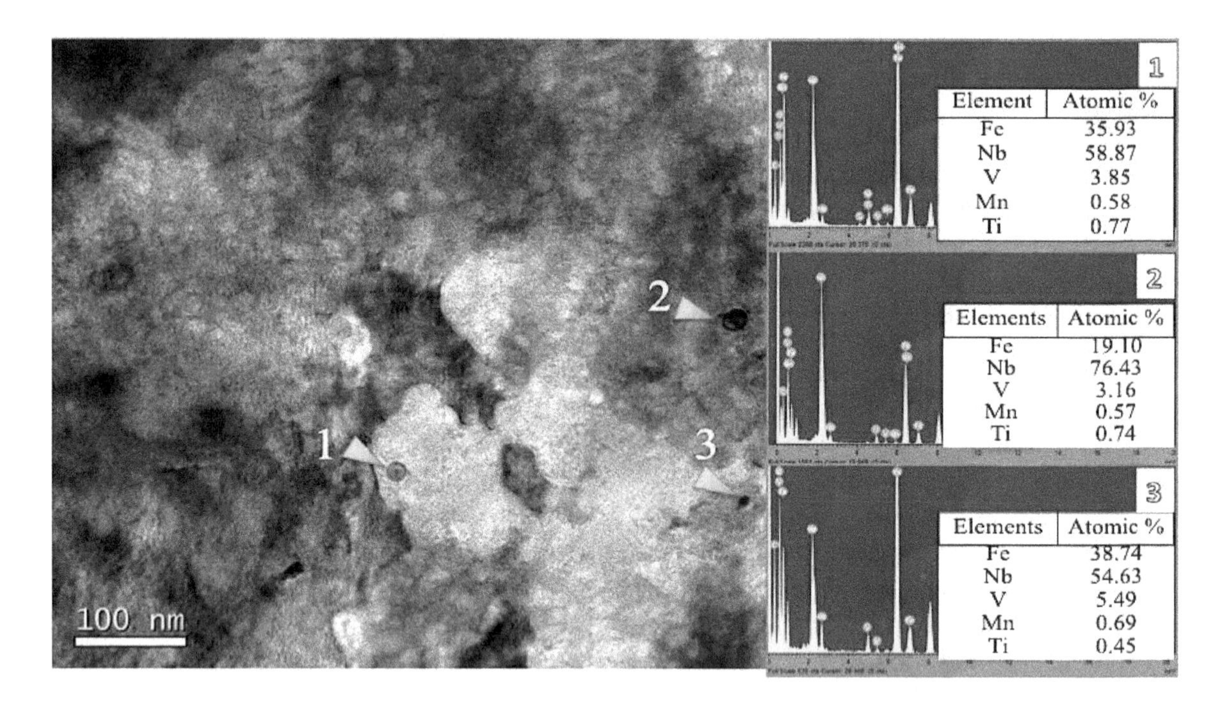

Figure 12. TEM micrograph of steel RE3 presenting Nb-rich precipitates along with their EDS microanalysis.

It is known that the difference between the thermal contraction of inclusions and the matrix during cooling can create stress fields around the inclusions leading to the adjacent matrix deformation or discontinuity between the matrix and inclusion [37–39]. Figure 13 illustrates the thermal expansion coefficients of conventional inclusions and includes data from different references [40–42]. This figure has been copied, with permission, from Figure 10 in reference [43]. If the thermal expansion coefficient of an inclusion is lower than that of the matrix (ferrite), like that of Al_2O_3, stress fields appear around the inclusions; however, in the case of higher contraction coefficient than ferrite e.g. MnS, vacancies and subsequently gaps could appear [37]. It can be seen that MnS has one of the highest values among the typical inclusions while Al_2O_3 and other oxides have lower values. It is worth mentioning that the thermal expansion coefficient of ferrite lies between the values for MnS and Al_2O_3, which is reported to be around 11×10^{-6} (1/°C) [44]. In addition, based on the ThermoCalc predictions (Figure 10), the investigated steels are hyper-peritectic type; i.e., delta ferrite is the first phase that solidifies from molten steel during cooling. In addition, MnS has been found to form in the lower range of the pasty region. It is known that the solubility of sulfur in molten steel is higher than the solid state, so, as solidification proceeds, the sulfur concentration would build up in the remained molten steel, resulting in higher sulfide formation [26]. In addition, non-equilibrium condition/heterogeneous nucleation can alter the formation temperature of MnS to higher temperature. Both phenomena would lead to the formation of MnS in the temperature range where delta ferrite coexists with the molten steel. In other words, the proposed mechanism considers the difference between the thermal expansion coefficient of MnS and the delta ferrite, while when it is compared with that of austenite, such a difference does not exist. For RE-based inclusions, this factor has been reported to be similar to that of ferrite [38,45]. Eventually, due to the considerable difference between the thermal contraction of MnS and delta ferrite, MnS creates stress fields at its interface with matrix as well as losing its solid continuity, which is likely to remain even after transformation of delta ferrite to austenite. Thus, these areas could provide preferential sites for Nb accumulation/precipitation at high temperature, as it has been experimentally observed in this investigation.

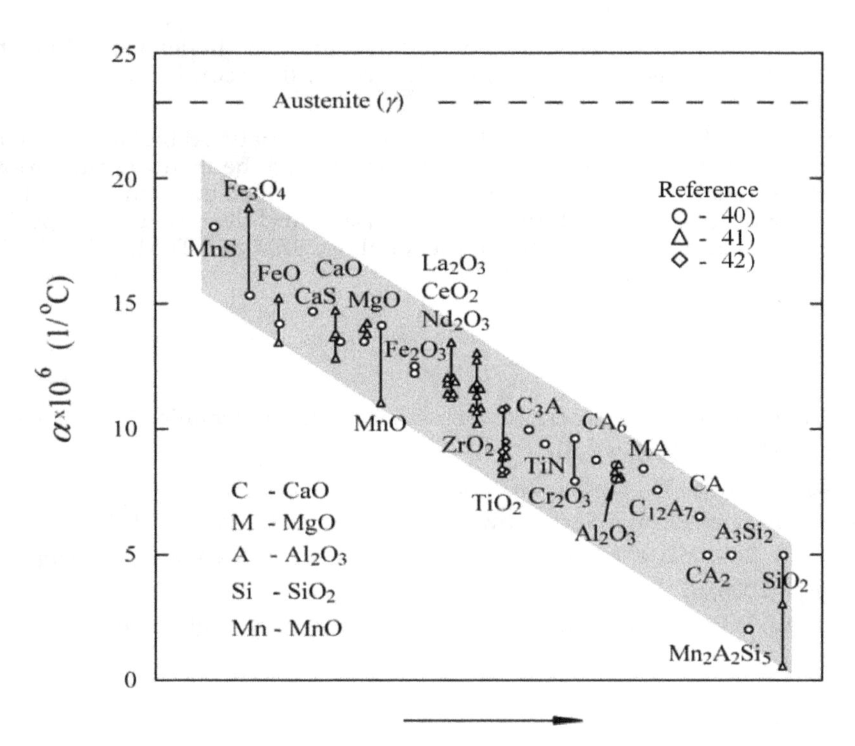

Figure 13. Thermal expansion coefficient of conventional inclusions found in steels; this Figure has been copied from Figure 10 in reference [43].

5. Conclusions

The major findings of the present investigation have been highlighted as follows: according to SEM results, RE addition can change the nature of inclusions formed during casting. In the base and low RE-added steels, Al_2O_3 exists in the molten steel and MnS inclusions form in the pasty region at lower temperatures, sometimes nucleating at these alumina particles and forming complex inclusions. Higher level of RE additions to the base steel (RE3 and RE4) promotes the formation of inclusions with an RE-based matrix instead of Al_2O_3-MnS inclusions that can modify the Al_2O_3 cluster as well.

The results of image analyses showed that the inclusions observed in RE3 and RE4 are rounder than those Al_2O_3-MnS found in RE1 and RE2. The rest of parameters e.g., area fraction and size of the inclusions did not follow a clear trend; compared to RE3, inclusions in steel RE4 were larger with higher area fraction that can lead to poor mechanical properties.

Formation of MnS was suppressed in steels RE3 and RE4, which has been found to serve as preferential sites for the precipitation of Nb-rich phases. As a consequence, alloying the steel with more than 200 ppm of RE inhibited the formation of coarse Nb-based precipitates. Thus, Nb remains in solid solution and available for nanoprecipitation as NbCN.

The precipitation of Nb-rich phases on MnS inclusions would be due to the difference in the thermal expansion coefficient between the matrix and the MnS particles. This difference could cause stress fields as well as solid discontinuity at the interface of MnS with matrix during cooling providing nucleation sites for Nb-rich phases.

Acknowledgments: The authors from the University of Tehran gratefully acknowledge the financial support provided by the Office of International Affairs and the Office for Research Affairs, College of Engineering, for the project number 8107009.6.34. The authors from Centro Nacional de Investigaciones Metalúrgicas (CENIM) that belong to the Consejo Superior de Investigaciones Científicas (CSIC) would like to acknowledge the financial support from Comunidad de Madrid through the project Diseño Multiescala de Materiales Avanzados (DIMMAT-CM_S2013/MIT-2775). Javier Vivas acknowledges financial support in the form of a FPI (Formación de Personal Investigador) Grant BES-2014-069863. Authors are grateful to the Phase Transformations and Microscopy labs from CENIM-CSIC and to the Centro Nacional de Microscopia Electronica (CNME), located at Complutense

University of Madrid (UCM), for the provision of laboratory facilities. Mr. Javier Vara Miñambres from the Phase Transformations lab (CENIM-CSIC) is gratefully acknowledged for their continuous experimental support.

Author Contributions: Hadi Torkamani carried out the experiments, analyzed the data and wrote the manuscript; Shahram Raygan supervised the project, discussed the data and edited the manuscript; Carlos Garcia Mateo and David San Martin supervised, analyzed and discussed the results and edited the manuscript; Jafar Rassizadehghani supported the steelmaking process (production of the materials) and designed the experiments; Javier Vivas contributed to the experiments and characterization of materials by SEM and TEM; Yahya Palizdar contributed to analyzing the data and edited the manuscript.

References

1. Thornton, P.A. The influence of nonmetallic inclusions on the mechanical properties of steel: A review. *J. Mater. Sci.* **1971**, *6*, 347–356. [CrossRef]
2. Bytyqi, A.; Puksic, N.; Jenko, M.; Godec, M. Characterization of the inclusions in spring steel using light microscopy and scanning electron microscopy. *Mater. Tehnol.* **2011**, *45*, 55–59.
3. Zhang, L.; Thomas, B.G. State of the art in the control of inclusions during steel ingot casting. *Metall. Mater. Trans. B* **2006**, *37*, 733–761. [CrossRef]
4. Shi, G.; Zhou, S.; Ding, P. Investigation of nonmetallic inclusions in high-speed steels. *Mater. Charact.* **1997**, *38*, 19–23. [CrossRef]
5. Gao, J.; Fu, P.; Liu, H.; Li, D. Effects of rare earth on the microstructure and impact toughness of H13 steel. *Metals* **2015**, *5*, 383–394. [CrossRef]
6. Senberger, J.; Cech, J.; Zadera, A. Influence of compound deoxidation of steel with Al, Zr, rare earth metals, and Ti on properties of heavy castings. *Arch. Foundry Eng.* **2012**, *12*, 99–104. [CrossRef]
7. Opiela, M.; Kamińska, M. Influence of the rare-earth elements on the morphology of non-metallic inclusions in microalloyed steels. *JAMME* **2011**, *47*, 149–156.
8. Wang, L.M.; Lin, Q.; Yue, L.J.; Liu, L.; Guo, F.; Wang, F.M. Study of application of rare earth elements in advanced low alloy steels. *J. Alloy. Compd.* **2008**, *451*, 534–537. [CrossRef]
9. Grajcar, A.; Kaminska, M.; Galisz, U.; Bulkowski, L.; Opiela, M.; Skrzypczyk, P. Modification of non-metallic inclusions in high-strength steels containing increased Mn and Al contents. *JAMME* **2012**, *55*, 245–255.
10. Paul, S.K.; Chakrabarty, A.K.; Basu, S. Effect of rare earth additions on the inclusions and properties of a Ca-Al deoxidized steel. *Metall. Trans. B* **1982**, *13*, 185–192. [CrossRef]
11. Opiela, M.; Grajcar, A. Modification of non-metallic inclusions by rare-earth elements in microalloyed steels. *Arch. Foundry Eng.* **2012**, *12*, 129–134. [CrossRef]
12. Kimanov, B.M. Removing oxide and sulfide inclusions from molten steel by filtration. *Steel Transl.* **2008**, *38*, 641–646. [CrossRef]
13. Vahed, A.; Kay, D.A.R. Thermodynamics of rare earths in steelmaking. *Metall. Trans. B* **1976**, *7*, 375–383. [CrossRef]
14. Wu, Y.; Wang, L.; Du, T. Thermodynamics of rare earth elements in liquid iron. *J. Less Common Met.* **1985**, *110*, 187–193. [CrossRef]
15. Garrison, M.W., Jr.; Maloney, L.J. Lanthanum additions and the toughness of ultra-high strength steels and the determination of appropriate lanthanum additions. *Mater. Sci. Eng. A* **2005**, *403*, 299–310.
16. Akila, R.; Jacob, K.T.; Shukla, A.K. Gibbs energies of formation of rare earth oxysulfides. *Metall. Trans. B* **1987**, *18*, 163–168. [CrossRef]
17. Ma, Q.; Wu, C.; Cheng, G.; Li, F. Characteristic and formation mechanism of inclusions in 2205 duplex stainless steel containing rare earth elements. *Mater. Today Proc.* **2015**, *2*, 300–305. [CrossRef]
18. Handerhan, K.J.; Garrison, W.M. Effects of rare earth additions on the mechanical properties of the secondary hardening steel AF1410. *Scr. Metall.* **1988**, *22*, 409–412. [CrossRef]
19. Belyakova, A.F.; Kryankovskii, Y.V.; Paisov, I.V. Effect of rare earth metals on the structure and properties of structural steel. *Met. Sci. Heat Treat.* **1965**, *7*, 588–593. [CrossRef]
20. Liu, H.-L.; Liu, C.-J.; Jiang, M.-F. Effect of rare earths on impact toughness of a low-carbon steel. *Mater. Des.* **2012**, *33*, 306–312. [CrossRef]
21. Liu, H.L.; Liu, C.J.; Jiang, M.F. Effects of rare earths on the austenite recrystallization behavior in X80 pipeline steel. *Adv. Mater. Res.* **2010**, *129–131*, 542–546. [CrossRef]

22. Muro, P.; Gimenez, S.; Iturriza, I. Sintering behaviour and fracture toughness characterization of D2 matrix tool steel, comparison with wrought and PM D2. *Scr. Mater.* **2002**, *46*, 369–373. [CrossRef]
23. Porter, D.A.; Easterling, K.E.; Sherif, M. *Phase Transformations in Metals and Alloys*, 3rd ed.; Revised Reprint; CRC Press: London, UK, 2009.
24. Torkamani, H.; Raygan, S.; Rassizadehghani, J. Comparing microstructure and mechanical properties of AISI D2 steel after bright hardening and oil quenching. *Mater. Des.* **2014**, *54*, 1049–1055. [CrossRef]
25. Kasińska, J. Influence of rare earth metals on microstructure and inclusions morphology G17CrMo5–5 cast steel. *Arch. Metall. Mater.* **2014**, *59*, 993–996. [CrossRef]
26. Muan, A.; Osborn, E.F. *Phase Equilibria Among Oxides in Steelmaking*; Addison-Wesley: Boston, MA, USA, 1965.
27. Davis, J.R. *Alloying: Understanding the Basics*; ASM International: Novelty, OH, USA, 2001; p. 647.
28. Najafi, H.; Rassizadehghani, J.; Asgari, S. As-cast mechanical properties of vanadium/niobium microalloyed steels. *Mater. Sci. Eng. A* **2008**, *486*, 1–7. [CrossRef]
29. Najafi, H.; Rassizadehghani, J.; Norouzi, S. Mechanical properties of as-cast microalloyed steels produced via investment casting. *Mater. Des.* **2011**, *32*, 656–663. [CrossRef]
30. Rassizadehghani, J.; Najafi, H.; Emamy, M.; Eslami-Saeen, G. Mechanical properties of V-, Nb-, and Ti-bearing as-cast microalloyed steels. *J. Mater. Sci. Technol.* **2007**, *23*, 779–784.
31. Sosnin, V.V.; Longinov, A.M.; Barantseva, I.V.; Povkova, N.A.; Lyasotskii, I.V. Distribution of niobium and titanium carbonitrides in continuous-cast microalloy steels. *Steel Transl.* **2010**, *40*, 590–594. [CrossRef]
32. Chen, C.Y.; Yen, H.W.; Kao, F.H.; Li, W.C.; Huang, C.Y.; Yang, J.R.; Wang, S.H. Precipitation hardening of high-strength low-alloy steels by nanometer-sized carbides. *Mater. Sci. Eng. A* **2009**, *499*, 162–166. [CrossRef]
33. San Martin, D.; Caballero, F.G.; Capdevila, C.; Garcia de Andres, C. Austenite grain coarsening under the influence of niobium carbonitrides. *Mater. Trans.* **2004**, *45*, 2797–2804.
34. Vivas, J.; Celada-Casero, C.; San Martín, D.; Serrano, M.; Urones-Garrote, E.; Adeva, P.; Aranda, M.M.; Capdevila, C. Nano-precipitation strengthened G91 by thermo-mechanical treatment optimization. *Metall. Mater. Trans. A* **2016**, *47*, 5344–5351. [CrossRef]
35. Mousavi Anijdan, S.H.; Rezaeian, A.; Yue, S. The effect of chemical composition and austenite conditioning on the transformation behavior of microalloyed steels. *Mater. Charact.* **2012**, *63*, 27–38. [CrossRef]
36. Vivas, J.; Capdevila, C.; Jimenez, J.; Benito-Alfonso, M.; San-Martin, D. Effect of ausforming temperature on the microstructure of G91 steel. *Metals* **2017**, *7*, 236. [CrossRef]
37. Sohaciu, M.; Predescu, C.; Vasile, E.; Matei, E.; Savastru, D.; Berbecaru, A. Influence of MnS inclusions in steel parts on fatigue resistence. *Dig. J. Nanomater. Biostruct.* **2013**, *8*, 367–376.
38. Drar, H. Metallographic and fractographic examination of fatigue loaded PM-steel with and without MnS additive. *Mater. Charact.* **2000**, *45*, 211–220. [CrossRef]
39. Enomoto, M. Nucleation of phase transformations at intragranular inclusions in steel. *Met. Mater.* **1998**, *4*, 115–123. [CrossRef]
40. Samsonov, G.V. *Physico-Chemical Properties of Oxides. Handbook*; Metallurgiya: Moscow, Russia, 1978; p. 130. (In Russian)
41. Brooksbank, D.; Andrews, K.W. Stress fields around inclusions and their relation to mechanical properties. *J. Iron Steel Inst.* **1972**, *210*, 246–255.
42. Touloukian, Y.S. *The Thermo-Physical Properties of High Temperatures of Solids Materials*; Macmillan Co.: New York, NY, USA, 1967; p. 462.
43. Sarma, D.S.; Karasev, A.V.; Jonsson, P.G. On the role of non-metallic inclusions in the nucleation of acicular ferrite in steels. *ISIJ Int.* **2009**, *49*, 1063–1074. [CrossRef]
44. Cverna, F. *Asm Ready Reference: Thermal Properties of Metals*; ASM International: Materials Park, OH, USA, 2002.
45. Pan, F.; Zhang, J.; Chen, H.L.; Su, Y.H.; Kou, C.L.; Su, Y.H.; Chen, S.H.; Lin, K.J.; Hsieh, P.H.; Hwang, W.S. Effects of rare earth metals on steel microstructures. *Materials* **2016**, *9*, 417. [CrossRef] [PubMed]

7

Influence of Vanadium on the Microstructure and Mechanical Properties of Medium-Carbon Steels for Wheels

Pengfei Wang [1,2], Zhaodong Li [2,*], Guobiao Lin [1], Shitong Zhou [2,3], Caifu Yang [2] and Qilong Yong [2]

1 School of Materials Science and Engineering, University of Science and Technology, Beijing 100083, China; g20168309@xs.ustb.edu.cn (P.W.); lin571@163.com (G.L.)
2 Department of Structural Steels, Central Iron and Steel Research Institute, Beijing 100081, China; zhoushitong19@outlook.com (S.Z.); yangcaifu@cisri.com.cn (C.Y.); yongqilong@cisri.com.cn (Q.Y.)
3 Department of Materials Science and Engineering, Kunming University of Science and Technology, Kunming 650093, China
* Correspondence: lizhaodong@cisri.com.cn

Abstract: Steels used for high-speed train wheels require a combination of high strength, toughness, and wear resistance. In 0.54% C-0.9% Si wheel steel, the addition of 0.075 or 0.12 wt % V can refine grains and increase the ferrite content and toughness, although the influence on the microstructure and toughness is complex and poorly understood. We investigated the effect of 0.03, 0.12, and 0.23 wt % V on the microstructure and mechanical properties of medium-carbon steels (0.54% C-0.9% Si) for train wheels. As the V content increased, the precipitation strengthening increased, whereas the grain refinement initially increased, and then it remained unchanged. The increase in strength and hardness was mainly due to V(C,N) precipitation strengthening. Increasing the V content to 0.12 wt % refined the austenite grain size and pearlite block size, and increased the density of high-angle ferrite boundaries and ferrite volume fraction. The grain refinement improved the impact toughness. However, the impact toughness then reduced as the V content was increased to 0.23 wt %, because grain refinement did not further increase, whereas precipitation strengthening and ferrite hardening occurred.

Keywords: medium-carbon steel; grain refinement; precipitation strengthening; strength and toughness

---

1. Introduction

Steels for high-speed train wheels must have good strength, toughness, and wear resistance. Pearlite steel, which has a lamellar microstructure obtained by special processing, is exceptionally strong and is often used for this application. The performance of the microstructure mainly depends on the proeutectoid ferrite and pearlite content, and the size of the pearlite substructure.

Microalloying is used to improve the performance of wheel steel by increasing its strength and preventing crack propagation via obtaining a microstructure with good toughness. V has a high solubility in γ-Fe, and it exists in solution or as a precipitate below austenitizing temperatures. The addition of V results in solution strengthening, precipitation strengthening, grain refinement, and hardening. Thus, V has an important effect on the toughness of steel [1–3]. Adding 0.1 wt % V to eutectoid steel can delay the pearlite transformation and effectively refine the pearlite colony size and interlamellar spacing [4]. The dissolution and precipitation of V in medium-carbon steel has a beneficial effect on microstructure refinement and ferrite content, and increases the steel strength [5,6]. Studies have shown that ferrite volume fraction and grain size, pearlite colony/block size, pearlite interlemellar spacing, and austenite grain size all contribute to the toughness of hypoeutectic steel [7,8]. The addition of V increases the hardness of steel, while a high V content decreases the impact toughness [9]. A previous study reported that in steel containing 0.4 wt % C, the toughness of the steel dropped sharply when the V content reached 0.2 wt %, but a detailed explanation was not given [10]. Most research focuses on low-carbon or medium-carbon steels with a C content of 0.4 wt %, or steels with a low Si content, so that the volume fraction of proeutectoid ferrite is high. There has been limited research on wheel steel with a C content of around 0.54 wt %, a high Si content, and a low ferrite content. In 0.54% C-0.9% Si wheel steel, increasing the V content from 0.03 to 0.075 or 0.12 wt % can substantially refine the grains and increase the ferrite content and toughness, although it has a complex effect on strength [11]. The effect of the V content on the microstructure and toughness of medium-carbon wheel steel (0.54% C-0.9% Si) with a low ferrite content remains poorly understood. Thus, in this paper, we investigate the effect of V content on the microstructure and toughness of medium-carbon wheel steel with a low ferrite content.

2. Materials and Methods

2.1. Materials and Processing Technology

Three experimental steels were prepared by smelting in a vacuum induction furnace and rolling. The equilibrium phase-transformation temperatures of the steels were calculated by using Thermo-Calc (TCFE7) software (TCFE7 database, Stockholm, Sweden). The chemical composition of the steel samples and the A_{e3} (γ/α temperature) and A_{e1} ($\gamma+\alpha/\gamma+\alpha+\theta$ temperature) equilibrium temperatures are listed in Table 1. The size of the rolled specimen was 250 mm (length, rolling direction) $\times$ 125 mm (width) $\times$ 60 mm (thickness). The rolling process involved heating to 1280 °C for 2 h, several passes of continuous rolling, and finishing rolling above 900 °C, followed by air cooling to room temperature.

For the heat treatment experiments, samples of 65 mm (length)$\times$ 62 mm (width) $\times$ 15 mm (thickness) were used. The heat treatment process was heating to 860 °C for 1 h, air cooling to room temperature (an average cooling rate of about 3 °C/s in the range of 700–500 °C), tempering at 520 °C for 2 h, and air cooling to room temperature. The schematic of the thermo-mechanical treatment of the investigated steels is shown in Figure 1.

Table 1. Compositions of the investigated steels (wt %), A_{e3}, and A_{e1} (°C).

Sample	C	Si	Mn	P	S	Cr	V	N	A_{e3}	A_{e1}
0.03 V	0.54	0.88	0.78	0.0079	0.0077	0.17	0.03	0.0014	768.2	740.5
0.12 V	0.54	0.87	0.78	0.0072	0.0078	0.18	0.12	0.0016	769.3	740.2
0.23 V	0.54	0.95	0.80	0.0071	0.0071	0.18	0.23	0.003	779.0	740.0

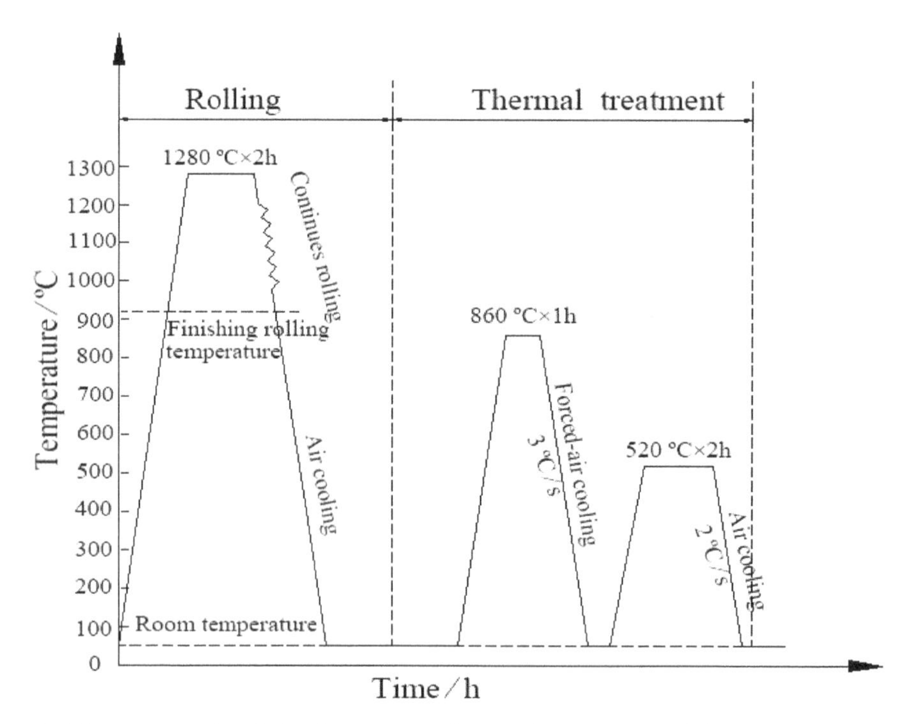

Figure 1. The schematic of the thermo-mechanical treatment of the investigated steels.

2.2. Mechanical Properties

Standard tensile samples with a diameter of 5 mm and a standard length of 25 mm were tested at room temperature on a tensile testing machine (WE-300, Jinan Kairui Machinery Equipment Co., Ltd., Jinan, China), to obtain the yield strength ($R_{p0.2}$), tensile strength (R_m), section shrinkage (ψ), and elongation (δ).

The heat-treated samples were mechanically ground and polished, and then etched with 4% nitric acid in alcohol. The hardness was measured using a Vickers hardness tester with a load of 5 kg and a loading time of 10 s. Microhardness measurements were performed made using a Vickers hardness tester (VH-5, Univer, Qingdao Fupida Electromechanical Technology Co., Ltd., Qingdao, China) with a load of 10 g and a loading time of 10 s. Before measuring the microhardness, the sample was polished and slightly etched with 2% nitric acid in alcohol to distinguish the proeutectoid ferrite and pearlite. The Charpy V impact tests were conducted on an impact tester (JBW-300N, Shanghai Zhujin Instrument Co., Ltd., Shanghai, China) using transverse specimens (10 mm × 10 mm × 55 mm) at test temperatures of −20 and 20 °C.

2.3. Microstructure Analysis

After mechanical grinding, polishing, and etching with 4% nitric acid in alcohol, the heat-treated samples were examined by optical microscopy (OM, GX51, Olympus, Tokyo, Japan) and scanning electron microscopy (SEM, S-4300, Hitachi, Tokyo, Japan). Since the pro-eutectoid ferrite is formed at the austenite grain boundary, we used the intercept method to measure the amount of pro-eutectoid ferrite per unit distance, to obtain the spacing of the pro-eutectoid ferrite, thereby estimating the austenite grain size. The volume fraction of the proeutectoid ferrite was measured from the OM images by the point count method, and there were at least 15 photographs with a magnification of 100×. The intercept method was used to measure the pearlite colony size and the lamellar spacing from the SEM images. The pearlite block size was measured from a large-angle interface diagram with an electron backscatter diffraction (EBSD, Nordlys F+, Oxford, London, UK) step size of 0.5 μm. The density of the high-angle ferrite interface was the ratio of the total length of the interface greater than 15° to the measured area in the ferrite interface diagram per unit area. The cementite thickness (t_c) was calculated from the relationship between the pearlite interlamellar spacing (S) and the carbon content (wt % C) [12].

$$t_c = \frac{S \times 0.15(wt\ \%\ C)}{V} \tag{1}$$

where V is the volume fraction of pearlite.

The samples were examined by SEM and EBSD after electrolytic polishing. Thin-film samples with a diameter of 3 mm were prepared using a double-spout electrolytic polishing device in 8% perchloric acid in ethanol, and the microstructure and precipitation of the samples were analyzed by transmission electron microscopy (TEM; H-800, Hitachi, Tokyo, Japan) at an acceleration voltage of 200 kV.

The precipitated phase in the steels was qualitatively and quantitatively investigated by physicochemical phase analysis. Six 20 mm × 60 mm samples were prepared, subjected to electrolysis, and then cleaned to remove residues. The samples were analyzed by X-ray diffraction (XRD), quantitative analysis, and X-ray small angle diffraction to determine the particle size. The residue after electrolysis was also analyzed by XRD (APD-10, Phillips, London, UK). The element mass fraction was determined by inductively coupled plasma-atomic emission spectrometry, and the particle size analysis of the MC phase was performed by Kratky small-angle X-ray scatterometry.

3. Results and Discussion

3.1. Microstructure

The microstructure of the heat-treated samples was lamellar pearlite with a small amount of proeutectoid ferrite. The OM (Figure 2a–c) and SEM (Figure 2d–f) images show the microstructure of the steels. The microstructural parameters of the steels are shown in Table 2. As the V content was increased from 0.03 to 0.12 wt %, the sizes of the austenite grains, pearlite colonies, and pearlite blocks decreased, and the proeutectoid ferrite volume fraction and the density of high-angle ferrite boundaries increased. The changes in these parameters were small when the V content was further increased from 0.12 to 0.23 wt %. In addition, the proeutectoid ferrite size decreased and the cementite thickness did not change significantly as the V was content increased.

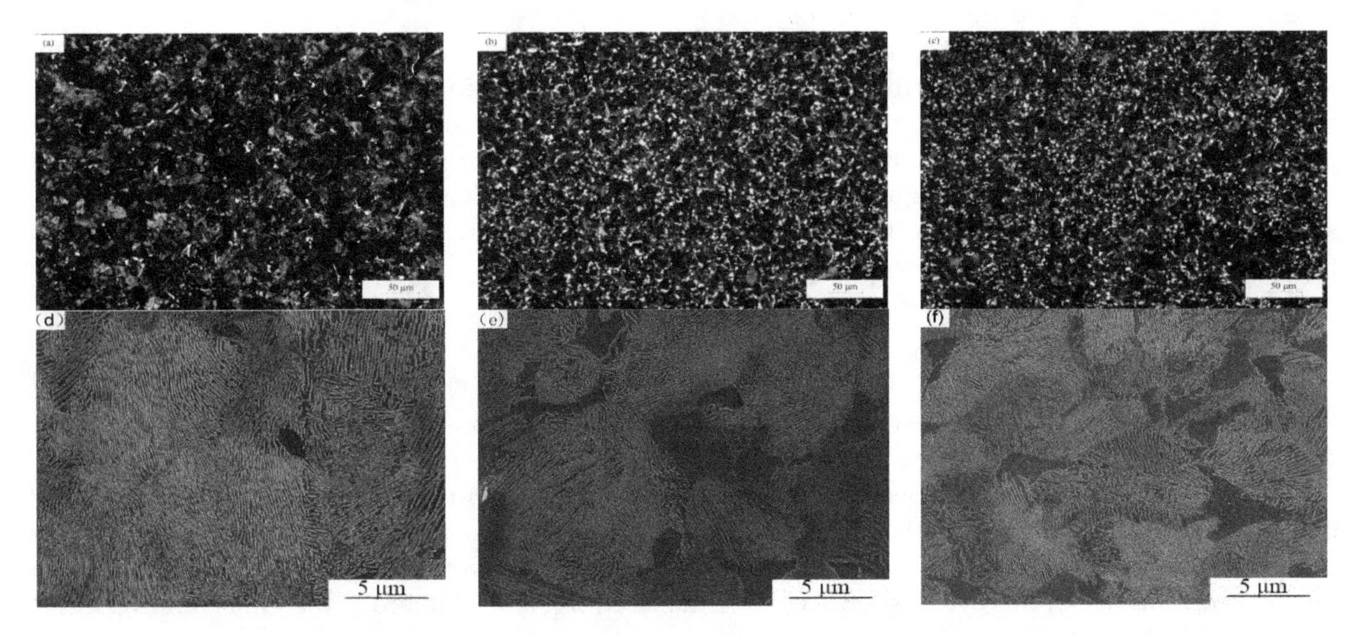

Figure 2. OM (optical microscopy) and SEM (scanning electron microscopy) images showing the microstructures of the investigated steels. (**a**,**d**) Sample 0.03 V; (**b**,**e**) sample 0.12 V; (**c**,**f**) sample 0.23 V.

Table 2. Microstructural parameters of the investigated steels.

Sample	AGS (μm)	f_{α} (%)	d_{α} (μm)	PS (μm)	NS (μm)	S (μm)	t_c (μm)	ρ_{α} (μm)
0.03 V	32.4 ± 1.8	2.0 ± 0.4	4.02 ± 1.5	9.3 ± 1.5	12.1 ± 0.4	0.158 ± 0.03	0.014	7.76
0.12 V	16.6 ± 0.8	11.1 ± 0.4	3.66 ± 0.4	5.5 ± 1.0	7.4 ± 0.3	0.143 ± 0.02	0.014	14.36
0.23 V	15.7 ± 0.7	11.2 ± 0.3	3.41 ± 0.2	5.2 ± 0.5	6.6 ± 0.3	0.131 ± 0.02	0.012	15.01

Austenite grain size (AGS); proeutectoid ferrite volume fraction (f_{α}); proeutectoid ferrite size (d_{α}); pearlite colony size (PS); pearlite block size (NS); interlamellar spacing (S); cementite thickness (t_c); density of high-angle ferrite boundaries (ρ_{α}).

Figure 3 shows the results of the Thermo-Calc calculations. V solid solution in austenite in steel samples 0.03 V and 0.12 V had a small amount of precipitate from 830 °C to 940 °C. In contrast, sample 0.23 V showed precipitate from 1010 °C, the amount of solid solution decreased from 0.23 to 0.11 wt % at 940 °C, and subsequently, the amount of solid solution was the same as that of sample 0.12 V.

Compared with samples 0.03 V and 0.12 V, V precipitation gradually increased with the increase in V content. The precipitated V(C,N) was formed at the austenite grain boundary, preventing the grain from growing, and thus refining the grain. Grain refinement provides more nucleation positions and facilitates proeutectoid ferrite transformation. Compared with samples 0.12 V and 0.23 V, as the V content continued to increase, a large amount of V was coarse precipitate, which had little effect on the austenite refinement as the temperature was decreased from 1010 to 940 °C, due to the high temperature and low N content. When the driving force of the grain growth is balanced with the resistance of the second-phase particles to grain growth, high-V steel can precipitate too many particles at a high temperature, and the resistance to grain growth exceeds the driving force for balancing grain growth. Thus, some precipitated particles do not produce a refinement effect [13].

The higher the V content in the experimental steels, the higher the precipitation temperature of VC; the initial precipitation temperatures of VC for samples 0.03 V, 0.12 V, and 0.23 V were 830 °C, 940 °C, and 1010 °C, respectively. In austenite at a temperature of 860 °C, V is completely dissolved in 0.03 V steel and only partially dissolved in 0.12 V and 0.23 V steels. Furthermore, the amount of undissolved precipitates (VC) present in 0.23 V steel was higher than that in 0.12 V steel, and the solution content of V was the same.

Comparing samples 0.03 V and 0.12 V, V precipitation gradually increased with increasing V content. The precipitates formed at the austenite grain boundary, preventing the grain from growing, and thus refining the grain. Grain refinement provides more nucleation sites and facilitates proeutectoid ferrite transformation. As shown in Figure 4, the amount of VC precipitated in 0.03 V steel was small, leading to weak grain refinement. The amount of VC precipitated in 0.12 V and 0.23 V steels was relatively higher, resulting in an obvious grain refinement effect. According to Zener's formula, the relationship between the precipitates and grain size is:

$$D_C = A\frac{d}{f} \tag{2}$$

where D_C is the critical grain size (μm), A is a proportionality factor, d is the S of the precipitate (μm), and f is the volume fraction of the precipitate.

Since the precipitation temperature of VC and the content of undissolved V are both higher for 0.23 V steel, the VC size precipitated at 860 °C will be greater than that for 0.12 V steel. Meanwhile, as can be seen from Figure 3, the volume fraction of VC in 0.23 V steel at 860 °C is also higher than that in 0.12 V steel, and the values of d/f should be essentially equivalent. Therefore, the grain-refining effect of VC on 0.23 V steel and 0.12 V steel is basically the same.

The dissolved V contents in 0.03 V, 0.12 V, and 0.23 V steels were 0.03, 0.045, and 0.045 wt %, respectively, while the solution contents of C were 0.54, 0.52, and 0.50 wt %, respectively. The solute drag effect of V plays a role in grain refinement. The solid solutions of V in samples 0.12 V and 0.23 V were the same, and there was no further refinement effect. Therefore, the austenite size of sample 0.12 V was substantially smaller than that of sample 0.03 V, whereas those of samples 0.12 V and 0.23 V were similar. The change in proeutectoid ferrite content was mainly caused by the austenite grain size and the C solid solution content. The austenite refinement of sample 0.12 V was considerably greater than that of sample 0.03 V, the C solid solution content was lower, and the proeutectoid ferrite content was higher. The austenite refinement of sample 0.23 V was not obvious compared with sample 0.12 V, the C solid solution was lower, and the proeutectoid ferrite content was similar.

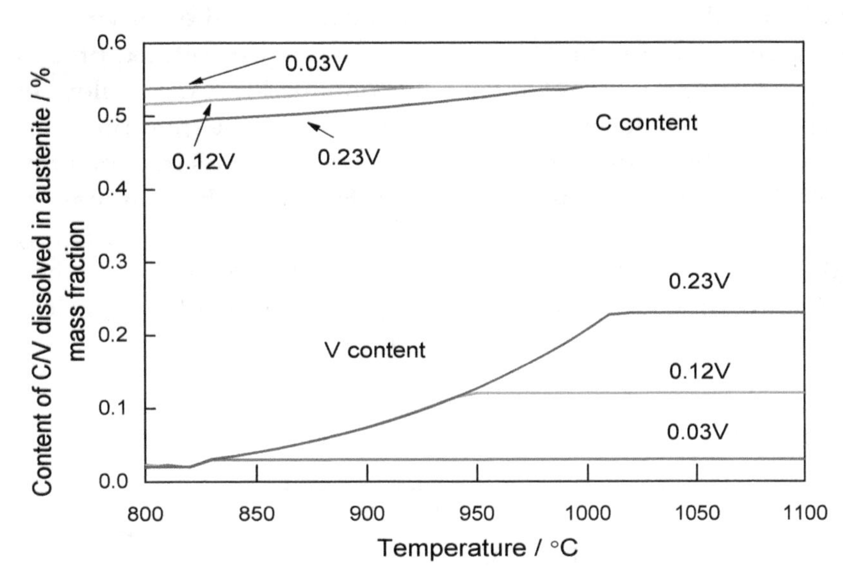

Figure 3. Effect of V content on dissolved C and V contents in austenite in steel samples 0.03 V (red), 0.12 V (green), and 0.23 V (blue).

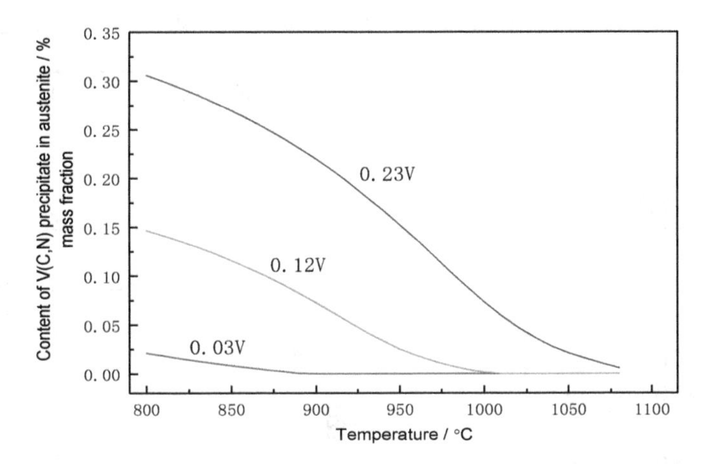

Figure 4. Content of V(C,N) precipitate in austenite in steel samples 0.03 V (red), 0.12 V (green), and 0.23 V (blue).

As the V content increases, the amount of proeutectoid ferrite at the austenite grain boundary increases. This destroys the continuity between the austenite grains, and reduces the size of the austenite grains that transformed to pearlite, thus refining the sizes of the pearlite colonies and blocks after the transformation. The nucleation and growth rates of pearlite colonies and blocks are also affected by the pearlite transformation temperature, and decreasing the transformation temperature refines the sizes of the pearlite colonies and blocks. The V solid solution increases the diffusion activation energy of C atoms, improving the stability of the austenite, and increasing the undercooling of the material at the same cooling rate. Owing to the differences in the austenite grain size and pearlite transformation temperature, sample 0.12 V had substantially smaller pearlite colonies and blocks than sample 0.03 V, whereas those of samples 0.12 V and 0.23 V were similar.

As the V content increased, the temperature of the pearlite transformation decreased. Because V is a strong carbide-forming element that hinders the diffusion of C atoms, the interlamellar spacing also decreased.

3.2. Analysis of the Precipitate Phase

XRD and TEM were used to examine the precipitated phases. The V in sample 0.03 V was almost completely dissolved at 860 °C, as Figure 4 shows. For the convenience of the following analysis, we ignore the precipitation in 0.03 V steel; only the precipitated phases in samples 0.12 V and 0.23 V were analyzed. Samples 0.12 V and 0.23 V contained the M_3C (alloy cementite) phase and the MC (V(C,N) phase, which includes a small amount of Cr. The quantitative analysis results are presented in Tables 3 and 4. The amount of V precipitate and the mass fraction of the M_3C and M(C,N) phases were obtained. The XRD patterns are shown in Figure 5. The nominal chemical formulas of the M(C,N) phase in samples 0.12 V and 0.23 V were calculated as $(V_{0.767}Cr_{0.233})(C_{0.937}N_{0.063})$ and $(V_{0.876}Cr_{0.124})(C_{0.943}N_{0.057})$, respectively. M(C,N) particles were precipitated in the proeutectoid ferrite and pearlitic ferrite (Figure 6). Figure 7 shows the size distribution of V(C,N) in samples 0.12 V and 0.23 V, which had average sizes of 48.9 and 40.7 nm, respectively.

Table 3. Elemental content of M_3C in steel samples 0.12 V and 0.23 V.

Sample	Elemental Content of M_3C (wt %)					
	Fe	Cr	Mn	V	C	Σ
0.12 V	5.65	0.027	0.137	0.012	0.418	6.244
0.23 V	5.323	0.023	0.129	0.02	0.394	5.889

Table 4. Elemental content of M(C,N) in steel samples 0.12 V and 0.23 V.

Sample	Elemental Content of M(C,N) (wt %)				
	V	Cr	N	C	Σ
0.12 V	0.071	0.022	0.0016	0.02	0.115
0.23 V	0.167	0.024	0.003	0.042	0.236

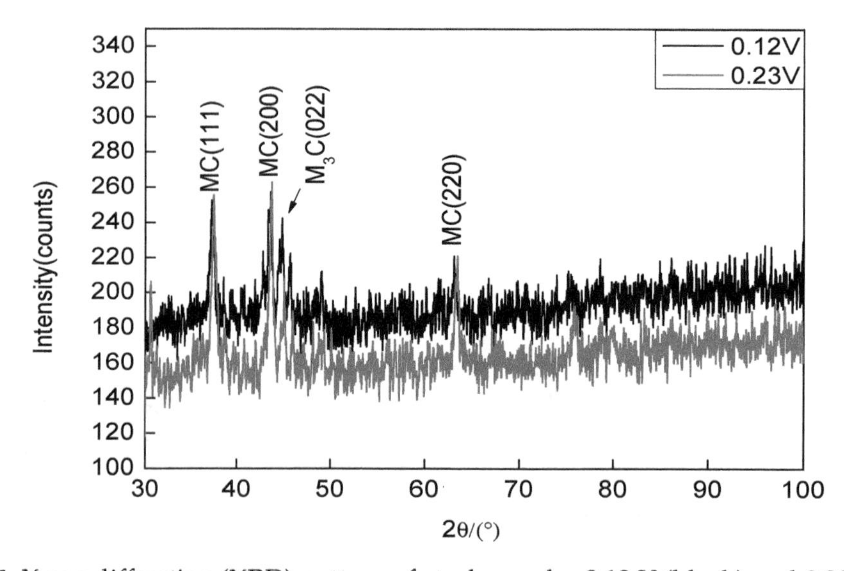

Figure 5. X-ray diffraction (XRD) pattern of steel samples 0.12 V (black) and 0.23 V (red).

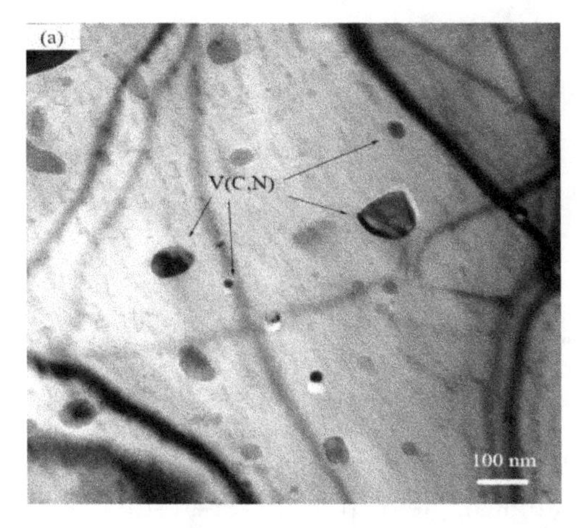

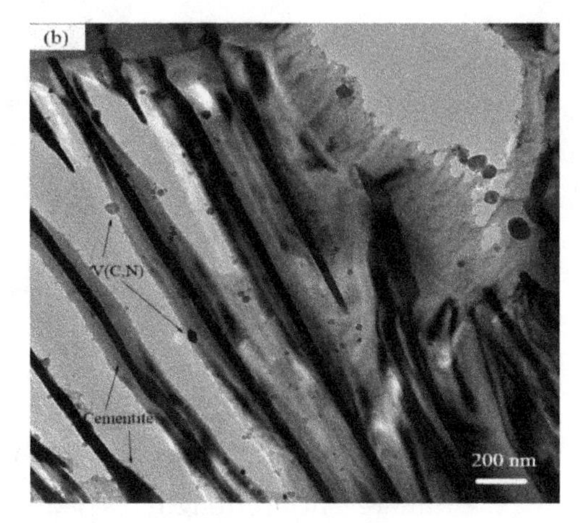

Figure 6. Transmission electron microscopy (TEM) images of sample 0.23 V showing (**a**) V(C,N) particles in proeutectoid ferrite, and (**b**) pearlitic ferrite.

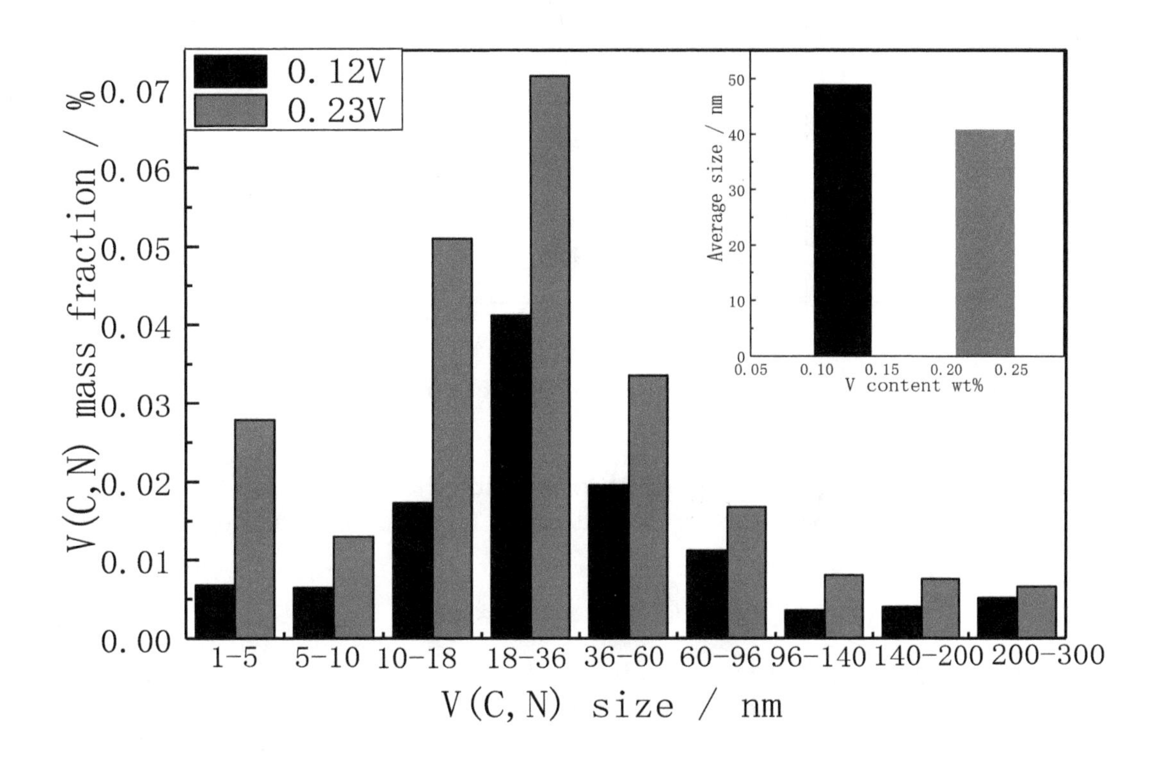

Figure 7. Average size and size distribution of V(C,N) in samples 0.12 V (red) and 0.23 V (black).

3.3. *Strength and Hardness*

The ultimate tensile strength (UTS), yield strength, Vickers hardness (HV), and microhardness of the proeutectoid ferrite and pearlite increased considerably with the increase in V content (Figures 8 and 9). However, the elongation and section shrinkage of the steels were similar. The microstructure of the steels were proeutectoid ferrite and pearlite. The strength of the multiphase mainly depends on the soft phase; thus, because the proeutectoid ferrite content was lower in the steel samples, the effect of pearlite on the strength was also considered.

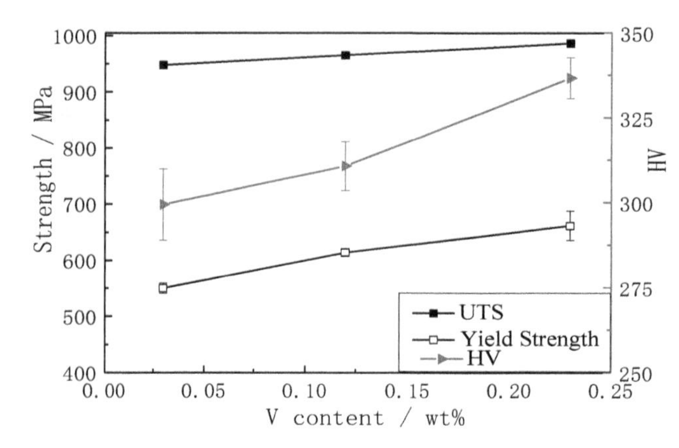

Figure 8. Ultimate tensile strength (closed squares, red triangles), yield strength (open squares), and Vickers hardness (red triangles) for the investigated steels (0.03, 0.12, and 0.23 wt % V).

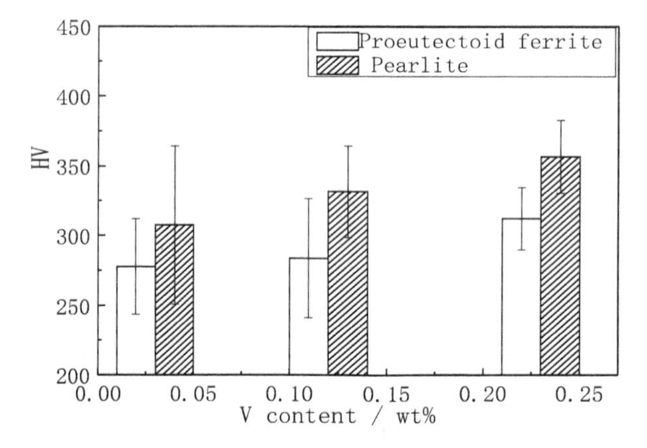

Figure 9. Microhardness of proeutectoid ferrite and pearlite in investigated steels (0.03, 0.12, and 0.23 wt % V).

Gladman et al. [7] used regression analysis to describe the relationship between steel strength and the strength and volume fraction of proeutectoid ferrite and pearlite, as expressed in the following equation:

$$\sigma_{ys} = f_{\alpha}^{\frac{1}{3}}\sigma_{\alpha} + \left(1 - f_{\alpha}^{\frac{1}{3}}\right)\sigma_{P}, \quad (2)$$

where σ_{ys}, σ_{α}, and σ_{P} are the yield strengths of the steel, proeutectoid ferrite, and pearlite, respectively. f_{α} is the volume fraction of proeutectoid ferrite. In contrast to low-carbon steel and eutectoid steel, in proeutectoid steel, the strength of proeutectoid ferrite increases with the volume fraction of pearlite, which is independent of the size of the proeutectoid ferrite grains. The strength and microhardness of pearlite also increases with the pearlite volume fraction, independent of the pearlite interlamellar spacing [14]. As the V content increased from 0.03 to 0.12 wt %, the volume fraction of pearlite decreased. Therefore, the reason for why the strength of 0.12 V steel is higher than that of 0.03 V steel is basically independent of the change in pearlite volume fraction; upon increasing the V content from 0.12 to 0.23 wt %, the pearlite volume fraction remained unchanged. Therefore, the yield strength change of sample 0.23 V was independent of the grain refinement, in contrast to sample 0.12 V. The yield strength was also affected by the solid solution elements and the precipitated phase. The V contents in the solid solutions in 0.03 V, 0.12 V, and 0.23 V steels of 0.03, 0.045, and 0.045 wt %, respectively, showed negligible differences at 860 °C. The contents of other elements were also similar. Therefore, the increase in strength was independent of solution strengthening.

Previous analysis has been based on the assumption that the strength of pearlite is independent of interlamellar spacing. However, Gladman et al. [7,15,16] concluded that the strength of pearlite is related to interlamellar spacing, and this is described by the following strength formula for proeutectoid steel [7]:

$$\sigma_{ys} = 15.4\left(f^{\frac{1}{3}}\left[2.3 + 3.8(\%Mn) + 1.13d_{\alpha}{}^{-\frac{1}{2}}\right] + 1 - f^{\frac{1}{3}}\left[11.6 + 0.25S^{-\frac{1}{2}}\right]\right) + 4.1(\%Si) + 27.6(\%N) \quad (3)$$

where f is the volume fraction of ferrite (%), d_{α} is the grain size of ferrite (mm), and S is the interlamellar spacing of pearlite (mm). We calculated that the yield strength increases of samples 0.12 V and 0.23 V compared with sample 0.03 V were −13 and 1 MPa, respectively. Thus, the increase of yield strength in this experiment was independent of fine crystal reinforcing and solution strengthening.

The precipitation strengthening of V in medium-carbon steel can be described by the Ashby–Orowan model [17]. Therefore, the increase in precipitation strengthening was calculated by the following equation obtained from the Ashby–Orowan model [13]:

$$\sigma_p = 8.995 \times 10^3 \frac{f^{1/2}}{d} \ln(2.417d) \quad (4)$$

where d is the average diameter of the precipitate particles (nm) and, f is the volume fraction of the precipitated phase (%),which is obtained from $f = f_{MC} \times \frac{\rho_{Fe}}{\rho_{MC}}$, and where f_{MC} is the mass fraction of the precipitated phase (wt %), ρ_{Fe} is the density of the α-Fe matrix, which is 7.875 g/cm^3, and ρ_{MC} is the theoretical density of the precipitated phase. Because the atomic weight of Cr is close to that of V, it was regarded as equivalent to V. The N content was negligible, the nominal formula of phase MC in the experimental steel was identified as VC, and the density of VC is 5.717 g/cm^3.

To further explain the calculated effect for overall strengthening, calculation results of precipitation hardening increments are shown in Table 5, and the effect of a precipitated phase of different sizes was superimposed on the root mean square [10]. The increases in precipitation strengthening for samples 0.12 V and 0.23 V were calculated as 83 and 147 MPa, respectively, and they were mainly produced by particles that were smaller than 60 nm in diameter.

Table 5. Calculations results of precipitation hardening increments.

Sample Size (nm)	1–5	5–10	10–18	18–36	36–60	60–96	96–140	140–200	200–300
0.12 V	57.4	32.7	34.9	33.1	14.6	6.8	3.0	2.4	0.8
0.23 V	116.3	46.5	60.0	43.8	19.1	8.4	4.5	3.2	2.2

The effect on solution strengthening, fine crystal strengthening, and precipitation strengthening was similar to the tensile strength and yield strength. Therefore, precipitation strengthening was the most important strengthening factor.

3.4. Impact Toughness

The Charpy impact energies of the steel samples are shown in Figure 10. As the V content increased, the impact energies initially increased, reached a maximum at a V content of 0.12 wt %, and then decreasing sharply. The impact specimens did not contain inclusions at the crack origins.

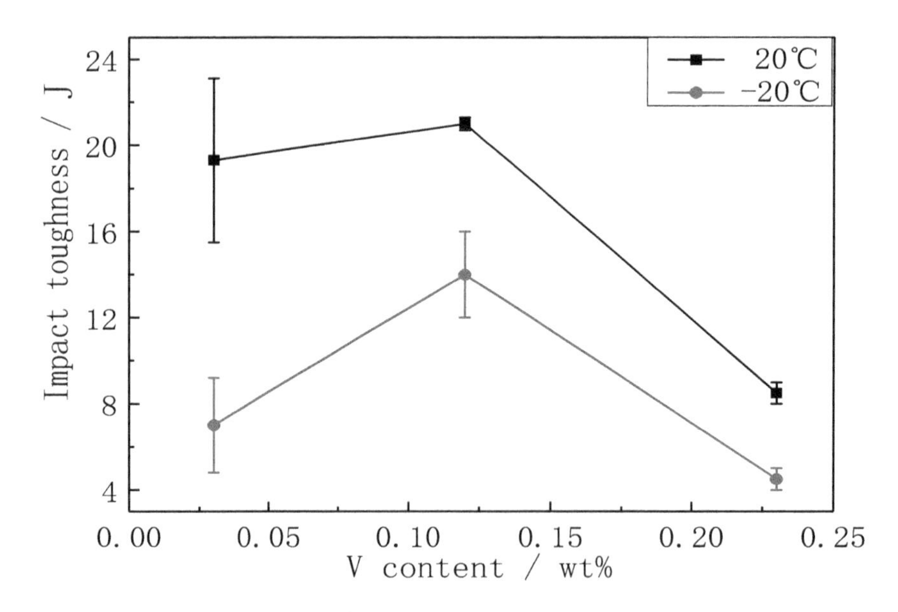

Figure 10. Charpy impact energies for the investigated steels (0.03, 0.12, and 0.23 wt % V).

The volume fraction of proeutectoid ferrite, the size of austenite grains and pearlite colonies and blocks, and pearlite interlamellar spacing affect the toughness of hypoeutectoid steel [5,8]. When the V content increased from 0.03 to 0.12 wt % (samples 0.03 V and 0.12 V), the volume fraction of proeutectoid ferrite increased, increasing the impact toughness. The effect of austenite grain refinement on toughness occurred via the pearlite substructure. The ferrite/cementite interface hardly changed the direction of crack propagation, which showed that although the cementite lamellae inhibited the dislocation slip, they did not inhibit the growth of cleavage cracks, unlike traditional grain boundaries [18]. Pearlite interlamellar spacing has little effect on toughness. Alexander and Bernstein [18,19], and Mishra and Singh [20] have shown that refinement of the interlamellar spacing increases both the cleavage fracture stress and the yield strength. The larger the ratio of cleavage fracture stress to yield strength, the more difficult it is for cleavage fracture to occur, and the higher the toughness. The cleavage fracture stress and yield strength are calculated by [19]:

$$\begin{aligned} \sigma_{fc} &= 156.5S^{-1} + 423.8, \\ \sigma_y &= 73.1S^{-1} + 99.3, \end{aligned} \tag{5}$$

where σ_{fc} is the cleavage fracture stress (MPa), σ_y is the yield strength (MPa), and S is the pearlite interlamellar spacing (μm). The ratios of cleavage fracture stress to yield strength were 2.52, 2.49, and 2.46 for samples 0.03 V, 0.12 V, and 0.23 V, respectively. Although the refinement of the interlamellar spacing in the experimental steel may slightly reduce the toughness, the overall effect was negligible.

The effect of austenite grain size on the size of the pearlite blocks is much greater than that on the size of the pearlite colonies, and the pearlite blocks control fractures [16,21,22]. The microstructures of the steels were observed by EBSD, and the pearlite block size and large-angle interface were analyzed (Figures 11 and 12). The pearlite block size was counted according to the large-angle interface (>15°), and the density of the high-angle ferrite boundaries was calculated. The average size of the pearlite blocks for the 0.03 V, 0.12 V, and 0.23 V steels were 11.2, 8.2, and 7.9 μm, respectively. Large-angle interfaces hinder cracks, and the greater the interface density, the greater the toughness. Table 2 shows the ferrite interface density (total interface length per unit area; >15°) in the steel samples. When the V content was increased from 0.03 to 0.12 wt % (samples 0.03 V and 0.12 V), the size of the pearlite blocks decreased, and the high-angle boundary density increased with the austenite grain refinement. Thus, the remarkable increase in the toughness was mainly due to the refinement of the microstructure, although the increase in the volume fraction of proeutectoid ferrite also increased the toughness.

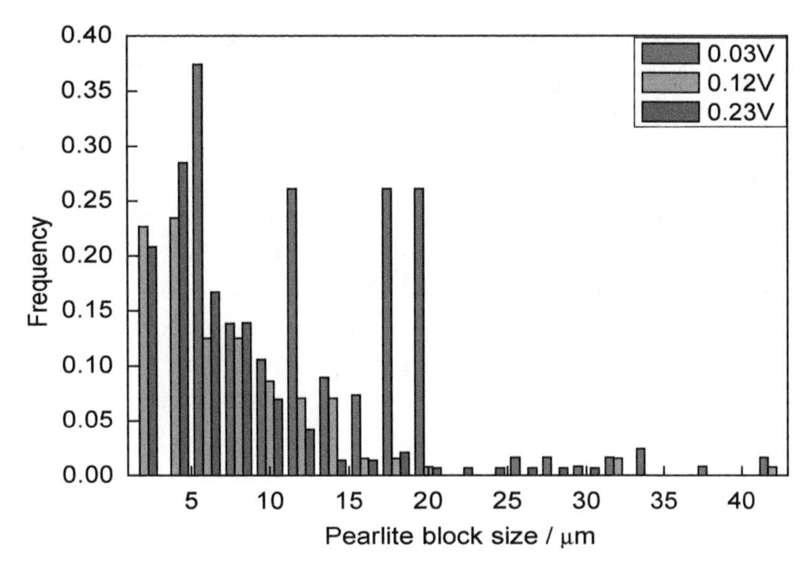

Figure 11. Distribution of the pearlite block size in samples 0.03 V (red), 0.12 V (green), and 0.23 V (blue).

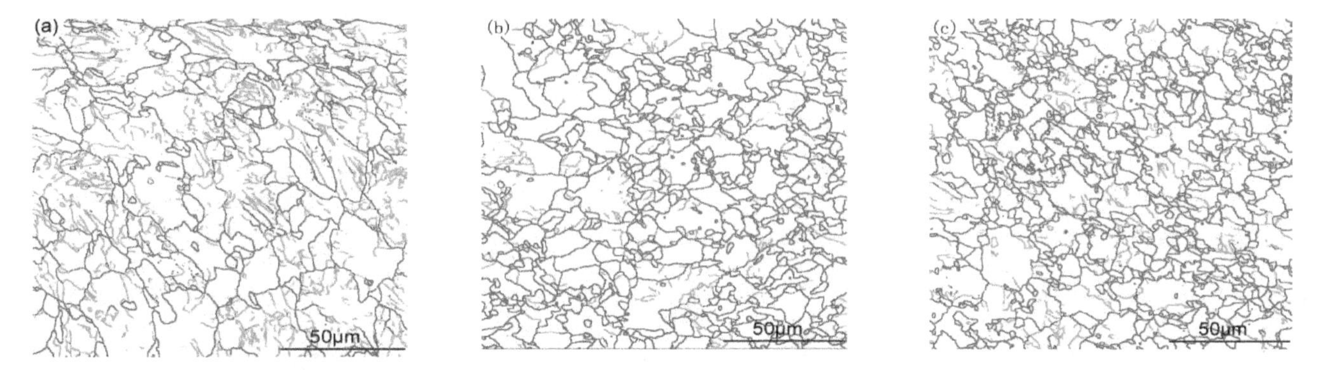

Figure 12. Ferrite orientation maps for samples (**a**) 0.03 V; (**b**) 0.12 V; (**c**) 0.23 V. Red lines represent high-angle boundaries (>15°), and green lines represent low-angle boundaries (5°–15°).

When the V content was increased from 0.12 to 0.23 wt %, V had little effect on the microstructure refinement, and the refinement of the pearlite blocks and the increase in the density of the high-angle boundaries were not important. Besides, the volume fraction of the proeutectoid ferrite is equivalent. Theoretically, the toughness values of the steels should be similar. However, the toughness of sample 0.23 V was substantially lower. As the V content increased, the microhardness of the proeutectoid ferrite and pearlite increased considerably (Figure 11).

The hardening of pearlite is mainly controlled by ferrite, although it is also affected by the interlamellar spacing. The effect of the interlamellar spacing on the strength and hardness was small. In hypoeutectoid steel, when the interlamellar spacing is less than 627 nm, the overlapping of the plastic deformation zone in ferrite will increase its hardness and reduce the toughness. As the interlamellar spacing decreases, the overlapping of the plastic deformation zone in pearlitic ferrite increases the strength and hardness, and reduces the toughness. When the spacing is smaller, the hardening reaches saturation, and the change in strength, hardness, and toughness tends to zero [13,22,23]. In this study, the interlamellar spacing of pearlite was small, and the hardening of the ferrite obtained by the refinement of the lamellar spacing may have reached the limit. Combined with the results obtained from Equation (5), the interlamellar spacing has little effect on the hardening of ferrite. There was a coherent or semi-coherent relationship between the precipitated V(C,N) and the matrix, which greatly increased the friction resistance between the crystal lattices and limited the plastic deformation of ferrite , thereby hardening the ferrite [13,24]. Combined with the microhardness, as the V content increases, the hardness of the proeutectoid ferrite and pearlite increases. In the analysis of strength and hardness, it can be seen that precipitation strengthening is the most important strengthening factor. The addition of V increased the volume fraction of ferrite and refined the structure, while hardening the ferrite. At a V content of 0.03 wt %, V was completely dissolved in the microstructure, resulting in a small effect on strengthening and toughening. When the V content was increased to 0.12 wt %, the microstructure refinement and increase in the volume fraction of ferrite were greater, and the precipitation strengthening was lower. Thus, the positive effect of grain refinement on toughness outweighed the negative effect of ferrite hardening, increasing the impact toughness substantially. When the V content was increased to 0.23 wt %, the effect on the grain refinement was smaller, but the precipitation strengthening continued to increase. The decrease in toughness caused by ferrite hardening was much greater than the positive effect of grain refinement and the increase in proeutectoid ferrite volume fraction, and the impact toughness was reduced.

4. Conclusions

In medium-carbon wheel steel, the effect of precipitation strengthening increased with increasing V content, and the grain refining effect and the volume fraction of proeutectoid ferrite initially increased, and then remained unchanged. The yield strength, tensile strength, and hardness of the steels increased substantially with increasing V content, owing to the precipitation strengthening by V(C,N) particles. As the V content was increased from 0.03 to 0.12 wt %, the size of the austenite grains and pearlite blocks decreased, and the density of the high-angle boundaries and the volume fraction of proeutectoid ferrite increased. The positive effect of grain refinement on toughness was greater than the negative effect of ferrite hardening caused by precipitation strengthening, and the impact toughness was improved. However, at 0.23 wt % V, the grain refining effect did not increase, whereas the precipitation strengthening increased. Thus, the negative effect of the ferrite hardening on toughness was greater than the positive effect of the grain refinement, and the impact toughness was reduced.

Author Contributions: Conceptualization was completed by Z.L. and P.W.; visualization and methodology were completed by G.L.; investigation, software, data collation, formal analysis, and writing of the manuscript were completed independently by P.W.; verification was completed by P.W., Z.L., and S.Z.; resources, supervision, and capital acquisition were completed by Z.L.; the review and editing were completed by Q.Y.; and C.Y. was responsible for project management.

Acknowledgments: The authors would like to thank the professors, senior engineers, doctoral students, etc., for their guidance and help with these of this experiments, and also acknowledge the technical support provided by other laboratory staff.

References

1. Qi, J.J.; Huang, Y.H.; Zhang, Y. *Microalloyed Steels*; Metallurgical Industry Press: Beijing, China, 2006; p. 31, ISBN 7502439692.
2. Rune, L.; Bevis, H.; Tadeusz, S.; Stanislaw, Z. *The Role of Vanadium in Microalloyed Steels*; Metallurgical Industry Press: Beijing, China, 2015; pp. 19–31, ISBN 9787502470968.
3. Wu, D.Y.; Xiao, F.R.; Wang, B. Investigation on grain refinement and precipitation strengthening applied in high speed wire rod containing vanadium. *Mater. Sci. Eng. A* **2014**, *592*, 102–110. [CrossRef]
4. Li, J.; Yang, Z.M. The effects of V on phase transformation of high carbon steel during continuous cooling. *Acta Metall.* **2010**, *46*, 1501–1510. [CrossRef]
5. Sakamoto, H.; Toyama, K.; Hirakawa, K. Fracture toughness of medium-high carbon steel for railroad wheel. *Mater. Sci. Eng. A* **2000**, *285*, 288–292. [CrossRef]
6. Ishikawa, F. Intragranular ferrite nucleation in medium-carbon vanadium steels. *Metall. Mater. Trans. A* **1994**, *25*, 929–936. [CrossRef]
7. Gladman, T.; McIvor, I.; Pickering, F. Some aspects of the structure-property relationships in high-C ferrite-pearlite steels. *J. Iron Steel Inst.* **1972**, *210*, 916–930.
8. Ma, Y. Study on Toughening Mechanism and Production Process Optimization of Wheel Steel for High Speed Train. Ph.D. Thesis, Central Iron & Steel Research Institute, Beijing, China, May 2012.
9. Parsons, S.A.; Edmonds, D.V. Microstructure and mechanical properties of medium-carbon ferrite pearlite steel microalloyed with vanadium. *Mater. Sci. Technol.* **1987**, *3*, 894–904. [CrossRef]

10. Wu, Y.F.; Hui, W.J.; Chen, S.L. Influence of vanadium on microstructure and mechanical properties of medium carbon forging steel. *J. Iron Steel Res.* **2016**, *28*, 56–62. [CrossRef]
11. Zuo, Y.; Zhou, S.T.; Li, Z.D. Effect of V and Si on microstructure and mechanical properties of medium-carbon pearlitic steels for wheels. *Chin. J. Mater. Res.* **2016**, *30*, 401–408. [CrossRef]
12. O'Donnelly, B.E.; Reuben, R.L.; Baker, T.N. Quantitative assessment of strengthening parameters in ferrite-pearlite steels from microstructural measurements. *Metall. Sci. J.* **2013**, *11*, 45–51. [CrossRef]
13. Yong, Q.L. *The Second Phase in Steel*; Metallurgical Industry Press: Beijing, China, 2006; pp. 47–55, ISBN 7502440003.
14. Ray, K.K.; Mondal, D. The effect of interlamellar spacing on strength of pearlite in annealed eutectoid and hypoeutectoid plain carbon steels. *Acta Metall. Mater.* **1991**, *39*, 2201–2208. [CrossRef]
15. Mondal, D.P.; Ray, K.K.; Das, S. The strength of ferrite in annealed Armco iron and hypoeutectoid steels. *Z. fuer Metallkunde* **1998**, *89*, 635–641.
16. Hyzak, J.M.; Bernstein, I.M. The role of microstructure on the strength and toughness of fully pearlitic steels. *Metall. Trans. A* **1976**, *7*, 1217–1224. [CrossRef]
17. Miyamoto, G.; Hori, R.; Poorganji, B.; Furuhara, T. Interphase precipitation of VC and resultant hardening in V-added medium carbon steels. *ISIJ Int.* **2011**, *51*, 1733–1739. [CrossRef]
18. Park, Y.J.; Bernstein, I.M. The process of crack initiation and effective grain size for cleavage fracture in pearlitic eutectoid steel. *Metall. Trans. A* **1979**, *10*, 1653–1664. [CrossRef]
19. Alexander, D.J.; Bernstein, I. M. Cleavage fracture in pearlitic eutectoid steel. *Metall. Trans. A* **1989**, *20*, 2321–2335. [CrossRef]
20. Mishra, K.; Singh, A. Effect of interlamellar spacing on fracture toughness of nano-structured pearlite. *Mater. Sci. Eng. A* **2017**, *706*, 22–26. [CrossRef]
21. Nakase, K.; Bernstein, I.M. The effect of alloying elements and microstructure on the strength and fracture resistance of pearlitic steel. *Metall. Trans. A* **1988**, *19*, 2819–2829. [CrossRef]
22. Furuhara, T.; Kikumoto, K.; Saito, H.; Sekine, T.; Ogawa, T.; Morito, S.; Maki, T. Phase transformation from fine-grained austenite. *ISIJ Int.* **2008**, *48*, 1038–1045. [CrossRef]
23. Dollar, M.; Bernstein, I.M.; Thompson, A.W. Influence of deformation substructure on flow and fracture of fully pearlitic steel. *Acta Metall.* **1988**, *36*, 311–320. [CrossRef]
24. Modi, O.P.; Deshmukh, N.; Mondal, D.P.; Jha, A.K.; Yegneswaran, A.H.; Khaira, H.K. Effect of interlamellar spacing on the mechanical properties of 0.65% C steel. *Mater. Charact.* **2001**, *46*, 347–352. [CrossRef]

8

Property Optimization in As-Quenched Martensitic Steel by Molybdenum and Niobium Alloying

Hardy Mohrbacher [1,2]

[1] Department of Materials Engineering (MTM), KU Leuven, 3001 Leuven, Belgium; hm@niobelcon.ne

[2] NiobelCon bvba, 2970 Schilde, Belgium

Abstract: Niobium microalloying is the backbone of modern low-carbon high strength low alloy (HSLA) steel metallurgy, providing a favorable combination of strength and toughness by pronounced microstructural refinement. Molybdenum alloying is established in medium-carbon quenching and tempering of steel by delivering high hardenability and good tempering resistance. Recent developments of ultra-high strength steel grades, such as fully martensitic steel, can be optimized by using beneficial metallurgical effects of niobium and molybdenum. The paper details the metallurgical principles of both elements in such steel and the achievable improvement of properties. Particularly, the underlying mechanisms of improving toughness and reducing the sensitivity towards hydrogen embrittlement by a suitable combination of molybdenum and niobium alloying will be discussed.

Keywords: martensitic steel; direct quenching; microalloying; hardenability; toughness; grain refinement; Hall–Petch coefficient; microalloy precipitates; hydrogen embrittlement

---

1. Introduction

As-quenched martensite is the hardest and strongest microstructure of low-carbon steel, yet is often considered to have a tendency for brittleness. Tempering treatment mitigates the properties of martensite by reducing the strength and simultaneously increasing toughness and ductility [1,2]. Quenched and tempered steels find many applications for components subjected to demanding operating conditions.

As-quenched martensite is being used particularly for applications with extreme strength or hardness demands. Such applications traditionally appear in the mining industry where hardness and strength are fundamental assets providing good functionality and sufficiently long life of components subjected to wear conditions. More recently, as-quenched martensite has gained sizeable market potential in the automotive industry for making crash-resistant car body components [3,4]. State-of-the-art car bodies use these steels typically with a tensile strength of around 1500 MPa. They can represent up to 40 percent of the car body weight. Furthermore, hot-rolled martensitic steel has potential in structural applications where ultra-high strength is required for weight reduction, for instance in mobile hoisting equipment [5]. Since in all these applications impact loading must be expected, sufficient upper shelf toughness over the temperature range of operation is required, demanding a correspondingly matching ductile-to-brittle transition temperature. Another significant problem perceived in all ultra-high strength steels is the sensitivity to hydrogen embrittlement [6]. The presence of diffusible interstitial hydrogen in such steel can cause unexpected brittle fracture or leads to so-called "delayed cracking" [7]. In that respect, optimization of as-quenched martensitic steel is necessary.

Principally, two different processing routes are practiced for producing flat rolled steel with martensitic microstructure. Re-austenitizing & quenching is the traditional process by which the as-rolled steel, typically having ferritic–pearlitic microstructure, is austenitized at temperatures of 900 to 950 °C and subsequently quenched, generally using water as a cooling medium. This process is particularly used for plate products. Cold-rolled strip can be treated in this way using a continuous annealing line equipped with a water quenching section. In the so-called press hardening process, nowadays widely used by the automotive industry, quenching is achieved by solid-to-solid heat conduction between the hot sheet and the cold forming die. In the production of hot-rolled strip and plate, direct quenching from the rolling heat is increasingly often being practiced. This procedure makes the re-austenitizing process obsolete and thus allows for considerable energy savings and simpler logistics in the steel works. However, direct quenching from austenite, depending on rolling conditions and alloy concept, can lead to an anisotropic microstructure [8,9].

This paper focuses on property improvement of as-quenched martensite utilizing the metallurgical functionality of niobium and molybdenum as alloying elements. In this approach, established knowledge on the most significant relationships between microstructure and properties are briefly reviewed. Subsequently, successful examples for dedicated application of niobium and molybdenum alloying for property optimization in re-austenitized quenched as well as direct quenched flat rolled martensitic steels will be demonstrated.

2. Review of Microstructure–Property Relationships in Martensite

The crystal structure of low-carbon lath martensite site is bcc, but at carbon contents in excess of 0.20% the structure becomes slightly tetragonal [10]. Quenched steel containing carbon higher than 0.5% forms more complicated microstructures, as lath martensite is gradually replaced by plate martensite and the fraction of retained austenite increases. In the basic microstructure of lath martensitic steel [11–15] each prior austenite grain is subdivided into packets (Figure 1) separated by high-angle grain boundaries. Packets are sub-divided into blocks, which can be separated by either low-angle or high-angle grain boundaries [13–16]. Given the need to minimize elastic energy, the same microstructural pattern is favored for all prior austenite grains of reasonable size. Accordingly, the block and packet sizes tend to scale linearly with the parent austenite grain size until the grain size becomes so small that surface effects become dominant. Maki et al. [14] analyzed correlations of substructure size and parent austenite size in detail. The packet size can never be larger than the prior austenite grain size and, hence, decreases with the latter. Larger austenite grains transform into a substructure consisting of several packets. For a prior austenite grain size finer than 10 μm it is possible that some prior austenite grains consist of a single packet or even a single block. Yet, the effect of prior austenite grain size on the lath width appears to be small.

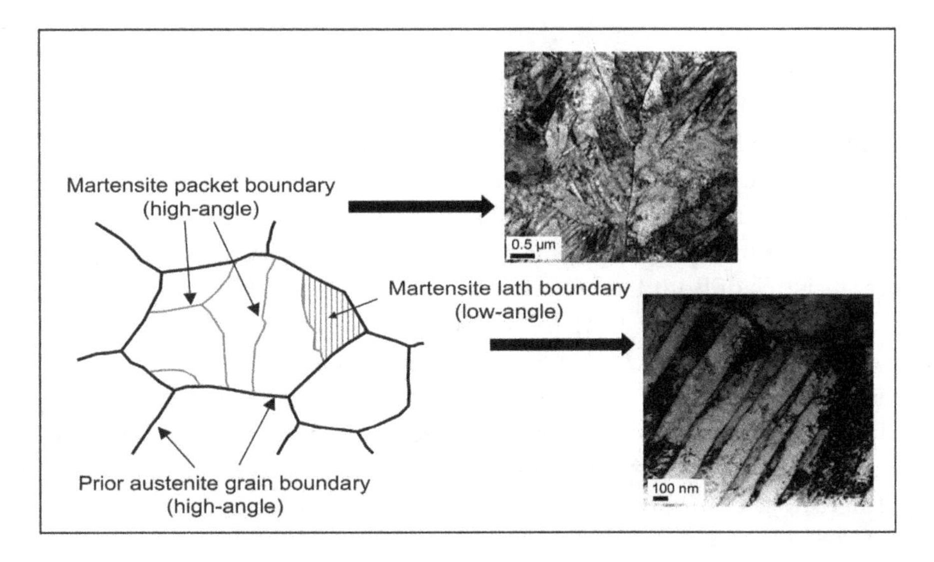

Figure 1. Microstructural features of low-carbon martensitic steel.

It is well established that many important properties of polycrystalline materials are related to the grain size. Smaller grain size in ferritic steels increases strength and toughness while it lowers the ductile-to-brittle transition temperature according to Hall–Petch type relationships. In polygonal ferritic steels the grain size can easily be determined by light optical microscopy. In martensite, however, determination of the grain size is not trivial and often requires sophisticated microstructural analysis tools. Yet, the parent austenite grain size can be determined quite easily and, hence, is of great practical interest. With respect to properties, the fundamental question that needs to be answered is "which is the relevant definition for grain size in martensitic steel?". Morris and co-workers [17] have approached this question by defining an "effective" grain size (d_{eff}) in martensite and pointed out that it may be different with respect to the various properties.

2.1. *Effective Grain Size for Strength*

Early work by Cohen [18,19] suggested the yield strength of carbon-free martensite to be inversely proportional to the square root of the parent austenite grain size as defined by

$$\sigma_y = \sigma_0 + K_y d_{eff}^{-1/2} \quad (1)$$

where K_y is the Hall–Petch coefficient for strength. Later studies by Tomita and Okabayashi on Fe-0.20C-Ni-Cr-Mo steel, considered the packet size as the effective grain size in lath martensite [20]. This correlation could be well repeated by Wang et al. [21] on a similar steel although correlation with the parent austenite grain size was similarly good. Work by Morito and Obha [22] showed that the replacement of packet size by block size in the Hall–Petch relation of Fe-0.2C and Fe-0.2C-2Mn steels produced a more consistent value of the Hall–Petch coefficient. Also Morris [23] argued that the block size could be the effective grain size for strength. However, Bain variants share common slip planes and, hence, such block boundaries may not effectively inhibit slip [24]. Data by Ohmura and Tsuzaki [25] suggest that strengthening by block or packet boundaries is largely due to boundary decoration by carbon or carbide films. A detailed analysis of the effective grain size on strengthening in direct quenched steel was performed by Hannula et al. [26]. In these steels, austenite has a pronounced pancake morphology prior to quenching. Reduced thickness of the austenite pancake leads to smaller effective grain size. The effective grain size was determined by high-angle boundary misorientations (>15°) via an electron backscatter diffraction (EBSD) based technique, and its square root correlates well with the measured yield strength.

2.2. *Effective Grain Size for Toughness*

Two fracture modes, quasi-cleavage and ductile fracture, were observed by Irani [27] in an early study on the fracture behavior of martensitic steel. Under impact loading, the former absorbs less energy than the latter. The absorbed impact energy increases with the amount of ductile fractures. Quasi-cleavage proceeds by the nucleation and growth of submerged cracks ahead of the advancing fracture front. The crack front advances in a stepwise manner as the cracks in front of the fracture tip grow until coalescence takes place. The advance of a quasi-cleavage fracture crack through a martensitic structure is transgranular with respect to parent austenite grains or packets. The quasi-cleavage fracture surface shows characteristics of both true cleavage and plastic rupture.

Nailor et al. [28,29] pointed out that micro-cracks formed within laths are too small to start cleavage due to the limited lath width. Micro-cracks in adjacent laths must join together to create a crack of critical length to propagate as cleavage. This crack experiences small deflections when crossing lath or sub-block boundaries and larger deflections or even crack arrest when passing block boundaries. The blocks are thought to coincide with the facets seen in fractography.

The cleavage fracture stress follows a relationship of the form:

$$\sigma_f = K_f d_{eff}^{-1/2} \quad (2)$$

where K_f is the Hall–Petch coefficient for cleavage. The microstructural mechanism of cleavage in bcc steels is well known; bcc steels cleave along {100} planes. Consequently, in the case of lath martensitic steels the effective grain size d_{eff} is the coherence length on {100} planes, which determines the cleavage crack length. The {100} coherence length is fixed by the "block" size, which is the basic crystallographic unit. Refining the block size in martensitic steel is hence an effective means of increasing its resistance to transgranular cleavage fracture, since Bain variant boundaries are crystallographic discontinuities in the {100} cleavage planes [30].

The dimpled topography of ductile fracture is due to the concave depressions formed by the growth and coalescence of spherical micro-voids with the advancing crack front. These micro-voids may be nucleated at any heterogeneity; hence, the size and distribution of heterogeneities has an important influence on the formation, growth, and coalescence of voids. The size to which a micro-void can grow depends partly on the work-hardened state of the matrix. Thus, the number of voids required for the propagation of a fracture front will increase with an increase in the work hardened condition of the matrix. Improving the cleanness of steel is an important means of increasing energy absorption in ductile fracture mode. In very clean steels voids can be generated at grain corners defining the limit to this improvement effort.

2.3. *Effective Grain Size for Ductile-to-Brittle Transition Temperature*

Reducing the carbon content and refining the grain size can efficiently lower the ductile-to-brittle transition temperature in bcc steel. Grain refinement compensates the loss of strength originating from the carbon reduction. This principle has found its culmination in modern thermo-mechanical processed HSLA steels. The ductile–brittle transition temperature (DBTT) T_B often obeys a constitutive equation of the form:

$$T_B = T_0 - K_B d_{eff}^{-1/2} \quad (3)$$

In martensitic steel the transition temperature from a ductile dimple-type fracture to brittle cleavage-type fracture depends on the block size and the strength. Reducing carbon is not always an option as carbon is needed for strength and the trade-off with Hall–Petch strengthening from grain refinement may be insufficient. The linear correlation between inverse square root of the packet size, thus, inherently block size and T_B according to Equation (3) was demonstrated for different alloys [29,30]. The Hall–Petch coefficient K_B was found to be larger in magnitude when the steel has higher carbon content [31].

The connection between DBTT and the cleavage fracture stress can be understood on the basis of a model that was originally suggested by the Russian physicist, Yoffee, in the early 20th century [32]. The ductile–brittle transition occurs at a temperature close to the crossover point. Accordingly, two generic ways of suppressing the ductile–brittle transition are feasible: raising the brittle fracture stress or lowering the yield strength. Since high yield strength is a prerequisite in many structural steels, most of the efforts have concentrated on raising the brittle fracture stress. Hanamura et al. [33] analyzed the impact of effective grain size on the cleavage fracture stress for various steel microstructures. The as-quenched martensitic microstructure according to their results has the largest Hall–Petch coefficient K_f in Equation (2). For as-quenched martensite, K_f was estimated to be around 160 $N{\cdot}mm^{-3/2}$ and is 4 times larger than that of ferritic–pearlitic steel.

2.4. *Effective Grain Size for Intergranular Embrittlement*

Under specific circumstances intergranular brittle fracture is observed. Parent austenite grains usually constitute the facets of that fracture type. Intergranular fracture occurs when the grain boundary cohesion is weaker than either yield strength and cleavage stress. There are three principal origins for such weakening of austenite grain boundaries:

(1) Segregation of cohesion reducing solutes and impurities to the austenite grain boundary;
(2) Precipitation of particles (carbides, nitrides) at the austenite grain boundary;

(3) Hydrogen embrittlement.

Solute atoms and impurities tend to segregate to the austenite grain boundary. In as-quenched martensite, segregation or precipitation can only take place in the austenite phase. Thus, an increased total grain boundary area as a consequence of grain refinement or austenite pancaking should lead to a "geometrical dilution" of segregated elements. Solute hydrogen has a high diffusibility even at low temperatures and can thus segregate to the parent austenite grain boundary after martensite transformation. Hydrogen-induced fracture separates parent austenite grain boundaries by intergranular decohesion at rather low stress intensity [34–38]. Increasing yield strength enhances the sensitivity for hydrogen embrittlement, because it increases the local hydrogen concentration at the tip of a stressed crack or notch and also facilitates reaching the local cohesive stress. Transgranular cracking due to hydrogen embrittlement occurs at higher stress intensities than intergranular decohesion and appears to propagate along planes of maximum shear stress [39]. The fracture appearance resembles that of quasi-cleavage.

Based on the previous considerations, a strategy for optimizing as-quenched martensite basically relies on microstructural refinement, clean steel, and strong parent austenite grain boundaries as schematically indicated in Figure 2. Furthermore, trapping of diffusible hydrogen by nano-sized particles should be considered as a means of counteracting hydrogen induced embrittlement and delayed cracking.

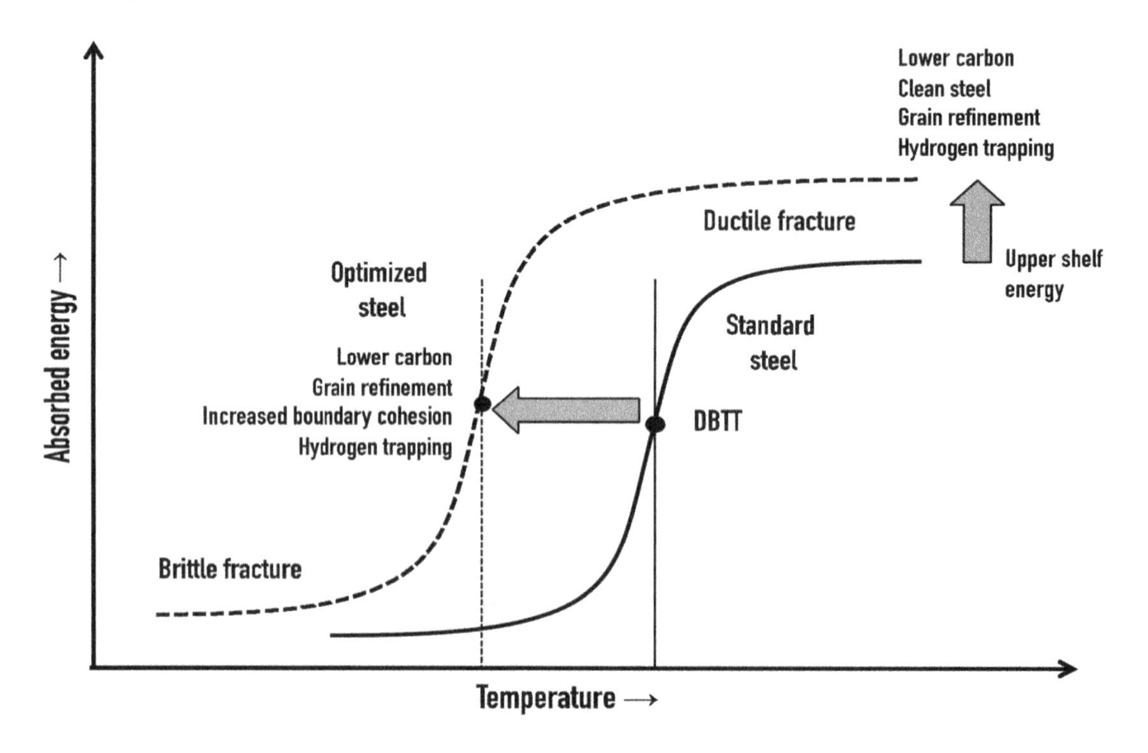

Figure 2. Strategies for optimizing the toughness characteristics of as-quenched martensitic steels.

3. Alloy Concepts

Conventional martensitic steels are usually based on a rather simple carbon–manganese alloy concept, often using boron microalloying for boosting hardenability. In such steel, hardness is determined directly by the carbon content [1,40]. Figure 3 demonstrates a good linear correlation of measured Vickers hardness (HV) data with the carbon content for steels [41]. Also shown are non-linear relationships between carbon content and strength as proposed by Takaki et al. [42]. The tensile strength well fits the measured hardness data over a wide carbon range if a hardness-to-strength ratio of 1:3 is adjusted. The impact of carbon content on the yield strength is, however, significantly weaker so that the yield-to-tensile ratio decreases with increasing carbon content. However, with increasing hardness the toughness of as-quench martensite decreases to rather low values. Therefore, alloyed steel types are often chosen for applications requiring superior toughness.

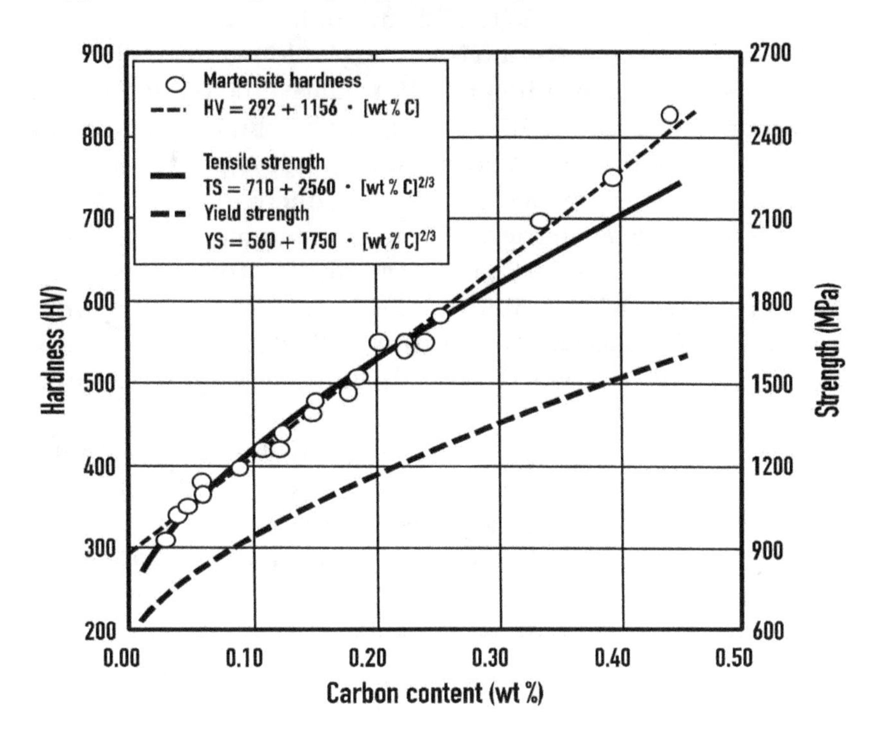

Figure 3. Effect of carbon content on measured hardness (HV0.5) [41] and calculated strength [42] in unalloyed CMn steel with as-quenched martensitic microstructure.

3.1. Alloy Design for Hardenability

A fundamental aspect in designing alloy concepts for martensitic steel is hardenability. The alloy design in combination with the applied quenching method must achieve sufficient hardness (strength) over the entire strip or plate thickness. In conventional carbon–manganese steel the hardening depth is limited to a few millimeters only. However, for many applications, much heavier gauges of hardened material are required. Thus, a well-adjusted addition of alloying elements in combination with high cooling rates is needed to achieve a high hardness value in the core of thicker components. Adding alloying elements such as molybdenum, manganese, chromium, copper, and nickel can significantly enhance hardenability. The hardenability effect of molybdenum, manganese, and chromium is related to a reduced diffusivity of carbon when these alloying elements are present in solid solution. The reduced carbon diffusivity strongly retards the formation of ferrite and pearlite. In this respect molybdenum has the strongest hardenability effect. Molybdenum additions of 0.2 to 0.5 mass percent are sufficient for through hardening of heavy gauges under typical quenching conditions. Besides its excellent hardenability effect, molybdenum has the additional benefit of increasing parent austenite grain boundary cohesion, and thus enhancing the resistance against intergranular fracture. This potency of molybdenum has long been noticed empirically by its effect of counteracting tempering embrittlement of martensitic steel. Recent first-principle calculations by Geng et al. [43] confirmed that molybdenum is a strong boundary cohesion enhancer, whereas chromium has no effect in this respect and manganese decreases boundary cohesion.

3.2. Microalloying in Martensite

Vanadium microalloying has been standardly applied in martensitic steels for a long time as it provides precipitation strengthening during tempering treatment. Yet, it is rather ineffective in as-quenched martensite due to its good solubility in austenite. Vanadium precipitates providing grain size control during re-austenitizing or hydrogen trapping in the as-quenched martensitic microstructure are consequently absent.

Boron is added in very small amounts (10 to 50 ppm) for its significant hardenability effect. Solute boron strongly segregates to the austenite grain boundary at lower austenite temperature. The high concentration of solute boron in the austenite grain boundary efficiently obstructs the nucleation of ferrite grains below equilibrium transformation temperature, thus preserving metastable austenite down to martensite-start temperature. Since boron tends to precipitate with free nitrogen at lower austenite temperatures, it is usually protected by titanium microalloying [44]. Titanium has a much higher affinity to nitrogen. Over-stoichiometric addition of titanium (wt % Ti $> 3.4 \times$ wt % N), however, tends to form primary nitride particles in the liquid steel. These particles are often several micrometers in size and deteriorate toughness [45,46]. With regard to achieving an optimum combination of strength and toughness, the Ti/N mass ratio should be adjusted to around 4. Simultaneously one should aim for the lowest possible nitrogen level, which might require vacuum degassing during steel making.

Historically, the use of niobium microalloying in martensitic steels has not been very common. Its strong grain refining effect in austenite has been considered to reduce hardenability. This concern has meanwhile been discarded [47]. If at all, it is only relevant to simple low-carbon carbon–manganese steels when quenched directly after strong austenite conditioning. Furthermore, it was assumed that there is insufficient solubility of niobium at typical slab reheating temperature considering the relatively high carbon content of typical martensitic grades. Recently, however, niobium microalloying is being increasingly used for improving toughness behavior in as-quenched martensitic strip and plate steels [26,45,48]. Applying typical slab soaking practice, the amount of Nb necessary for deploying its key metallurgical effects, usually up to 0.05 mass percent, can be efficiently brought into solid solution.

3.3. Alloy Design for Grain Refinement

According to Section 2 of this paper, refining the "effective" grain size (defined as either parent austenite grains, packets, blocks) is the essential means of improving the properties of as-quenched martensite. Under the conditions of large-scale industrial production, however, only the parent austenite grain size (PAGS) can be efficiently and reproducibly controlled. Boron appears to have an effect on the sub-structure development within the parent austenite grain. Hannula et al. [26] demonstrated that a finer and more homogeneous substructure can be achieved by omitting boron.

Considering the various processing routes for producing martensitic steel the following steps have an influence on the final grain size.

During slab soaking (T > 1100 °C), austenite grains tend to become rather large in size. A successful way of limiting the austenite grain size in this process step is the dispersion of fine TiN particles which are stable at this high temperature. In order to have this pinning effect maximized it is advisable to adjust the titanium addition in a near-stoichiometric ratio to the residual nitrogen in the steel, as mentioned before.

Hot rolling in the recrystallizing regime leads to moderate grain refinement and homogenization. Combining a short inter-pass time (strip mill) and grain boundary drag by solute atoms such as molybdenum and niobium provides additional refinement. The grain morphology is globular.

Finish hot rolling below the recrystallization-stop temperature, so-called thermo-mechanical rolling, causes pancaking of the austenite grain. Niobium is the most effective microalloying element suppressing recrystallization at acceptable finish rolling temperatures. This effect is caused by precipitation of niobium to NbC particles that pin austenite grain boundaries.

In direct quenched steel either a recrystallized or a pancaked austenite grain structure is converted into martensite, depending on the finish rolling temperature and microalloy concept. In the re-austenitizing quenching route, the hot rolled austenite firstly cools down to ambient temperature, usually transforming into ferrite–pearlite microstructure. This microstructure is finer grained than the original austenite microstructure, particularly after thermo-mechanical rolling. When alloying elements like niobium or molybdenum are present as solutes after finish rolling, they will delay the phase transformation to a lower temperature, which additionally refines the ferrite–pearlite microstructure. Depending on the cooling rate after finish rolling, remaining solute niobium will at least partially precipitate during or after phase transformation. These particles are typically in the lower nano-meter size range. Molybdenum, however, stays in solution for typical martensitic steel alloy concepts. Upon re-heating the ferritic–pearlitic microstructure to austenite, any remaining solute niobium precipitates very quickly. The newly formed austenite grains are globular and grow in size depending on the temperature and holding time. Dispersed niobium precipitate particles and, to some extent, solute drag by molybdenum can efficiently impede the austenite grain growth at that stage and thus condition the austenite microstructure before the quenching process.

4. Examples of Optimized Alloy Concepts and Property Improvement

Quenchable steel grades with carbon contents from 0.08 to 0.33 wt % were used as listed in Table 1, covering a tensile strength range from 1000 to over 2000 MPa. The alloy effects of molybdenum and niobium were systematically investigated by varying these elements in the range of 0 to 0.5 wt % and 0 to 0.08 wt %, respectively. In selected cases the effect of boron and manganese was additionally considered. From a processing point of view, direct quenching was applied after hot rolling. Re-heat quenching was executed after hot as well as cold rolling. The study initially analyzes the effect of alloy concept and processing conditions on the microstructure. Subsequently, microstructural influences on the steel properties are being verified and discussed in the background provided in Section 2 of this paper.

Table 1. Chemical composition in wt % of hot rolled (HR) and cold rolled (CR) direct quenched (DQ) and re-heat quenched (RHQ) steels analyzed in this study.

Material	C	Si	Mn	Cr	Ni	Ti	B	Mo	Nb
DQ grades									
0.08% C HR	0.08	0.2	1.8	1.1	-	0.02	0 or 0.0025	0.15	var. 0–0.05
0.16% C HR	0.16	0.2	1.1	0.5	0.5	-	-	var. 0–0.5	var. 0–0.04
RHQ grades									
HB450 HR	0.18	0.2	1.2	0.2	-	0.03	0.0020	0.25	0.03
22MnB5 HR	0.22	0.2	1.2	0.2	-	0.03	0.0025	var. 0–0.15	0.05
16MnB5 CR	0.16	0.4	2.3	-	-	0.03	0.0025	-	var. 0–0.08
22MnB5 CR	0.23	0.3	1.2	0.2	-	0.03	0.0025	-	var. 0–0.08
32MnB5 CR	0.33	0.1	1.2 or 2.5	-	-	0.02	0.0020	var. 0–0.5	var. 0–0.05

4.1. Microstructure of Low-Carbon Direct Quenching Steel

In a series of 0.08% C-1.8% Mn-1.1% Cr-0.15% Mo steels with boron microalloying, the niobium content was systematically varied from zero to 0.05% Nb. The alloy containing 0.02% Nb was also produced without boron microalloying. All steels were solution treated at 1250 °C and hot rolled in six passes to 6 mm gauge as described by Hannula et al. [26]. Two finish rolling temperatures, namely 920 and 820 °C, have been applied in individual rolling schedules, the latter leading to stronger austenite conditioning. The finish rolled strips were then directly quenched to room temperature using water providing a cooling rate of around 90 °C/s. All steels were fully hardenable, except the boron-free alloy which was finish rolled at 920 °C. Analyzing the parent austenite grain morphology, only the niobium-free alloy finish rolled at 920 °C shows a nearly globular grain shape. However, the same alloy finish rolled at 820 °C is clearly pancaked. In the absence of niobium microalloying, this can be due to solute drag effects acting on the austenite grain boundary caused by segregated molybdenum and boron. All niobium microalloyed steels exhibit even more pronounced austenite pancaking under both finishing temperatures. Austenite pancake thicknesses are in the range of 5 to 6 μm and 8 to 10 μm for the lower and higher finishing temperatures, respectively. A synergy between niobium and boron is noticed in the sense that boron further suppresses recrystallization above the level already provided by niobium alone. Based on EBSD analysis [26], effective grain and lath sizes were determined as equivalent circle diameter (ECD) values with low-angle (3–15°) and high-angle boundary (>15°) misorientation, respectively. As proposed before, the effective grain size in the as-quenched material is determined by the austenite pancake thickness. Figure 4 correlates the effective grain size with the parent austenite pancake thickness. Higher niobium addition and lower finishing temperatures result in more severe pancaking and hence finer effective grain size. Wang et al. [21] demonstrated a similar behavior for martensite having a globular parent austenite grain morphology.

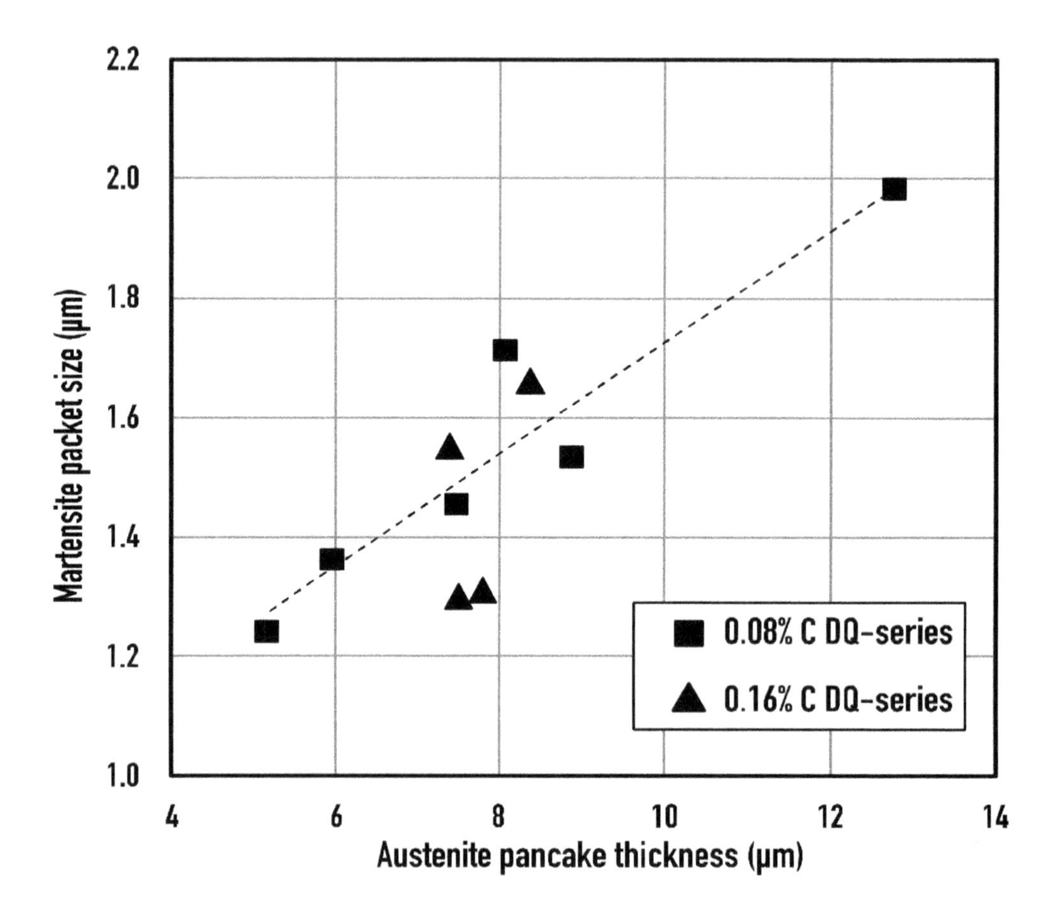

Figure 4. Correlation of martensite packet size with the parent austenite pancake thickness in direct quenched steel after non-recrystallizing hot rolling.

In a series of 0.16% C-1.1% Mn-0.5% Cr-0.5% Ni steels, the molybdenum content was varied from zero to 0.5% Mo. The reheating and rolling procedure was like that in the previous series of lower carbon steels; 900 °C was taken as finishing temperature for this series. Quenching was done using water directly after finish rolling resulting in fully martensitic microstructure. The parent austenite grain structure is shown in Figure 5. The Mo-free alloy exhibits globular shaped austenite grains, whereas adding 0.25% Mo results in pancaking under the same rolling conditions. Solute drag effects provided by molybdenum obstruct full recrystallization. With a further increased molybdenum content (0.5% Mo), austenite pancaking is marginally more pronounced. Adding however 0.04% Nb to the 0.25% Mo steel leads to significantly thinner austenite pancake thickness. The correlation of the effective grain size with the parent austenite thickness is shown in Figure 4.

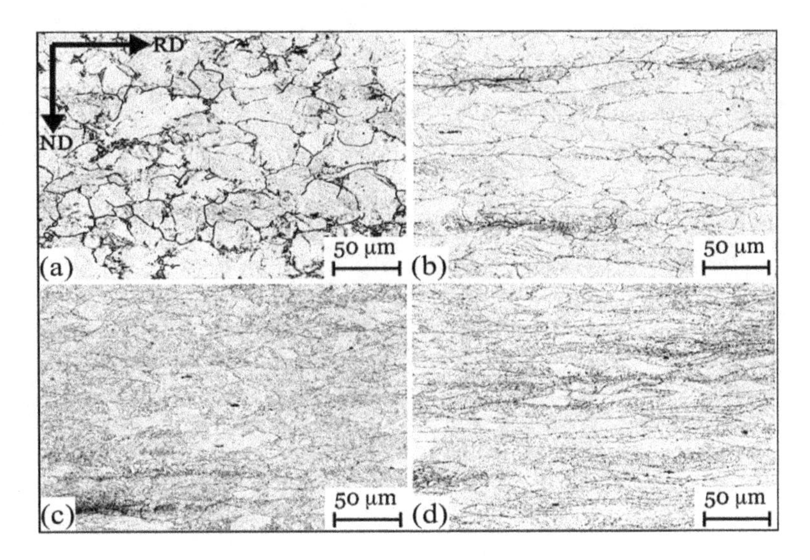

Figure 5. Parent austenite grain morphology of (**a**) 0% Mo; (**b**) 0.25% Mo; (**c**) 0.5% Mo; (**d**) 0.25% Mo + 0.04% Nb direct quenched steel (finish rolling temperature of 900 °C).

4.2. Microstructure of Re-Austenitize Quenching Steel

For this processing route, a series of medium carbon steels typically used for the automotive press hardening process is analyzed. These steels contain 0.20–0.35% C, 1.0–1.8% Mn, 0–0.50% Cr boron microalloying (20–40 ppm), molybdenum additions between zero and 0.25% Mo and niobium additions between zero and 0.08% Nb. Usually, there is no explicit austenite conditioning foreseen in the hot-rolling schedule of such steels. However, it is possible that for some of the niobium or molybdenum added variants a significant degree of austenite pancaking can be expected. Figure 6 shows microstructures of 22MnB5 with niobium (0.05% Nb) as well as combined niobium-molybdenum (0.05% Nb + 0.15% Mo) addition in the as-hot-rolled state (6 mm gauge). The ferritic–pearlitic microstructure indicates that the austenite indeed was pancaked (FRT: 850 °C). However, the transformed microstructure appears much finer in the Nb + Mo added steel. This can be due to a retarded transformation after finish rolling by solute molybdenum resulting in a bainitic microstructure. Parent austenite morphology analysis indicated that the Nb-only and Nb–Mo steel both had austenite pancake thicknesses between 5 and 6 μm. Figure 7 shows the pancake structure of the Nb–Mo steel. The pancake structure is erased and replaced by a globular austenite morphology after re-austenitizing (900 °C for 900 s). The average recrystallized parent austenite sizes are around 7 μm for both the Nb-only and Nb + Mo steel. Detailed microstructural analysis using an EBSD-based technique described by Hannula et al. [26] revealed that the effective grain size in these steels after re-austenitizing quenching is in the order of 1.5 μm. Accordingly, strength and toughness properties differ only marginally between the two steels.

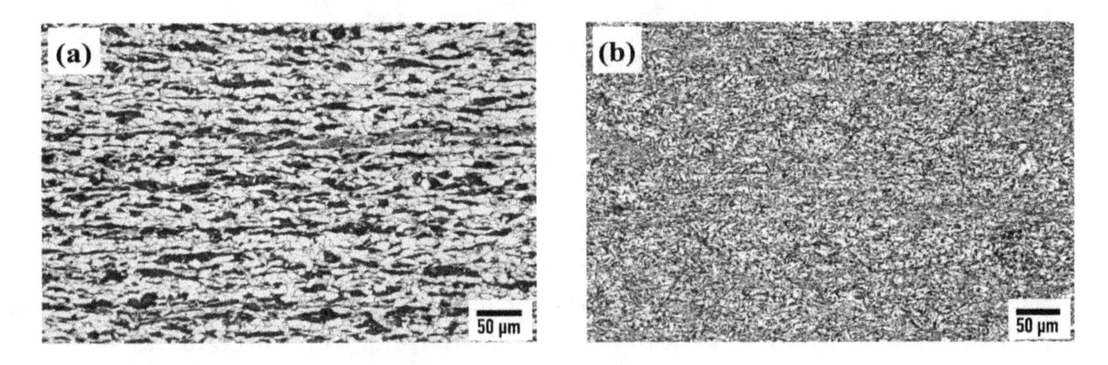

Figure 6. As-hot-rolled microstructure of 22MnB5 steel alloyed with Nb (**a**) and Mo + Nb (**b**).

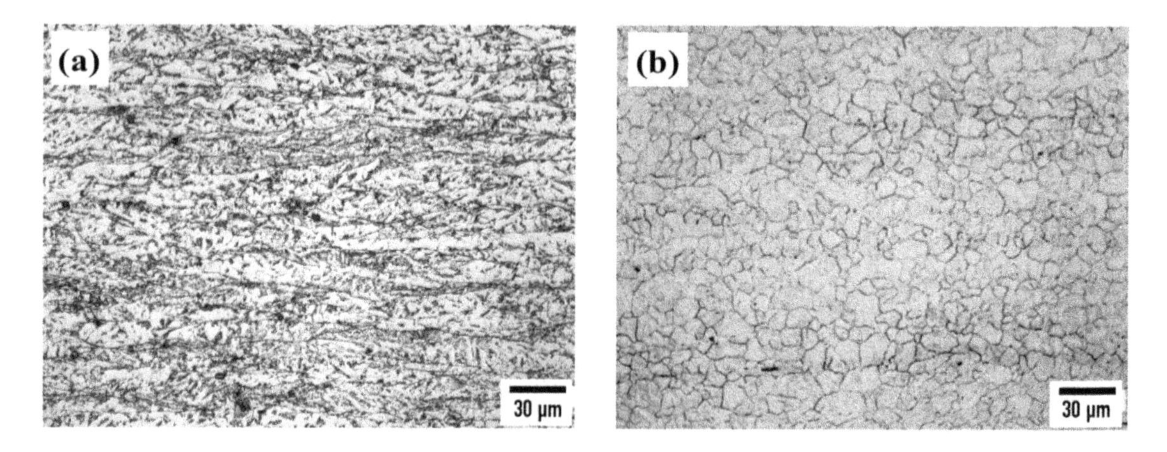

Figure 7. Parent austenite grain morphology of Mo + Nb alloyed 22MnB5 steel after hot rolling (**a**) and after re-austenitizing quenching (**b**).

The size of the recrystallized austenite grain with globular shape depends not only on the as-hot-rolled austenite grain size having globular or pancaked shape, but also on the heating conditions of the re-austenitizing treatment. Higher temperature or longer austenitizing duration impose growth of the recrystallized austenite grains so that these are usually bigger than the original ones (Figure 8). Therefore, it is of key importance to restrict this austenite growth. This can be achieved by optimizing the reheating schedule (low temperature, short time). However, under industrial circumstances over-heating cannot be always excluded and for process robustness one would likely not operate at the lower limit conditions. For that reason, it is favorable to obstruct austenite grain coarsening using solute drag or, more powerfully, by particle pinning effects.

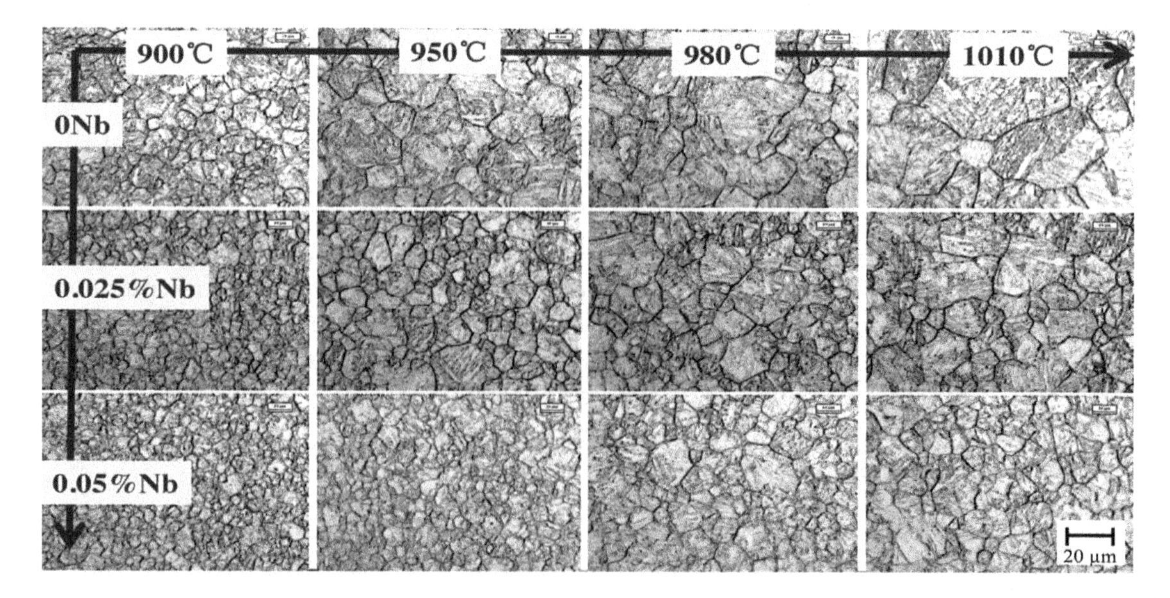

Figure 8. Effect of niobium microalloy content (rows) on parent austenite grain morphology in re-austenitized quenched 22MnB5 steel for various reheating temperatures (columns) for a treatment duration of 300 s.

Niobium carbide precipitate particles are thermodynamically stable at typical re-austenitizing temperatures. Hence, efficient grain boundary pinning can be expected when such particles are present in sufficient quantity and with a homogenous spatial distribution. In order to demonstrate this effect, 22MnB5 steel was alloyed with various additions of niobium. The as-hot-rolled ferritic–pearlitic material was further cold rolled to 1.5 mm final gauge. Subsequently, austenitizing treatments were done at temperatures between 900 and 1010 °C, each for 300 s. The parent austenite structure was developed after quenching as shown in Figure 8. The effect of NbC particles restricting austenite growth is evident at all re-heating temperatures. The higher niobium (0.05% Nb) addition provides a stronger growth-inhibiting effect, implying that the number of particles must be bigger with a more homogeneous spatial distribution. The addition of 0.05% Nb appears to be an optimum amount in this steel as higher niobium addition brings no further advantage (Figure 9). It is reasonable to conclude that more than 0.05% Nb cannot be brought into solution in this alloy during initial slab reheating. Only the dissolved niobium can precipitate to a particle size range providing efficient grain boundary pinning. The optimum niobium addition particularly shows its advantage at higher re-austenitizing temperatures (Figure 8). It allows a combination of a fine final parent austenite grain size with a robust re-austenitizing processing window.

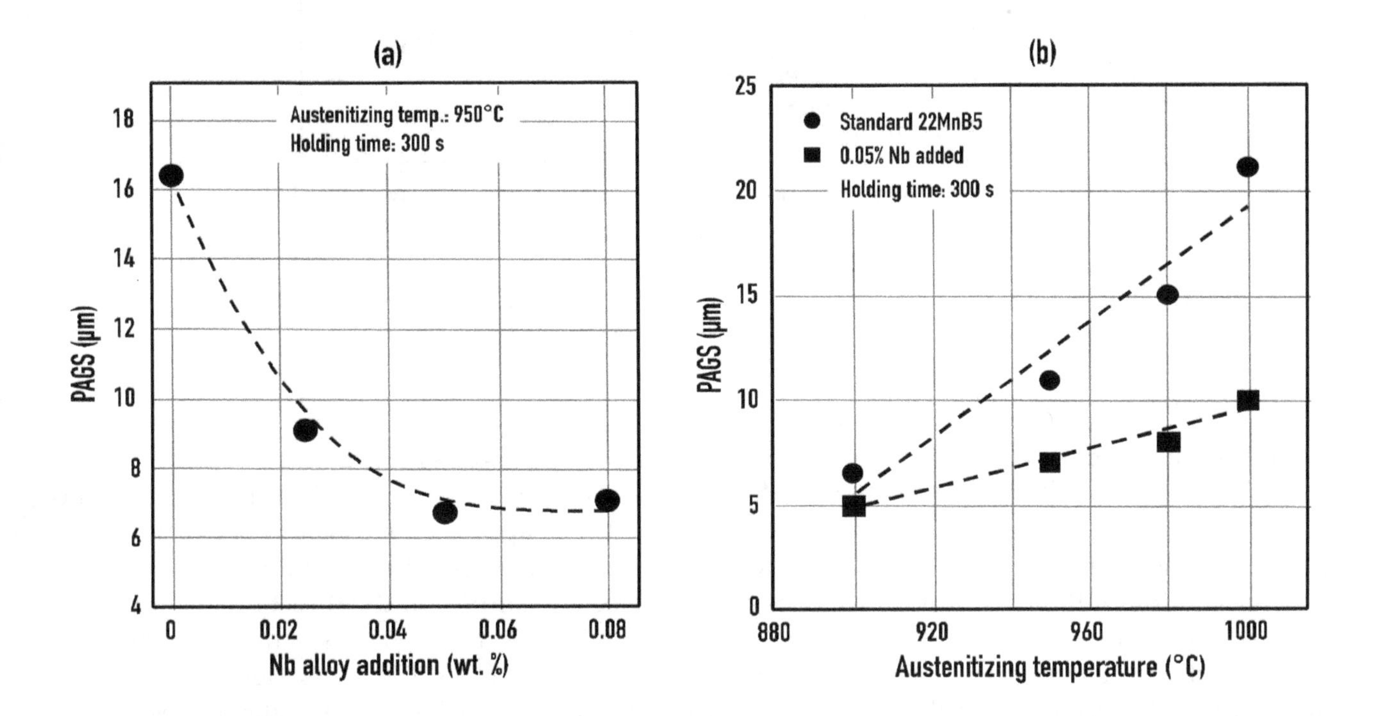

Figure 9. (**a**) Effect of niobium on parent austenite grain size control under typical re-austenitizing conditions used in the automotive press hardening process; (**b**) demonstration of robustness against overheating for a niobium alloyed variant.

4.3. Effective Grain Size and Strength

Strength data measured on various direct quenched and re-austenitized quenched steels are plotted against the inverse square root of the parent austenite grain size in Figure 10. The data sets show good linear correlation as expected by the Hall–Petch relationship (Equation (1)). The niobium microalloyed steel grades show yield strength Hall–Petch coefficients of similar magnitude, and irrespective of the actual strength level (carbon content). The Hall–Petch coefficients for the tensile strength are also similar but somewhat smaller in magnitude. The strength of the niobium-free steel with variable molybdenum content shows significantly larger Hall–Petch coefficients. The tensile strength has a larger Hall–Petch coefficient than the yield strength in these steels.

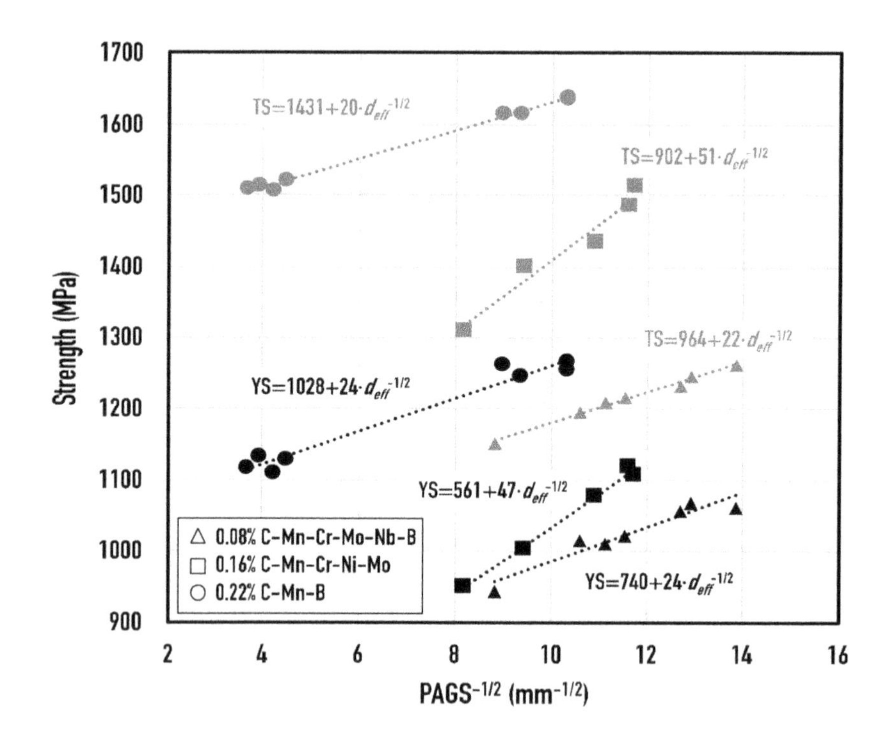

Figure 10. Hall–Petch type plot indicating the effect of parent austenite grain size (pancake thickness for direct quenched steels) on yield (black symbol) and tensile (grey symbol) strength.

4.4. Effective Grain Size and Toughness

Using the Hall–Petch coefficients for yield strength of around 24 N·mm$^{-3/2}$ for the Mn-B steel types and 47 N·mm$^{-3/2}$ for the Mo steel types, and the previously quoted Hall–Petch coefficient of 160 N·mm$^{-3/2}$ for the fracture stress in martensite in the Yoffee approach, it can be expected that grain refinement should clearly lower the ductile-to-brittle transition temperature in these steels.

Verification of this assumption was done using selected samples of the direct quenching steels. Toughness was measured using Charpy V-notch specimens. Due to the thickness of the rolled products (6 mm), sub-size samples (5 × 10 × 55 mm^3) had to be used. From the Mo-series (0.16% C-1.1% Mn-0.5% Cr-0.5% Ni-Mo variable), four samples were taken having a fully martensitic microstructure, that is, without detectable traces of other phases. Charpy tests were performed over the temperature range from 20 °C down to −150 °C. The upper shelf energy is always higher in the rolling direction (around 50 J) than in the transverse direction (around 30 J) reflecting the anisotropy present in direct quenched steels. The transition temperatures are shown Figure 11a for the test done in rolling direction taking 28 J as the defining criterion. It is obvious that samples with finer PAGS have a lower transition temperature and the data correlate according to a Hall–Petch-type relationship. The Hall–Petch coefficient according to (Equation (3)) is for these steels approximately 12 K·mm$^{1/2}$. These data become more meaningful when plotting the transition temperature versus the room temperature yield strength as in Figure 11b. It is obvious that despite a significant strength increase (170 MPa) the transition temperature is drastically lowered (−40 K). Also these data are in a linear relationship, since both quantities are originally linearly related to the square root of the PAGS.

The toughness behavior of the Nb-B series of steels (0.08% C-1.8% Mn-1.1% Cr-0.15% Mo-Nb-Ti, B variable) was in detail discussed by Hannula et al. [26] so that here only the most relevant observations are resumed. In the rolling direction, the upper shelf energy is between 40 and 50 J and thus somewhat lower than that observed for the Mo-series. Likewise, the lower shelf energy is below that seen in the Mo-series. The upper shelf energy in the transverse direction is much lower, reflecting the anisotropy present in direct quenched steels (Figure 12). The upper shelf energy in these steels decreases with the degree of austenite conditioning. A significant difference in the ductile-to-brittle transition behavior is observed between the boron-containing and the boron-free alloys, the latter performing much better. Hannula et al. [26] explained this result by a finer grained and more homogeneous microstructure of the boron-free steel. Microstructural refinement is generally more pronounced at lower finish rolling temperatures, yet so is the difference in ductile-to-brittle transition temperature. The fact that the microstructural inhomogeneity appears to be stronger at lower finish rolling temperature can be due to a retarding influence of boron on the recrystallization behavior in addition, and likely in synergy to that of niobium. Consequently, thorough homogenization of the austenite structure by multiple recrystallization at higher temperatures may be more severely obstructed. Mixed parent austenite grain size also results in enhanced quench distortion, which is due to the influence of austenite grain size on the martensite transformation temperature and accommodating residual stress between the grains [49].

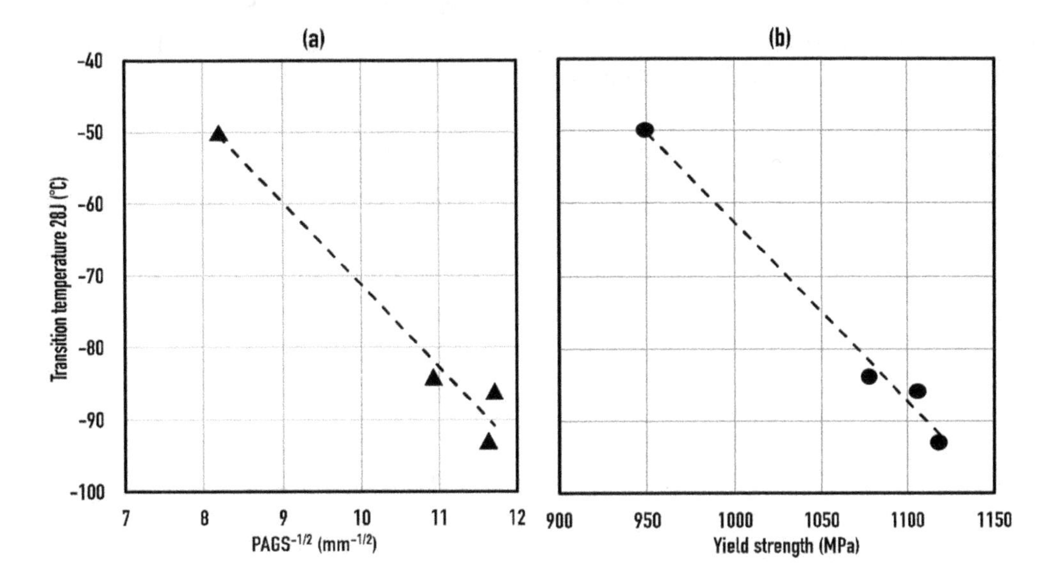

Figure 11. (**a**) Hall–Petch type plot indicating the effect of parent austenite grain size in direct quenched 0.16% C-1.1% Mn-0.5% Cr-0.5% Ni-Mo steels on ductile-to-brittle transition temperature (28 J criterion); (**b**) correlation between yield strength and ductile-to-brittle transition temperature for the same steels indicating simultaneous improvement of both properties.

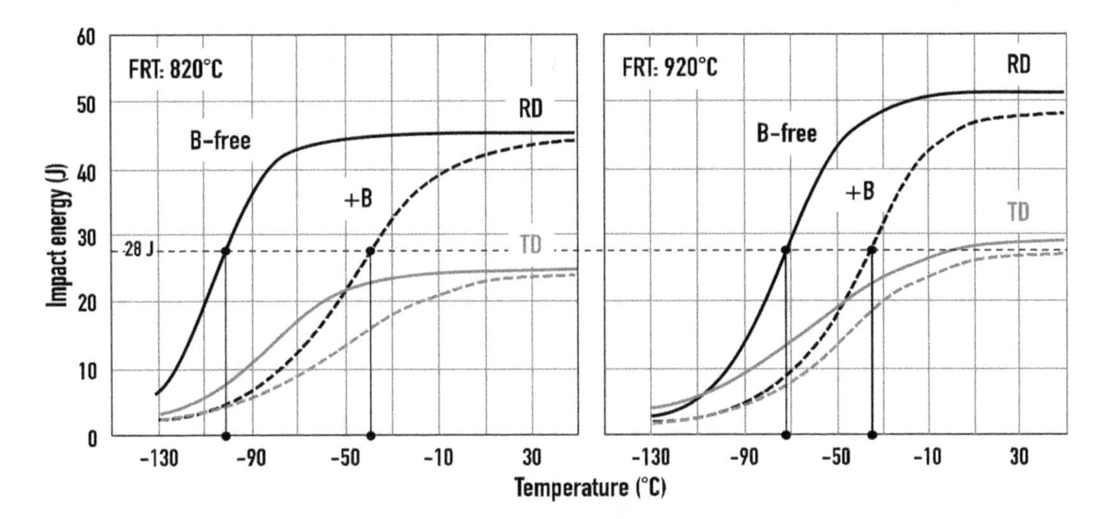

Figure 12. Anisotropy (black curve: rolling direction (RD), gray curve: transverse direction (TD)) of ductile-to-brittle transition behavior of direct quenched 0.08% C-1.8% Mn-1.1% Cr-0.15% Mo-Nb steel and influence of boron addition (solid line B-free, dashed line B-added) at two finish-rolling temperatures (FRT).

In re-austenitized-quenched steel the parent austenite microstructure is rather isotropic as a result of the normalization caused by the phase transformations occurring prior to quenching. In these steels, refinement of the parent austenite grain size increases the upper-shelf energy as shown in Figure 13 (data refer to the steels shown in Figure 9a). Another example (Figure 14) demonstrates the improvement that can be achieved by niobium microalloying to HB450 abrasion resistant steel plate. Due to the relatively low slab reheating temperature practiced in plate mills, niobium solubility limits the addition to a maximum of 0.03 wt % Nb. This amount is sufficient for restricting austenite coarsening at a re-austenitizing temperature of 900 °C as compared to a niobium-free alloy. It achieved refinement from an average parent austenite grain size of 18 μm in the Nb-free to 12 μm in the 0.03% Nb-added steel as well as a more homogeneous size distribution in the latter. This optimization results in a significant increase in upper-shelf energy by approximately 30 J and a reduction of ductile-to-brittle transition temperature by around 20 °C for the niobium added steel. Similar improvements using the same metallurgical approach were reported earlier by Kern et al. [46] and Ishikawa et al. [48] for optimizing toughness of wear plate in industrial production.

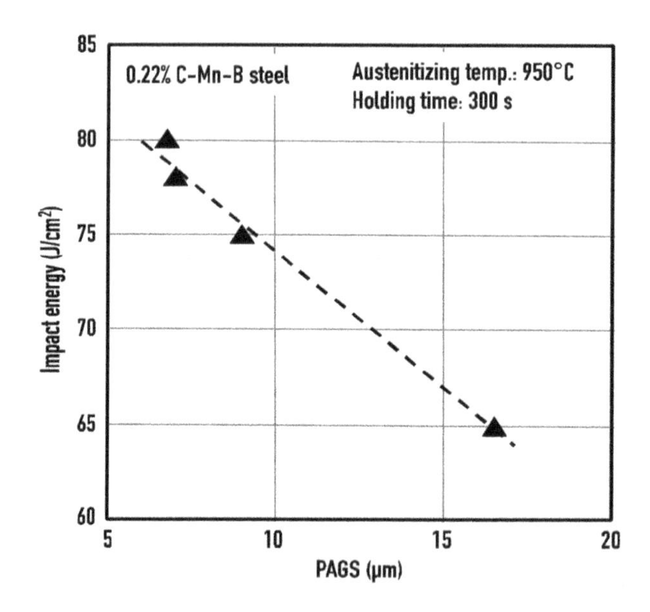

Figure 13. Effect of parent austenite grain size on upper-shelf energy for re-austenitized quenched 22MnB5 steel tested at ambient temperature (20 °C).

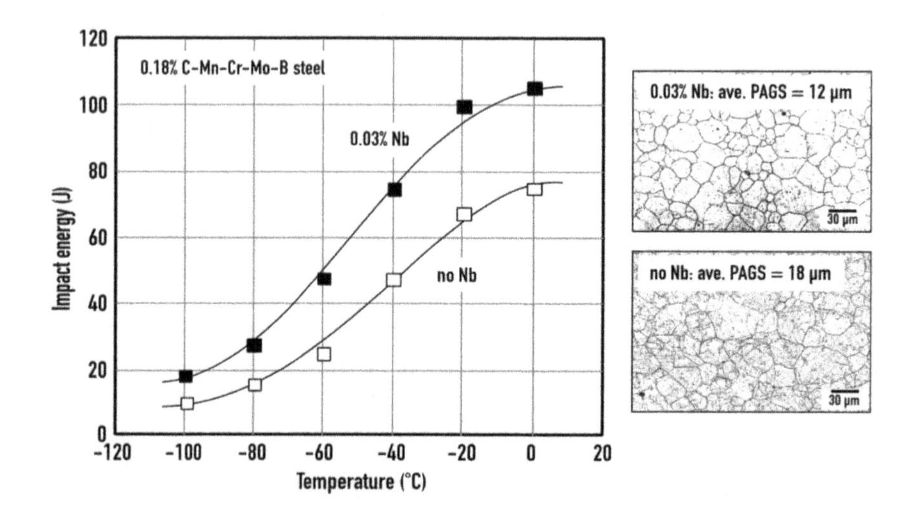

Figure 14. Influence of parent austenite grain refinement on ductile-to-brittle transition curves of HB450 abrasion resistant plate steel (0.18% C-Mn-Cr-Mo-B) produced by re-austenitizing quenching.

5. Optimization against Hydrogen Embrittlement

The presence of hydrogen in steel can severely limit the performance of high strength steels as detailed in a concise review by Gangloff [50]. Two types of non-ductile fracture can be observed in martensitic steel when hydrogen is present in small amounts. Intergranular fracture (Figure 15) is seen when hydrogen aggregates at parent austenite grain boundaries, weakening the grain boundary cohesion so much that failure occurs at a stress level far below the yield strength. Already the presence of around 0.4 ppm diffusible hydrogen is sufficient to cause intergranular fracture if microstructural optimization is not foreseen. Means of optimization are refinement of the parent austenite grain size, dispersion of particles with hydrogen trapping capacity, and metallurgical reinforcement of parent austenite grain boundaries. Depending on the efficiency of these means as well as the actual hydrogen content the fracture surface becomes of quasi-cleavage type (Figure 15) or even dimple type.

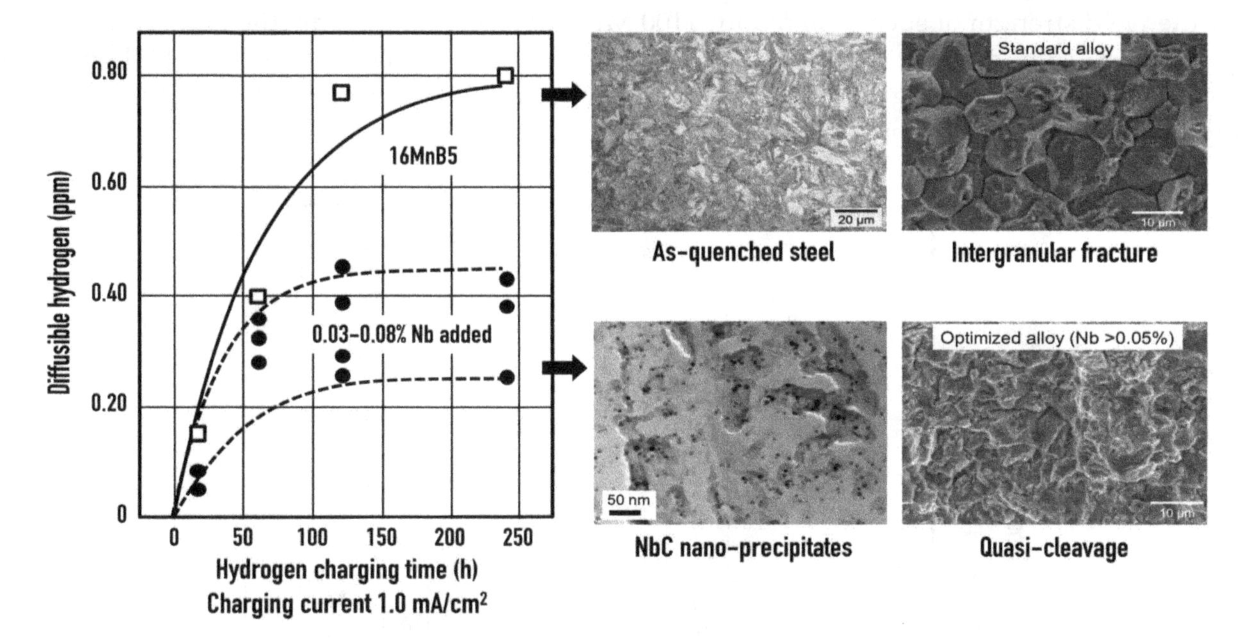

Figure 15. Amount of diffusible hydrogen in 16MnB5 measured by thermal desorption analysis (TDA) and influence of niobium microalloying; microstructural features and fracture surface after constant load testing.

5.1. *Effect of Grain Refinement*

For screening the hydrogen sensitivity of martensitic steels, in this case automotive press hardening sheet steels of 1.4 mm gauge, double notched tensile samples were immersed in 20% ammonium thiocyanate solution under constant load conditions with an acting stress of 1000 MPa at the notch [51]. Failure at this stress level is not instant but occurs with a certain time delay (time-to-fracture). Figure 16 indicates that the time to fracture becomes much shorter when the tensile strength level being 1500 MPa in grade 22MnB5 is raised to 2000 MPa in grade 32MnB5. These steel grades are standard alloys containing two different levels of carbon (0.22% and 0.32%), manganese (1.2 wt %), as well as boron (30 ppm) for hardenability. In a variant, the manganese level was set higher (2.5 wt %). In this variant, time-to-fracture becomes notably shorter indicating that manganese increases the sensitivity for hydrogen embrittlement and should therefore be kept as low as possible.

In order to show the influence of the parent austenite grain size and to exclude that of other effects such as hydrogen trapping, the PAGS of the base steels was varied between 4 and 60 μm by using different re-austenitizing temperatures. In all steels, smaller parent austenite grain size results in clearly longer time-to-fracture. Momotani et al. [52] demonstrated by hydrogen micro-print technique that hydrogen, which was originally distributed evenly within the matrix, aggregates on the parent austenite grain boundaries upon application of stress. Since this mechanism is diffusion controlled, it requires a certain time for hydrogen to redistribute. Substantial concentration of hydrogen at the parent austenite grain boundary leads to intergranular fracture. Reducing the parent austenite grain size has two effects with respect to that mechanism:

(1) The average diffusion distance from the grain bulk to the parent austenite grain boundary becomes shorter, hence reducing the time delay for hydrogen to aggregate on the boundary.
(2) The total grain boundary area becomes substantially larger so that for a given amount of hydrogen its average concentration per unit grain boundary area will be lower.

Similar delayed cracking tests by other researchers [53–55] applying constant load conditions to 22MnB5 under hydrogen charging conditions revealed that the failure stress after severe grain boundary embrittlement by hydrogen is only in the range of 400–600 MPa. This range is much below the yield strength of such steel (about 1100 MPa) and accordingly, ductile failure mechanisms are prevented. Higher addition of manganese reduces grain boundary cohesion [43] enhancing the negative effect of hydrogen embrittlement, whereas molybdenum has the opposite effect of counteracting hydrogen embrittlement.

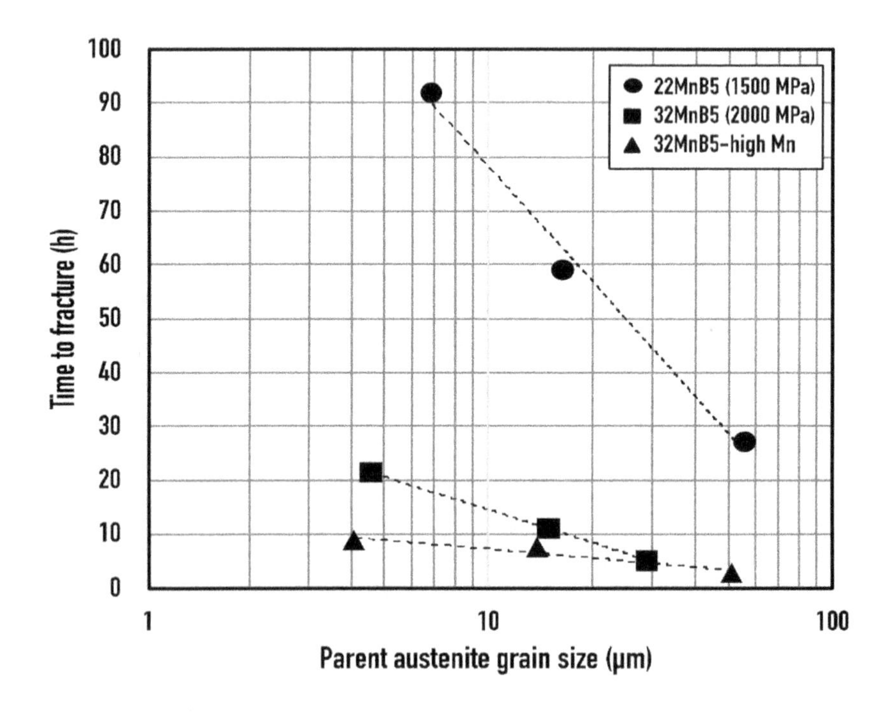

Figure 16. Delayed fracture behavior of as-quenched manganese–boron steels (without niobium microalloying) as a function of parent austenite grain size modified by variation of the re-austenitizing temperature.

5.2. Effect of Hydrogen Trapping by Precipitate Particles

Microalloy carbides are known to act as hydrogen traps in martensitic steels [50,56]. Wei et al. [57] identified that the hydrogen trapping potential depends on the microalloying element forming the carbide according to the following ranking: NbC > TiC > VC. Without going into the details of trapping mechanisms in this paper, evidence is presented that hydrogen trapping by NbC precipitates works efficiently for optimizing as-quenched martensitic steel and addition of molybdenum provides further improvement.

The effect of niobium was investigated in the same 22MnB5 steels [55], which were already presented in Figure 9a. Using hydrogen permeation tests, the hydrogen diffusivity was measured for different niobium additions (Figure 17a). The hydrogen diffusion coefficient is lowered by the addition of niobium reaching a minimum when 0.05% Nb is added. Niobium present in this steel completely precipitates after a re-austenitizing quenching treatment. Thorough transmission elcectron microscopy (TEM) precipitate size distribution analysis (Figure 17b) revealed that these precipitates have diameters in the range of up to 40 nanometers. Larger precipitate sizes are likely formed early in the process chain, for example, during hot rolling. The difference between the three niobium addition levels is that for 0.05% Nb the highest fraction of ultra-fine particles of sizes below 5 nanometers is observed. This size distribution consequently results in an increased particle density and high particle surface-to-volume ratio. It reasonable to assume that these features enhance the interaction of hydrogen with particles, leading to more efficient trapping. If NbC particles act as irreversible trap, the amount of diffusible hydrogen responsible for causing damage in the steel will be effectively lowered (Figure 15).

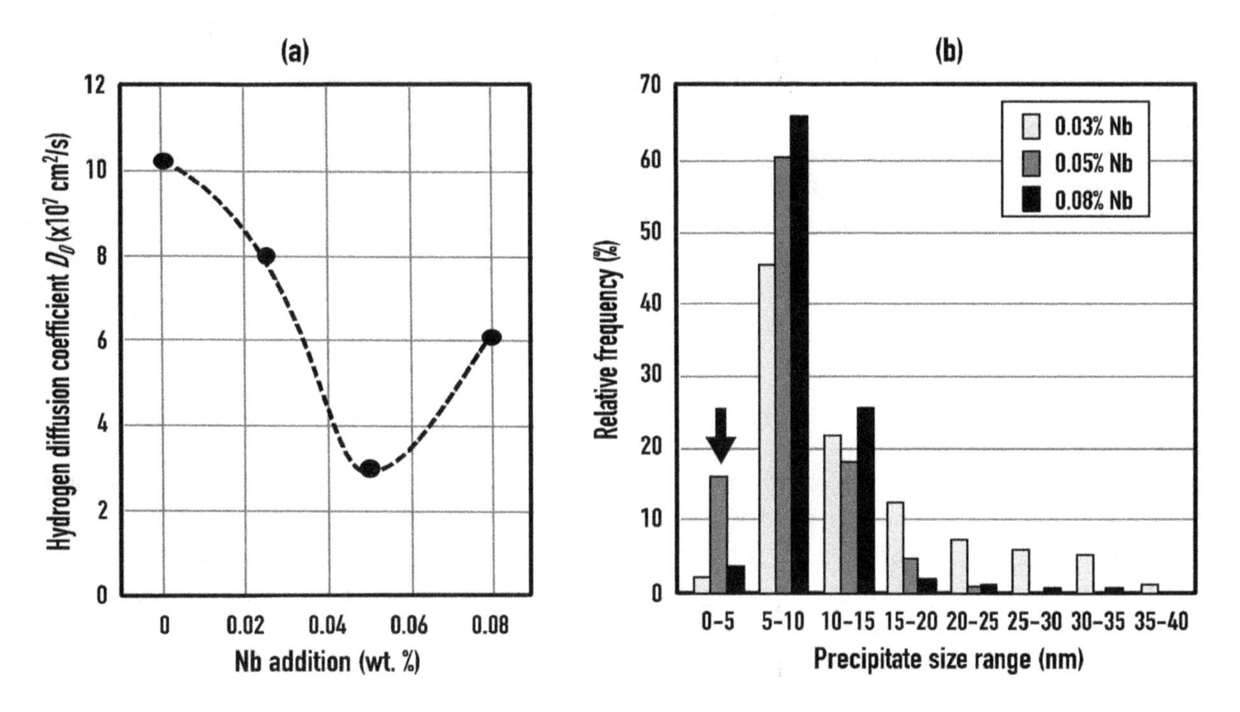

Figure 17. (**a**) Effect of niobium alloying on the hydrogen diffusion coefficient measured by hydrogen permeation tests; (**b**) size distribution of NbC precipitates measured by transmission elcectron microscopy analysis.

5.3. Combined Approach against Hydrogen Embrittlement

With increasing strength level of as-quenched martensitic steels, it will become necessary to combine several means of optimizing the resistance against hydrogen embrittlement. Parent austenite grain refinement, dispersion of ultra-fine precipitate particles, reinforcement of grain boundaries, and, generally, clean steelmaking practice are prerequisites for good performance at the highest strength level. Niobium microalloying provides both parent austenite grain refinement and particle dispersion. The particle size distribution can further be improved by the addition of molybdenum as it reduces the size of particle and thus increases the particle density. Simultaneously, molybdenum enhances grain boundary cohesion. It is further advisable to keep manganese additions low and to tightly limit phosphorous and sulphur residuals. Research in this respect is ongoing, yet first screening results indicate that the combination of individual approaches indeed brings about significant improvements. Figure 18 represents measured time-to-fracture improvements based after adding niobium and molybdenum, either single or combined, to grade 32MnB5 as shown before in Figure 16. The combined alloy effect is stronger than the sum of each individual effect, indicating a synergy between niobium and molybdenum in the alloy.

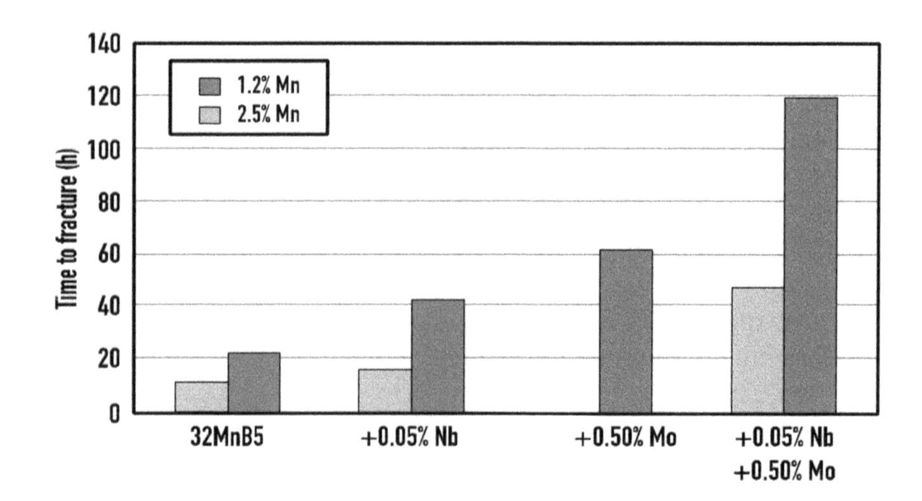

Figure 18. Effect of metallurgical optimization strategies on the delayed fracture resistance of a 2000 MPa tensile strength steel (32MnB5) using Mo and Nb alloying (double-notched tensile sample immersed in 20% ammonium thiocyanate solution under constant load conditions with an acting stress of 1000 MPa at the notch).

6. Conclusions

The presented results, originating from a large number of laboratory and industrial trials, confirmed that microstructural refinement is very efficient in optimizing the properties of as-quenched martensitic steels such as strength, toughness, ductile-to-brittle-transition temperature, and hydrogen embrittlement resistance. Although the definition of the effective grain size for each of these properties may refer to different microstructural features, the presented results indicated that in most cases the parent austenite grain size shows good correlation to respective properties in a Hall–Petch type relationship. This applies to both re-austenitizing quenched as well as direct quenched steels. In direct quenched steels with pancaked austenite morphology the grain thickness in the direction normal to the sheet surface defines the relevant parent austenite grain size.

The experimental data showed that the Hall–Petch coefficient for cleavage fracture stress appears to be significantly larger than the one for yield strength. Hence sufficiently strong grain refinement enables ductile fracture even at very low operating temperature. It was demonstrated that, by grain refinement, a significant strength increase can go along with a large drop in ductile-to-brittle transition temperature, contrary to all other strengthening mechanisms. Particularly in re-austenitizing quenched steels, parent austenite grain refinement effectively increases the upper-shelf impact energy. In direct quenched steels, toughness behavior is strongly anisotropic, having better properties in the rolling direction.

Because of their high yield strength, as-quenched martensitic steels are sensitive to hydrogen embrittlement causing delayed fracture phenomena under applied or residual stress. Parent austenite grain refinement was shown to be an efficient means of increasing the time-to-fracture by lowering the concentration of hydrogen on the grain boundary. Furthermore, hydrogen trapping by dispersed ultra-fine sized carbide precipitates can reduce the amount of diffusible hydrogen causing delayed fracture. Increasing the cohesion of parent austenite grain boundaries also counteracts decohesion caused by segregated hydrogen.

Molybdenum and niobium were demonstrated to be key alloying elements in this optimization strategy. Niobium provides efficient grain size control, actively in TMCP-based hot rolling processes and passively during re-austenitizing processes. Molybdenum not only provides high hardenability but adds to grain refinement by solute drag effects, optimizes the precipitation behavior of niobium, and increases grain boundary cohesion. The synergy between both alloying elements results in significantly improved delayed fracture resistance in martensitic steels of the highest strength level.

In quenched steels of highest strength level, it is advisable to opt for a low manganese level. High manganese additions increase the sensitivity for hydrogen embrittlement.

Acknowledgments: Parts of this work have been financially supported by the International Molybdenum Association (IMOA), London, UK and CBMM, Sao Paulo, Brazil. The collaboration and experimental support by Okayama University (Takehide Senuma), National Taiwan University (Jer-Ren Yang), Oulu University (Jukka Kömi, David Porter), and University of Science and Technology Beijing (Yunhua Huang) is gratefully acknowledged.

Author Contributions: Hardy Mohrbacher co-designed the experimental work, contributed to the metallurgical interpretation of the experimental results, prepared the literature review, and wrote the paper.

References

1. Leslie, W. *The Physical Metallurgy of Steels*; McGraw Hill: New York, NY, USA, 1981; pp. 216–226.
2. Krauss, G. Martensite in steel: Strength and structure. *Mater. Sci. Eng.* **1999**, *273*, 40–57. [CrossRef]
3. Belanger, P.J.; Hall, J.N.; Coryell, J.; Singh, J.P. Automotive Body Press-Hardened Steel Trends. In Proceedings of the International Symposium on the New Developments of Advanced High-Strength Steel, Vail, CO, USA, 23–27 June 2013; pp. 239–250.
4. Bian, J.; Mohrbacher, H. Novel Alloying Design for Press Hardening Steels with Better Crash Performance. In Proceedings of the International Symposium on the New Developments of Advanced High-Strength Steel, Vail, CO, USA, 23–27 June 2013; pp. 251–262.
5. Olsson, R.; Haglund, N.I. Cost effective fabrication of submarines and mobile cranes in high performance steels. *Int. J. Join. Mater.* **1991**, *3*, 120–128.
6. Sugimoto, K.-I. Fracture strength and toughness of ultra high strength TRIP aided steels. *Mater. Sci. Technol.* **2009**, *25*, 1108–1117. [CrossRef]
7. Nagumo, M. Function of Hydrogen in Embrittlement of High-strength Steels. *ISIJ Int.* **2001**, *41*, 590–598. [CrossRef]
8. Kaijalainen, A.J.; Suikkanen, P.; Karjalainen, L.P.; DeArdo, A.J. Effect of Austenite Conditioning in the Non-Recrystallization Regime on the Microstructures and Properties of Ultra High Strength Bainitic/Martensitic Strip Steel. In Proceedings of the 2nd International Conference on Super-High Strength Steels, Peschiera del Garda, Italy, 17–19 October 2010; Paper 115.
9. Schneider, A.S.; Cayla, J.L.; Just, C.; Schwinn, V. The Role of Niobium for the Development of Wear Resistant Steels with Superior Toughness. In Proceedings of the International Symposium on Wear Resistant Alloys for the Mining and Processing Industry, Campinas, Brazil, 4–7 May 2015; Mohrbacher, H., Ed.; TMS: Pittsburgh, PA, USA, 2018; pp. 173–186.
10. Speich, G.R. Tempering of low-carbon martensite. *Trans. AIME* **1969**, *245*, 2553–2564.
11. Guo, Z.; Lee, C.S.; Morris, J.W. On coherent transformations in steel. *Acta Mater.* **2004**, *52*, 5511–5518. [CrossRef]
12. Kinney, C.C.; Pytlewski, K.R.; Khachaturyan, A.G.; Morris, J.W. The microstructure of lath martensite in quenched 9Ni steel. *Acta Mater.* **2014**, *69*, 372–385. [CrossRef]
13. Morito, S.; Tanaka, H.; Konishi, R.; Furuhara, T.; Maki, T. The morphology and crystallography of lath martensite in Fe-C alloys. *Acta Mater.* **2003**, *51*, 1789–1799. [CrossRef]
14. Maki, T.; Tsuzaki, K. Tamura, K. The Morphology of Microstructure Composed of Lath Martensite. *Trans. Iron Steel Inst. Jpn.* **1980**, *20*, 207–214.
15. Morito, S.; Saito, H.; Ogawa, T.; Furuhara, T.; Maki, T. Effect of Austenite Grain Size on the Morphology and Crystallography of Lath Martensite in Low Carbon Steels. *ISIJ Int.* **2005**, *45*, 91–94. [CrossRef]
16. Qi, L.; Khachaturyan, A.G.; Morris, J.W. The microstructure of dislocated martensitic steel: Theory. *Acta Mater.* **2014**, *76*, 23–39. [CrossRef]
17. Morris, J.W.; Lee, C.S.; Guo, Z. The Nature and Consequences of Coherent Transformations in Steel. *ISIJ Int.* **2003**, *43*, 410–419. [CrossRef]
18. Cohen, M. The Strengthening of Steel. *Trans. AIME* **1962**, *224*, 638–657.
19. Cohen, M. On the Development of High Strength in Steel. *JISI* **1963**, *201*, 833–841.

20. Tomita, Y.; Okabayashi, K. Effect of microstructure on strength and toughness of heat-treated low alloy structural steels. *Metall. Trans. A* **1986**, *17*, 1203–1209. [CrossRef]
21. Wang, C.; Wang, M.; Shi, J.; Hui, W.; Dong, H. Effect of Microstructure Refinement on the Strength and Toughness of Low Alloy Martensitic Steel. *J. Mater. Sci. Technol.* **2007**, *23*, 659–664.
22. Morito, S.; Ohba, T. *Crystallographic Analysis of Characteristic Sizes of Lath Martensite Morphology. Fundamentals of Martensite and Bainite toward Future Steels with High Performance*; Furuhara, T., Tsuzaki, K., Eds.; ISIJ: Tokyo, Japan, 2007; pp. 57–62.
23. Morris, J.W. On the Ductile-Brittle Transition in Lath Martensitic Steel. *ISIJ Int.* **2011**, *51*, 1569–1575. [CrossRef]
24. Guo, Z.; Lee, C.S.; Morris, J.W. Grain Refinement for Exceptional Properties in High Strength Steel by Thermal Mechanisms and Martensitic Transformation. In *Proceedings Workshop on New Generation Steel*; Chinese Society for Metals: Beijing, China, 2001; pp. 48–54.
25. Ohmura, T.; Tsuzaki, K. A New Aspect of the Strengthening Factors of Fe-C Martensite through Characterization of Nanoindentation-induced Deformation Behavior. In *Fundamentals of Martensite and Bainite toward Future Steels with High Performance*; Furuhara, T., Tsuzaki, K., Eds.; ISIJ: Tokyo, Japan, 2007; pp. 35–46.
26. Hannula, J.; Kömi, J.; Porter, D.A.; Somani, M.C.; Kaijalainen, A.; Suikkanen, P.; Yang, J.R.; Tsai, S.P. Effect of Boron on the Strength and Toughness of Direct-Quenched Low-Carbon Niobium Bearing Ultra-High-Strength Martensitic Steel. *Metall. Mater. Trans. A* **2017**, *48*, 5344–5356. [CrossRef]
27. Irani, J.J. *Physical Properties of Martensite and Bainite*; Special Report 93; The Iron and Steel Institute: Scarborough, UK, 1965; pp. 193–203.
28. Naylor, J.P.; Blondeau, B. The Respective Roles of the Packet Size and the Lath Width on Toughness. *Metall. Trans.* **1976**, *7*, 891–894. [CrossRef]
29. Naylor, J.P.; Krahe, P.R. Cleavage Planes in Lath Type Bainite and Martensite. *Metall. Trans.* **1975**, *6*, 594–598. [CrossRef]
30. Morris, J.W.; Kinney, C.; Pytlewski, K.; Adachi, Y. Microstructure and cleavage in lath martensitic steels. *Sci. Tech. Adv. Mater.* **2013**, *14*, 041208. [CrossRef] [PubMed]
31. Matsuda, S.; Inoue, T.; Mimura, H.; Okamura, Y. Toughness and Effective Grain Size in Heat-Treated Low-Alloy High-Strength Steels. *Trans. ISIJ* **1972**, *12*, 325–333.
32. Morris, J.W.; Guo, Z.; Krenn, C.R.; Kim, Y.-H. The Limits of Strength and Toughness in Steel. *ISIJ Int.* **2001**, *41*, 599–611. [CrossRef]
33. Hanamura, T.; Yin, F.; Nagai, K. Ductile-Brittle Transition Temperature of Ultrafine Ferrite/Cementite Microstructure in a Low Carbon Steel Controlled by Effective Grain Size. *ISIJ Int.* **2004**, *44*, 610–617. [CrossRef]
34. McMahon, C.J. *Effects of Hydrogen on Plastic Flow and Fracture in Iron and Steel. Hydrogen Effects in Metals*; Bernstein, I.M., Thompson, A.W., Eds.; TMS: Pittsburgh, PA, USA, 1981; Volume 219.
35. Kim, Y.H. A Study of Hydrogen Embrittlement in Lath Martensitic Steels. Ph.D. Thesis, Department of Materials Science and Engineering, University of California, Berkeley, CA, USA, 1985.
36. Kim, Y.H.; Morris, J.W. The nature of quasicleavage fracture in tempered 5.5Ni steel after hydrogen charging. *Metall. Trans.* **1983**, *14*, 1883–1888. [CrossRef]
37. Kim, Y.H.; Kim, H.J.; Morris, J.W. The influence of precipitated austenite on hydrogen embrittlement in 5.5Ni steel. *Metall. Trans.* **1986**, *17*, 1157–1164. [CrossRef]
38. Yusa, S.; Hara, T.; Tsuzaki, K.J. Grain boundary carbide structure in tempered martensitic steel with serrated prior austenite grain boundaries. *Jpn. Inst. Met.* **2000**, *64*, 1230–1238. [CrossRef]
39. Takeda, Y.; McMahon, C.J. Strain controlled vs stress controlled hydrogen induced fracture in a quenched and tempered steel. *Met. Trans. A* **1981**, *12*, 1255–1266. [CrossRef]
40. Grange, R.A.; Hibral, C.R.; Porter, L.F. Hardness of Tempered Martensite in Carbon and Low-alloy Steels. *Metall. Trans. A* **1977**, *8*, 1775–1785. [CrossRef]
41. Mohrbacher, H. Laser welding of modern automotive high strength steels. In Proceedings of the 5th International Conference on HSLA Steels (2005), Sanya, Hainan, China, 8–10 November 2005; pp. 582–586.
42. Takaki, S.; Ngo-Huynh, K.-L.; Nakada, N.; Tsychiyama, T. Strengthening Mechanism in Ultra Low Carbon Martensitic Steel. *ISIJ Int.* **2012**, *52*, 710–716. [CrossRef]
43. Geng, W.T.; Freeman, A.J.; Olson, G.B. Influence of alloying additions on grain boundary cohesion of transition metals: First-principles determination and its phenomenological extension. *Phys. Rev. B* **2001**, *63*, 165415. [CrossRef]

44. Lin, H.-R.; Cheng, G.-H. Analysis of hardenability effect of boron. *Mater. Sci. Technol.* **1990**, *6*, 724–729. [CrossRef]
45. Kern, A.; Schriever, U. Niobium in Quenched and Tempered HSLA-Steels. In *Recent Advances of Niobium Containing Materials in Europe*; Verlag Stahleisen: Düsseldorf, Germany, 2005; pp. 107–120.
46. Kern, A.; Müsgen, B.; Schriever, U. Effect of Boron in Quenched and Tempered Steels. *Thyssen Tech. Ber.* **1990**, *1*, 43–52.
47. Nowill, C.A.; Speer, J.G.; De Moor, E.; Matlock, D.K. Effect of Austenitizing Conditions on Hardenability of Boron-Added Microalloyed Steel. *AIST Iron Steel Technol.* **2012**, *10*, 111–120.
48. Ishikawa, N.; Ueda, K.; Mitao, S.; Murotav, Y.; Sakiyama, T. High-Performance Abrasion-Resistant Steel Plates with Excellent Low-Temperature Toughness. In Proceedings of the International Symposium on the Recent Developments in Plate Steels, Winter Park, CO, USA, 19–22 June 2011; pp. 82–91.
49. Tobie, T.; Hippenstiel, F.; Mohrbacher, H. Optimizing Gear Performance by Alloy Modification of Carburizing Steels. *Metals* **2017**, *7*, 415. [CrossRef]
50. Gangloff, R.P. *Critical Issues in Hydrogen Assisted Cracking of Structural Alloys, in Environment Induced Cracking of Metals (EICM-2)*; Shipilov, S., Ed.; Elsevier Science: Oxford, UK, 2008; pp. 2–24.
51. Senuma, T.; Takemoto, Y. Influence of Nb Content on Delayed Fracture and Crash Relevant Properties of 2000 MPa class hot stamping steel sheets. In Proceedings of the International Conference on Steels in Cars and Trucks, Amsterdam, The Netherlands, 2017.
52. Momotani, Y.; Shibata, A.; Terada, D.; Tsuji, N. Effect of strain rate on hydrogen embrittlement in low-carbon martensitic steel. *Int. J. Hydrog. Energy* **2017**, *42*, 3371–3379. [CrossRef]
53. Lee, S.J.; Ronevich, J.A.; Krauss, G.; Matlock, D.K. Hydrogen Embrittlement of Hardened Low-carbon Sheet Steel. *ISIJ Int.* **2010**, *50*, 294–301. [CrossRef]
54. Lovicu, G.; Barloscio, M.; Bottazzi, M.; D'Aiuto, F.; De Sanctis, M.; Dimatteo, A.; Federici, C.; Maggi, S.; Santus, C.; Valentini, R. Hydrogen Embrittlement of Advanced High Strength Steels for Automotive Use. In Proceedings of the 2nd International Conference on Super High Strength Steels, Verona, Italy, 17–20 October 2010.
55. Zhang, S.; Huang, Y.; Sun, B.; Liao, Q.; Lu, H.; Jian, B.; Mohrbacher, H.; Zhang, W.; Guo, A.; Zhang, Y. Effect of Nb on hydrogen-induced delayed fracture in high strength hot stamping steels. *Mater. Sci. Eng. A* **2015**, *626*, 136–143. [CrossRef]
56. Pressouyre, G.M. Current Solutions to Hydrogen Problems in Steels. In Proceedings of the First International Conference on Current Solutions to Hydrogen Problems in Steels, Washington, DC, USA, 1–5 November 1982; pp. 18–36.
57. Wei, F.-G.; Hara, T.; Tsuzaki, K. Nano-Preciptates Design with Hydrogen Trapping Character in High Strength Steel. *ASM Int.* **2009**, 448–455. [CrossRef]

9

Effect of Microstructure on Post-Rolling Induction Treatment in a Low C Ti-Mo Microalloyed Steel

Gorka Larzabal [1,2], Nerea Isasti [1,2], Jose M. Rodriguez-Ibabe [1,2] and Pello Uranga [1,2,*]

1 CEIT, Materials and Manufacturing Division, 20018 San Sebastian, Basque Country, Spain; glarzabal@outlook.com (G.L.); nisasti@ceit.es (N.I.); jmribabe@ceit.es (J.M.R.-I.)
2 Universidad de Navarra, Tecnun, Mechanical and Materials Engineering Department, 20018 San Sebastian, Basque Country, Spain
* Correspondence: puranga@ceit.es

Abstract: Cost-effective advanced design concepts are becoming more common in the production of thick plates in order to meet demanding market requirements. Accordingly, precipitation strengthening mechanisms are extensively employed in thin strip products, because they enhance the final properties by using a coiling optimization strategy. Nevertheless, and specifically for thick plate production, the formation of effective precipitation during continuous cooling after hot rolling is more challenging. With the aim of gaining further knowledge about this strengthening mechanism, plate hot rolling conditions were reproduced in low carbon Ti-Mo microalloyed steel through laboratory simulation tests to generate different hot-rolled microstructures. Subsequently, a rapid heating process was applied in order to simulate induction heat treatment conditions. The results indicated that the nature of the matrix microstructure (i.e., ferrite, bainite) affects the achieved precipitation hardening, while the balance between strength and toughness depends on the hot-rolled microstructure.

Keywords: plate rolling; strengthening; precipitation; induction; titanium; molybdenum; microalloyed steels; EBSD; mechanical properties

---

1. Introduction

In the recent years, suitable thermomechanical sequences combined with advanced microalloying concepts have been developed to fulfill the demanding market requirements in terms of tensile and toughness properties. The addition of microalloying elements, such as Nb, Mo, Ti and V ensures the improvement of final mechanical properties. For the conventional High Strength Low Alloy (HSLA) steels, yield strength values of approximately 400–500 MPa can be achieved through a combination of strengthening contributions from solid solution, grain size, dislocation density and fine precipitation properties [1,2]. However, the hardening due to fine precipitation is not relevant for conventional HSLA steels. In recent years, combinations of Ti and Mo have been proposed for when higher yield strength values are required (higher than 700 MPa). Ti-Mo microalloyed steels offer excellent tensile properties and ductility balance, due to the formation of nanometer-sized carbides in the ferritic soft matrix [3]. A strategy for effective precipitation during cooling can promote an enhancement of approximately 300 MPa [4,5]. Therefore, significant effort has been dedicated to optimizing the cooling strategy, which ensures the formation of fine precipitates with a considerable hardening effect. Several works have investigated the influence of coiling strategies in the final tensile properties for thin strip products with the aim of selecting a coiling temperature that promotes additional precipitation hardening [2,5,6]. Conversely, the formation of fine precipitates during continuous cooling after plate hot rolling is not as effective as in coiled products [7]. In the current study, in order to take advantage of the microalloying elements that are available after hot rolling and cooling steps, a post-hot rolling induction heat treatment cycle after hot rolling is proposed for different microstructures. Even though the tensile property improvement caused by induction heat treatment for bainitic microstructures is clear, the benefits of using induction heating for ferritic microstructures is still unknown [8–10]. For the purpose of generating different pre-treatment microstructures (ferritic, ferritic/bainitic, and bainitic), the reproduction of plate hot rolling was carried out by plane compression tests. Finally, induction heat treatment was simulated for each microstructure. Specimens for tensile and Charpy tests were machined for each condition. This allowed us to conclude that a considerable strengthening contribution occurs after induction heat treatment for pre-treatment microstructures that is associated with the formation of nanosized precipitates during heat treatment. In the current analysis, the interactions between microstructure, tensile properties, and induction heating were evaluated. Furthermore, the influence of the mentioned tensile property enhancement on toughness properties was also studied for ferritic and bainitic microstructure types obtained before induction treatment.

2. Materials and Methods

In the current study, a low carbon steel microalloyed with titanium and molybdenum is selected. Its chemical composition is shown in Table 1. Plane strain compression tests were carried out following the thermomechanical schedule presented in Figure 1. The cycle included a reheating step of 5 min at 1200 °C, followed by a multipass deformation sequence. The first two deformations (ε = 0.4) were applied at 1100 and 1000 °C, with the aim of obtaining a fine recrystallized austenite. Then, a deformation was applied at 900 °C (ε = 0.4), below the non-recrystallization temperature, in order to ensure the accumulation of deformation in the austenite before transformation. Next, the specimens were cooled down at a rate of 15 °C/s to temperatures designated as the "Fast Cooling Temperatures" (FCTs; 790, 720, and 650 °C) with the purpose of generating different types of microstructures (ferritic, ferritic/bainitic, and bainitic, respectively). Then, the samples were cooled down slowly to room temperature at 1 °C/s. After the simulation of plate hot rolling, an induction heat treatment was applied by fast heating up to a treatment temperature of 710 °C with no holding time. Finally, the samples were cooled down to room temperature at 1 °C/s. The samples obtained after the plate hot rolling simulation (without heat treatment) were designated as FCT, while the induction heated specimens were designated as FCT-HT.

Table 1. Chemical compositions of the steels (weight percent).

Steel	C	Mn	Si	P	S	Ti	Mo	Al	N
Ti-Mo	0.048	1.61	0.20	0.020	0.006	0.09	0.20	0.02	0.0040

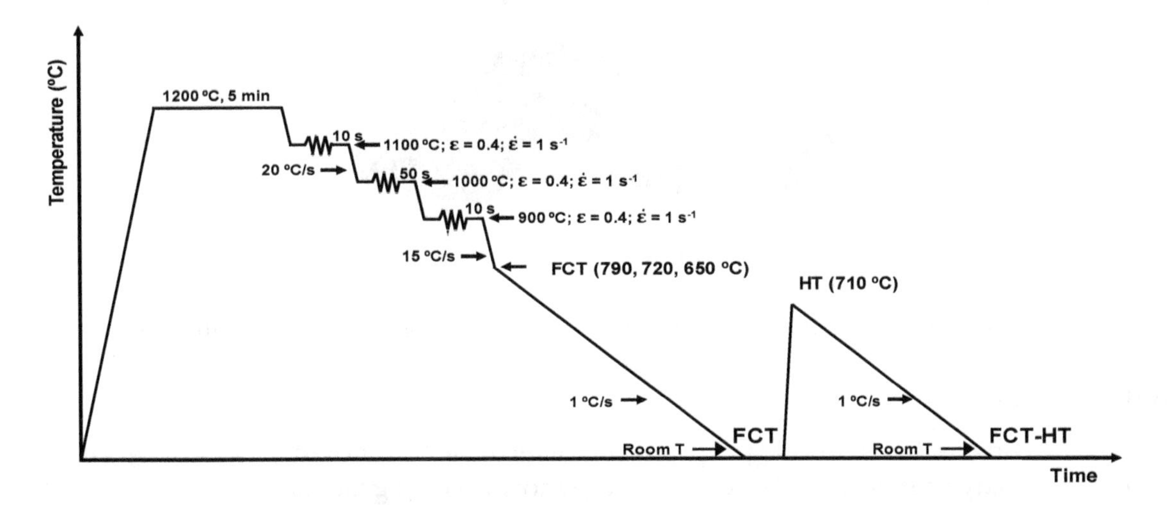

Figure 1. Schematic of the thermomechanical schedule performed in the plane strain compression machine.

In the plane compression specimens, the strain was heterogeneously distributed throughout the thickness owing to friction and sample geometry. In order to minimize the strain gradients, the characterization of the microstructure was performed at the central parts of the plane compression specimens [11]. In addition to the microstructural sample, specimens for tensile and Charpy tests were machined for each condition. The general microstructural characterization was performed by optical microscopy (OM, LEICA DMI5000 M, Leica Microsystems, Wetzlar, Germany) after etching in 2% Nital. More detailed analyses of microstructural features were carried out by field-emission gun scanning electron microscopy (FEGSEM, JEOL JSM-7000F, JEOL Ltd., Tokyo, Japan). In order to quantify the crystallographic unit size and evaluate dislocation densities, electron backscattered diffraction (EBSD) analysis was performed under each condition by means of a Philips XL 30CP SEM with W filament (TexSEM Laboratories, Draper, UT, USA). The EBSD sample preparation was based on a polishing down to 1 μm (using diamond liquids of 6, 3 and 1 μm), followed by a final polishing using colloidal silica. A step size of 0.4 μm was employed in the EBSD analysis, and an area of 200 × 200 μm^2 was scanned. TSL OIM^{TM} Analysis 5.31^{TM} software (EDAX, Mahwah, NJ, USA) was used for data processing. Finally, the study of fine precipitates was carried out by transmission electron microscopy (TEM, JEOL 2100, JEOL Ltd., Tokyo, Japan), characterized by a voltage of 200 kV and a thermionic filament of LaB_6. In order to obtain accurate information regarding fine precipitation, several electropolished thin foils were characterized for each condition. Precipitate size measurement was performed based on at least 25 TEM images and resulting in the formation of a number of precipitates between 120 and 600, depending on the precipitate density for each case.

Besides the characterization of the central part of the plane compression sample, two cylindrical tensile specimens (4 mm in diameter and a gauge length of 17 mm) were also machined under each condition (Schematics are shown in Figure 2). The 2% proof stress and the tensile strength were calculated as the average of two tensile tests for each condition. The tensile tests were carried out at room temperature, on an Instron testing machine (Instron, Grove City, PA, USA). In these tests, a strain rate of 10^{-3} s^{-1} was employed. Furthermore, Charpy tests were carried out (Tinius Olsen Model Impact 104 pendulum impact tester, Horsham, PA, USA), after machining sub-size Charpy specimens (~4 × 10 × 55 mm^3) from compression specimens. The impact transition curves were defined in accordance with the modified hyperbolic tangent fitting algorithm reported by Wallin [12].

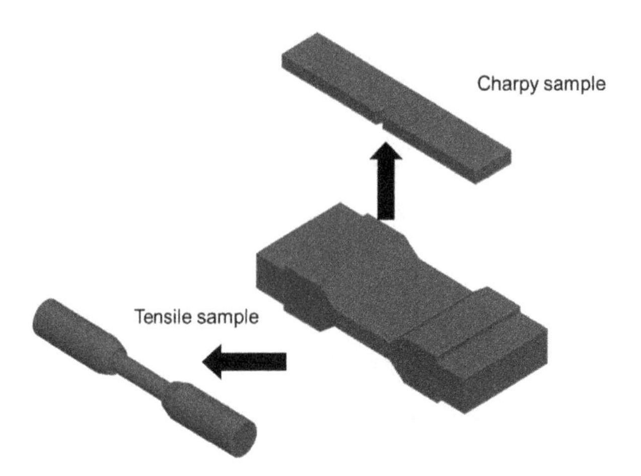

Figure 2. Schematics of the mechanical property sample extraction from the plane strain compression sample.

3. Results and Discussion

In the following sections, tensile property characterization is shown, followed by an exhaustive microstructural analysis in terms of microstructure morphology, grain size, dislocation density, and fine precipitation. In order to gain a better understanding of the benefits of applying induction technologies, an analysis of the contributing factors of the different strengthening mechanisms on tensile properties is presented. Finally, the effects of tensile property modification on the toughness properties are evaluated.

3.1. Tensile Properties

Tensile tests were performed for all the conditions before and after induction heat treatment. Figure 3a shows the stress-strain curves obtained for all the Fast Cooling Temperatures (FCTs). The stress-strain curves correspond to the samples without treatment, as well as the induction heated samples. Regarding the curves corresponding to the samples obtained before heat treatment, the results plotted in Figure 3a suggest that the modification of FCT has a considerable effect on tensile behavior. Tensile property improvement was observed as the Fast Cooling Temperature increased. In terms of the influence of the induction heat treatment, a clear tensile property improvement is observed after induction heating. This trend was seen for all of the FCTs. Higher yield strength and tensile strength values were measured in the induction heated samples in comparison to the specimens without heat treatment.

Concerning the shape of the stress-strain curve, the curves obtained in the non-heated samples showed continuous yielding behavior. As will be shown later, for the FCT790 condition, this behavior could be due to the formation of martensite/austenite (MA) islands within a ferritic matrix. This trend is in agreement with previously published studies [5]. The increment of the concentration of secondary phases leads to the increment of mobile dislocations located in the boundaries between the MA constituent and ferritic soft grains [13]. For the lowest, FCT650, a continuous shape was also noticed, which is associated with the formation of more bainitic phases. Conversely, a slightly different trend was distinguished when the tensile curves corresponding to the FCT-HT samples were analyzed. The curve obtained from the FCT790-HT condition exhibits a discontinuous behavior that is characteristic of ferritic-pearlitic microstructures. This is attributed to the modification of the secondary phase from MA islands to cementite. This aspect is analyzed in detail in the following section.

The yield and tensile strength values obtained from the stress-strain curves are plotted in Figure 3b as a function of the Fast Cooling Temperature. Tensile properties corresponding to the samples obtained before (FCT) and after induction heat treatment (FCT-HT) are shown in Figure 3b. Besides YS and TS values, the elongation and area reduction measurements are listed in Table 2. Looking at the results obtained in the samples without heat treatment, the increment of FCT from 650 to 790 °C promoted an increase in yield strength. Yield strength values of 495, 525 and 544 MPa were obtained, for FCT650, FCT720 and FCT790, respectively. Conversely, no significant effect of FCT on tensile strength values was noticed. Similar TS values were measured over the entire range of Fast Cooling Temperatures.

Table 2. Yield strength (YS), tensile strength (TS), elongation (%), and area reduction (%) values for the all of the conditions: FCT650, FCT650-HT, FCT720, FCT720-HT, FCT790, and FCT790-HT.

Condition	Yield Strength (MPa)	Tensile Strength (MPa)	Elongation (%)	Area Reduction (%)
FCT650	495 ± 3	667 ± 3	25 ± 2	79 ± 0
FCT650-HT	632 ± 1	738 ± 2	22 ± 0	76 ± 0
FCT720	525 ± 1	707 ± 3	19 ± 1	79 ± 1
FCT720-HT	654 ± 4	758 ± 1	27 ± 0	80 ± 1
FCT790	544 ± 0	676 ± 0	20 ± 0	77 ± 0
FCT790-HT	646 ± 3	739 ± 4	25 ± 1	80 ± 1

As mentioned previously, the induction heat treatment has considerable effects on tensile properties. Higher yield strength and tensile strength values were clearly observed after rapid heating. These improvements were observed for every Fast Cooling Temperature. The results shown in Figure 3b suggest that the observed improvement is more relevant for the yield strength. Enhancements of yield strength of 102, 129 and 137 MPa were promoted through the application of induction heating, for a FCTs of 790, 720 and 650 °C, respectively. The increment caused by induction treatment was shown to vary depending on the Fast Cooling Temperature and, therefore, the resulting microstructure (ferrite or bainite). A more relevant yield strength improvement due to induction heating was reached as the FCT decreased. Even though the benefits of induction heating were shown to be more important in terms of yield strength, tensile strength is also improved after applying a fast heating up to 710 °C. For example, by using a Fast Cooling Temperature of 720 °C, a TS values of 707 and 758 MPa were measured for the FCT720 and FCT720-HT samples, respectively.

Regarding the effect of FCT on the tensile behavior of the heated samples, it was observed that FCT does not affect tensile properties noticeably. YS and TS values remained approximately constant across the entire range of Fast Cooling Temperatures. Close yield strength values of 646 and 632 MPa were obtained, decreasing the FCT from 790 to 650 °C. Therefore, it was possible to attain similar tensile properties for the different microstructures after applying the induction heating process.

Figure 4 shows the yield to tensile strength ratio as a function of the FCT for the samples prior to induction treatment and for the heated samples. The YS/TS ratio ranged between 0.74 and 0.87, which is in line with previously reported values for low carbon microalloyed steels [1]. In both the FCT and FCT-HT conditions, the YS/TS ratio increased as the FCT increased from 650 to 790 °C, and the increment was more relevant for the samples without induction heat treatment. This trend is attributed mostly to the improvement in yield strength that was observed after the increment in FCT. Regarding the effect of the induction treatment, when a Fast Cooling Temperature of 720 °C was used, the yield to tensile strength ratio increased considerably from 0.74 to 0.86 after applying the heat treatment. This significant increment in the YS/TS ratio can be explained by several factors—the modification of the microstructure and strengthening due to fine precipitation—which contribute more significantly to YS than to TS.

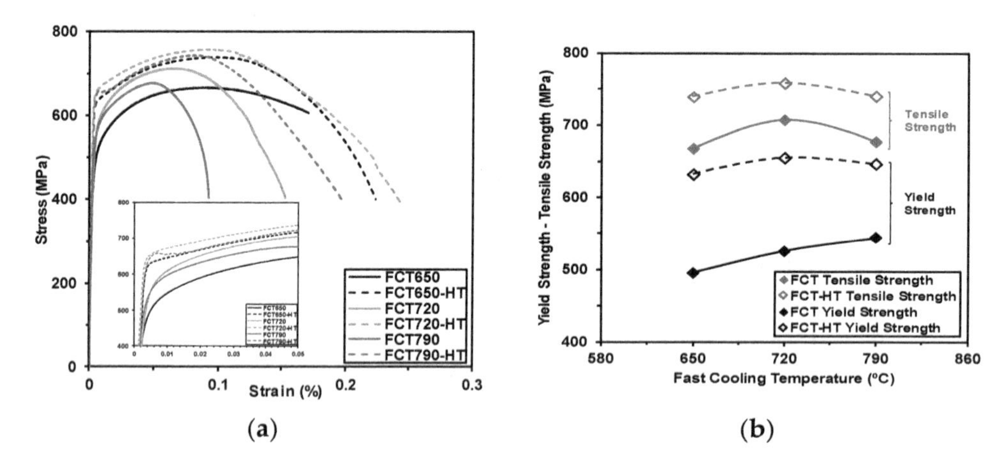

Figure 3. (**a**) Engineering tensile curves obtained for all of the Fast Cooling Temperatures (FCTs). Tensile curves measured before (FCT) and after induction heat treatment (FCT-HT) have been included. (**b**) Evolution of yield strength and tensile strength as a function of the Fast Cooling Temperature (samples obtained before and after induction treatment).

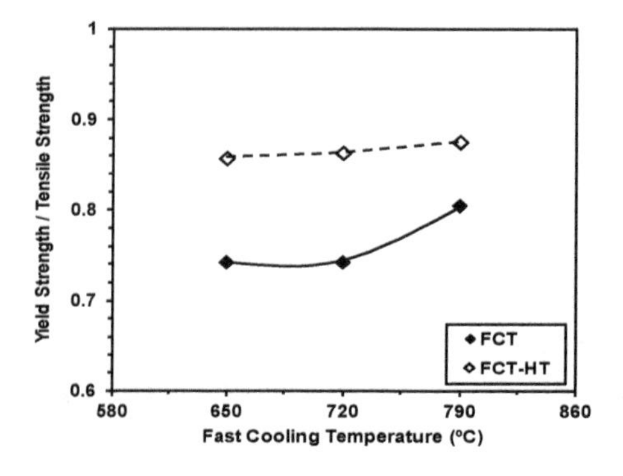

Figure 4. Evaluation of the effect of the induction heat treatment and the FCT on the yield/tensile strength ratio.

3.2. Microstructural Features and Unit Sizes

In order to understand the reasons for the observed improvement caused by induction heat treatment, a detailed microstructural characterization based on grain size, dislocation density, and precipitation was performed. In Figure 5, FEG-SEM micrographs related to the different FCTs (790, 720, and 650 °C) are presented. Figure 5a–c show the microstructures formed before (FCT samples) induction treatment, while Figure 5d–f illustrate the microstructural features of the samples obtained after rapid heating (FCT-HT samples). In the current work, the ISIJ Bainite Committee classification was used to designate the observed phases [14]. By analyzing the microstructures obtained after the continuous cooling to room temperature (FCT samples), completely different transformation products were distinguished depending on the FCT. For the highest FCT of 790 °C, mainly a ferritic microstructure was observed, in conjunction with martensite–austenite (MA) islands. When an intermediate FCT was applied (720 °C), the microstructure was composed of PF (polygonal ferrite), QF (quasipolygonal ferrite), GF (granular ferrite) and MA islands. For the lowest FCT, a mixture between the QF, GF, and MA islands was detected. The micrographs shown in Figure 5 clearly exhibit that microstructural refinement can be achieved by decreasing the Fast Cooling Temperature.

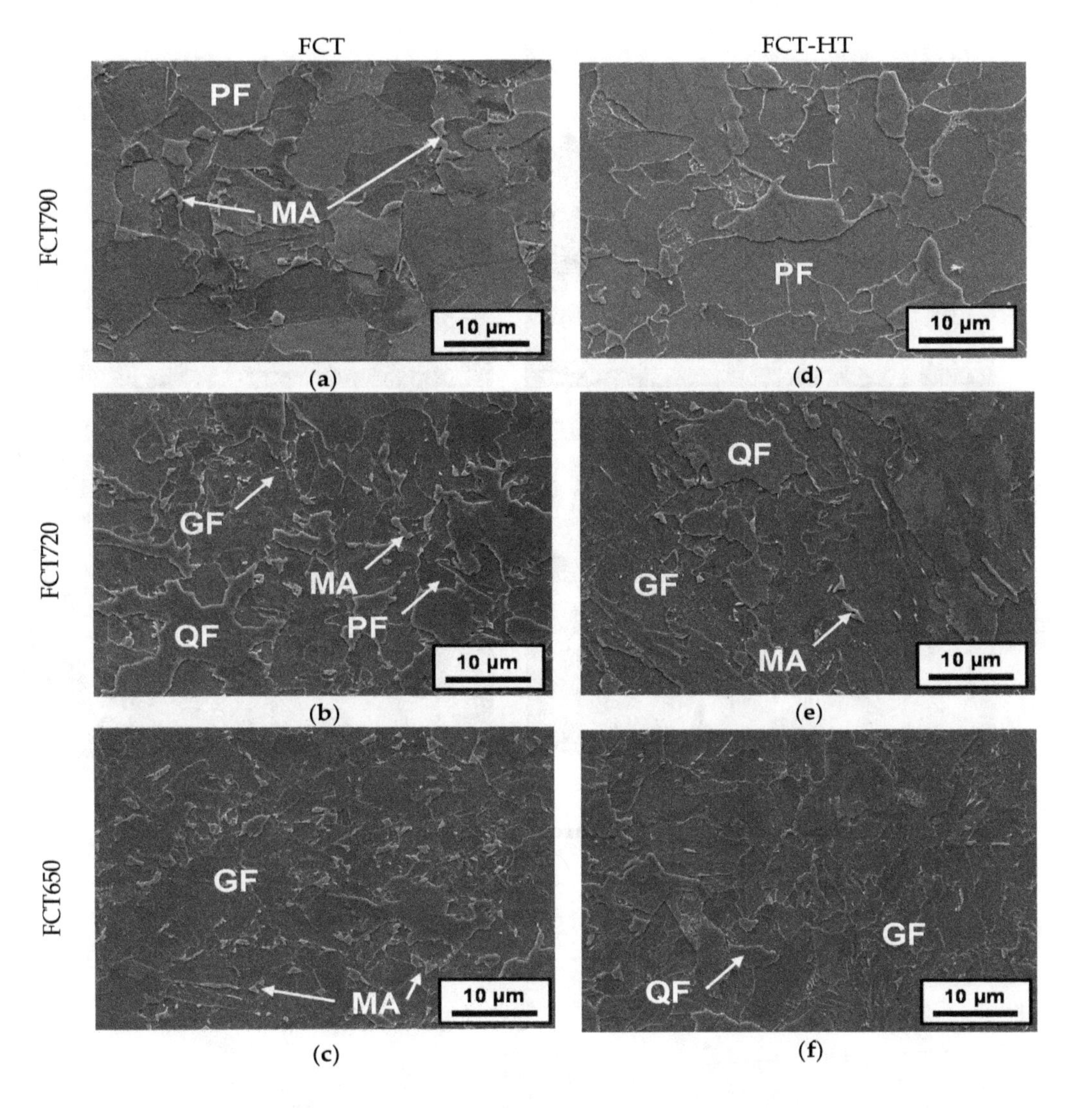

Figure 5. FEGSEM (field-emission gun scanning electron microscopy) micrographs corresponding to every Fast Cooling Temperature and both (**a–c**) FCT and (**d–f**) FCT-HT samples: (**a**,**d**) FCT790, (**b**,**e**) FCT720 and (**c**,**f**) FCT650.

When the microstructures obtained before and after induction treatment were compared, optical microscopy suggested that induction heating does not significantly modify the resulting microstructures. No considerable microstructural changes were observed during induction heat treatment. Either way, a more detailed analysis of secondary phases was required due to the presence of complex transformation products. In Figure 6, several FEGSEM micrographs obtained after FCTs of 790 °C (a,b), 720 °C (c,d), and 650 °C (e,f) can be compared. For the FCT790 condition, the images show that the secondary phase morphology was strongly modified during induction treatment. Prior to heat treatment, martensite–austenite (MA) islands were observed as the secondary phase, whereas in the induction heated sample, cementite particles were detected instead of MA islands. Similar phenomena can be noticed for the lowest Fast Cooling Temperature. Partial dissolution of the MA islands formed before treatment occurred during induction heat treatment, which is in line with previously published studies [9,15]. Xie et al. [15] claimed that after conventional tempering, the MA islands decomposed, and cementite was formed. Furthermore, they concluded that an increment in the tempering temperature promotes more relevant MA decomposition with smaller cementite particles, ensuring a significant enhancement of toughness properties. This modification also affects tensile curves, as observed in Figure 3b. The dissolution of MA islands affected the shape modification of tensile curves in the FCT790 condition, changing from a continuous yielding to a discontinuous one.

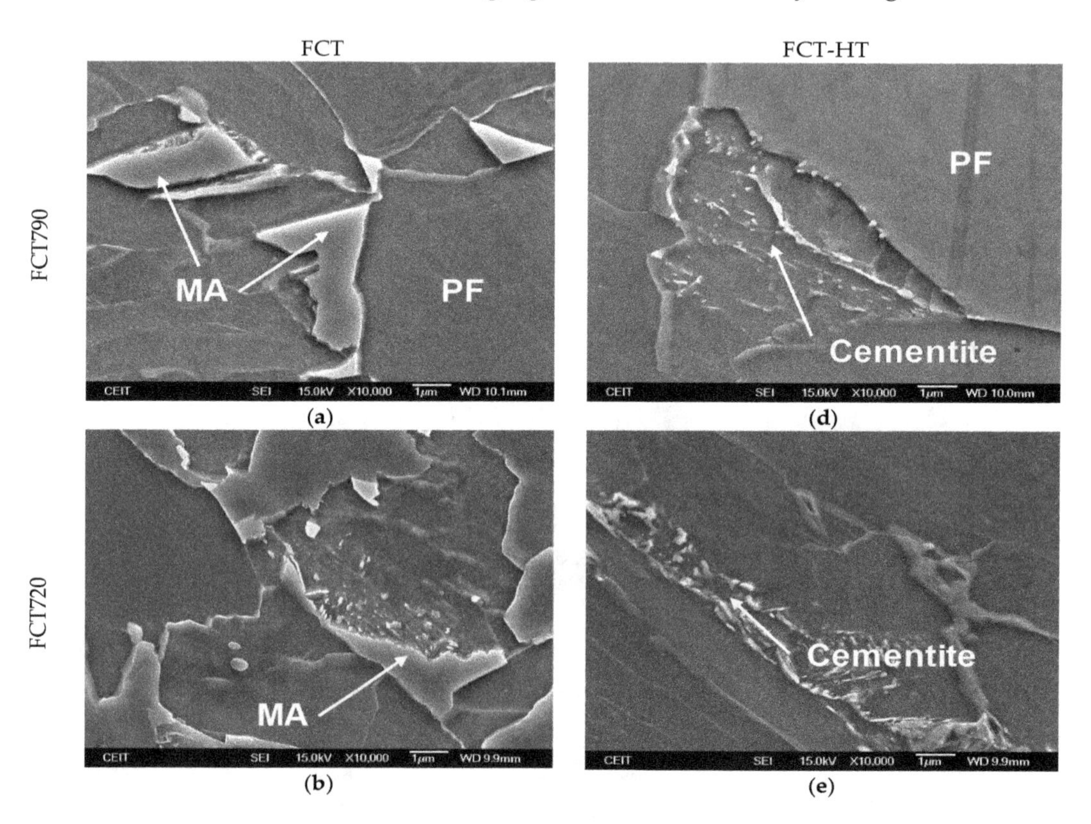

Figure 6. *Cont.*

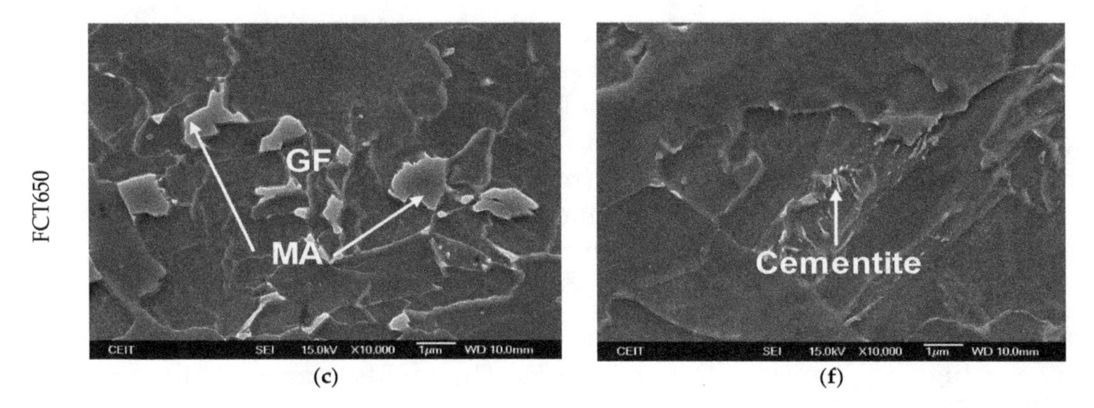

Figure 6. FEGSEM micrographs (higher magnifications) corresponding to every Fast Cooling Temperature and both (**a**–**c**) FCT and (**d**–**f**) FCT-HT samples: (**a**,**d**) FCT790, (**b**,**e**) FCT720 and (**c**,**f**) FCT650.

In order to resolve microstructural features that cannot be suitably resolved by conventional characterization procedures, such as optical and FEGSEM microscopy, an additional crystallographic characterization by means of EBSD technique was performed for all cases. As an example, Figure 7 presents EBSD maps corresponding to FCT790 and FCT650 conditions. To obtain further information concerning the phase morphology for each type of microstructure, the image quality and inverse pole figure are superimposed in both cases. By analyzing the results shown in Figure 7, completely different EBSD maps can be observed. When a high FCT is applied, polygonal ferrite grains were distinguished, whilst after a low FCT, non-polygonal, bainitic morphologies were detected. Moreover, a more pronounced substructure was shown for the FCT650 sample, a characteristic of mainly bainitic microstructures. In addition, the microstructural refinement due to FCT reduction was clearly evident. A significantly finer microstructure was achieved for the lowest FCT of 650 °C.

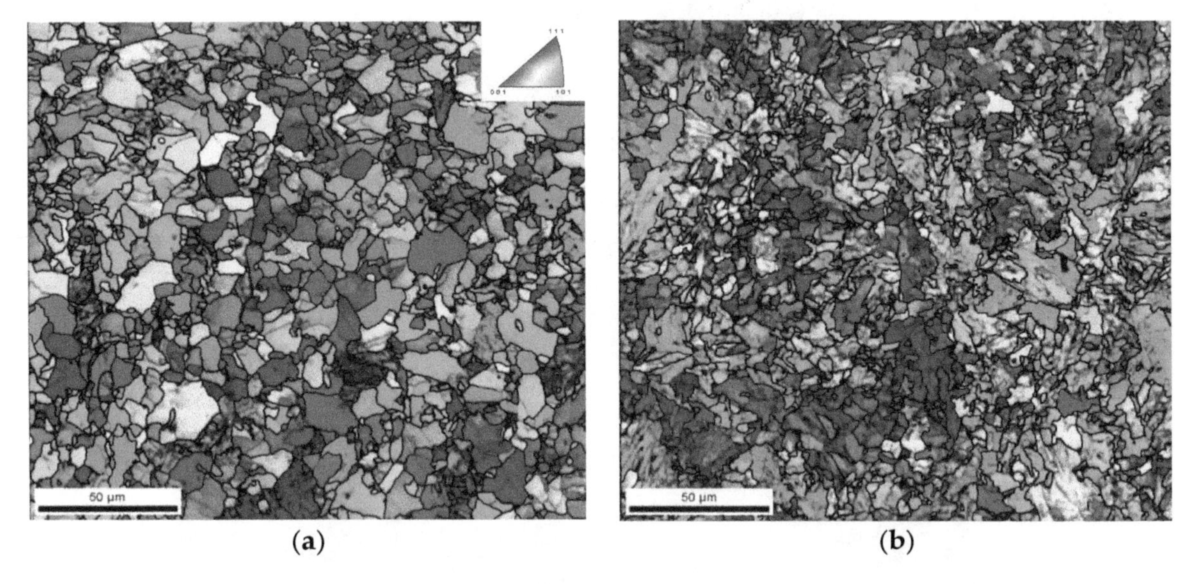

Figure 7. Image quality and inverse pole figure (IPF) maps corresponding to the samples without induction heat treatment: (**a**) FCT790 and (**b**) FCT650.

In the current work, low and high angle misorientation unit sizes were measured [16], considering 2° and 15° misorientation criteria, respectively. In the grain boundary maps, low and high angle boundaries are shown with red and black lines, respectively. Low angle boundaries are supposed to contribute to tensile properties owing to their opposition to dislocation movement, while high angle boundaries are assumed to be suitable for controlling crack propagation. Figure 8 illustrates several grain boundary maps for different types of microstructures (ferrite, ferrite/bainite, and bainite, for FCT790, FCT720, and FCT650, respectively). As expected, the variation in the Fast Cooling Temperature noticeably affected the microstructure morphology. When FCT decreased more, non-polygonal transformation products were detected and this was reflected in an increase in the content of low angle boundaries drawn in red. Moreover, finer microstructures were observed as the FCT decreased, which were associated with the formation of more bainitic phases. When the EBSD maps drawn for the samples obtained before and after induction treatment were compared, no significant differences were observed between the different types of microstructure (ferrite, ferrite/bainite, and bainite).

In Figure 9a, the effect of the Fast Cooling Temperature on the mean unit size was evaluated for both low and high angle misorientation criteria. The trends suggest that the reduction of FCT from 790 to 720 °C ensures microstructural refinement (considering both misorientation criteria). Taking into account high angle misorientation criteria, the mean unit size decreased from 6 to 2.9 μm when the FCT reduced from 790 to 720 °C. This behavior could be attributed to the modification of the microstructure from mainly ferritic phases to mixtures between polygonal ferrite and bainitic phases, such as quasipolygonal ferrite (QF) and granular ferrite (GF). Nevertheless, similar mean unit sizes were measured for FCTs of 720 °C and 650 °C.

Besides the evaluation of the influence of FCT on crystallographic unit sizes, an analysis of the potential of induction heating on the mean unit size was also carried out. Taking into account the low angle misorientation criterion, no significant variation in the mean unit size was observed. Similar mean unit sizes were measured prior to and after induction treatment. However, when the high angle misorientation criterion was analyzed, finer microstructures were measured after induction heat treatment. This trend was observed for every Fast Cooling Temperature. For example, mean grain sizes (considering the high angle misorientation criteria) of 6.21 and 4.67 μm were measured for the FCT720 and FCT720-HT samples, respectively. This effect could be related to the abovementioned decomposition of the secondary phases during rapid heating (see Figure 6).

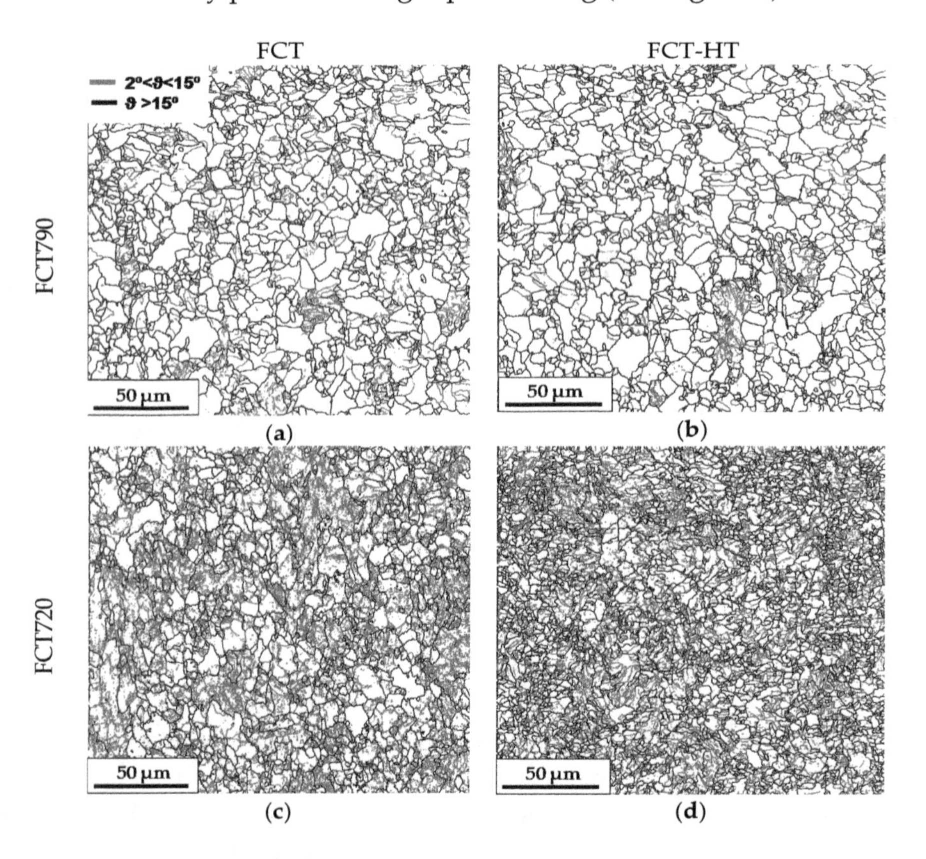

Figure 8. *Cont.*

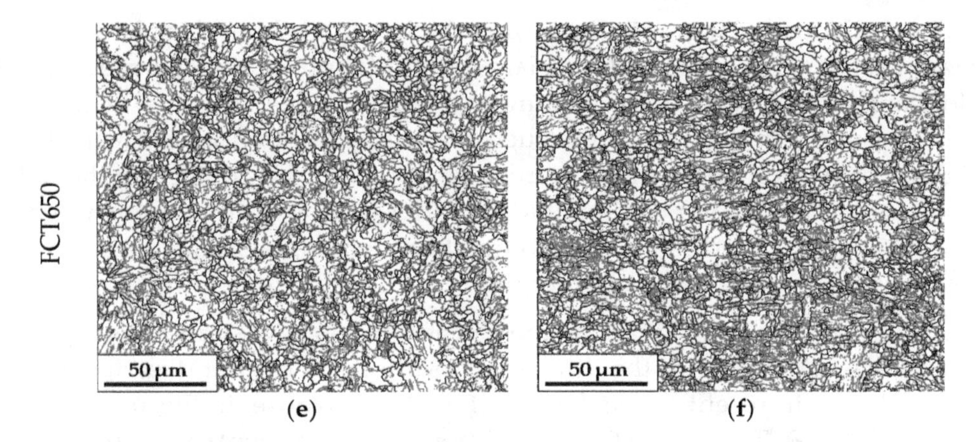

Figure 8. Grain boundary maps (red lines between 2° and 15°, black lines > 15°) corresponding to the samples obtained before (FCT in (**a**,**c**,**e**)) and after induction treatment (FCT-HT in (**b**,**d**,**f**)) and under different conditions: (**a**,**b**) FCT790 and (**c**,**d**) FCT720 and (**e**,**f**) FCT650.

In addition to analyzing the evolution of the mean grain size, an evaluation of microstructural heterogeneity was carried out. For that purpose, the ratio between the 20% critical grain size (Dc20%) and the high angle mean grain size (D15°) was estimated and plotted as a function of FCT in Figure 9b. In a grain size distribution, Dc20% is known as the cutoff grain size at the 80% area fraction and is able to evaluate the length of the grain size distribution tail [17]. Given that the presence of coarse grains affects the toughness properties, the evaluation of heterogeneity is crucial. Looking at Figure 9b, a clear effect of the microstructure on Dc20%/D15° can be observed. An increment in the Dc20%/D15° value occurred as the FCT decreased from 790 to 650 °C. This could be associated with the formation of more bainitic phases when a low Fast Cooling Temperature was used [18,19]. Nevertheless, no significant effect of rapid heating on heterogeneity was observed when the FCT and FCT-HT results were compared. Similarly, the Dc20%/D15° ratio was measured before and after induction treatment.

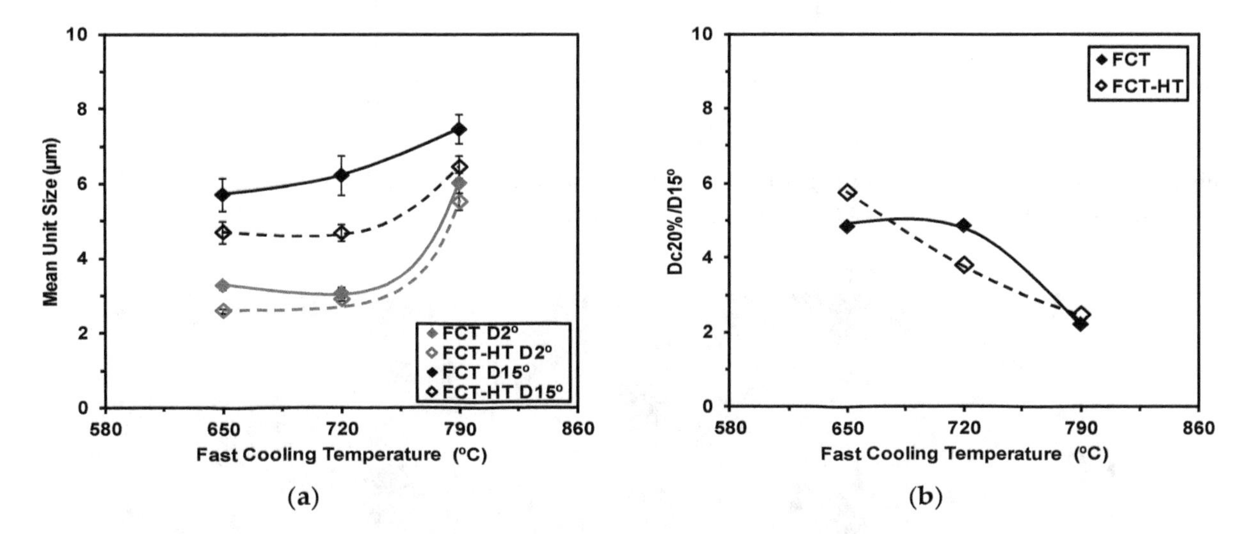

Figure 9. (**a**) Influence of the Fast Cooling Temperature and the induction heat treatment on the average unit size, using different threshold misorientation criteria: low angle (2°) and high angle (15°) boundaries. (**b**) Effect of FCT on the on Dc20%/D15°.

To further understand the differences observed in Figure 9a, an evaluation of the effect of induction treatment on the density of high and low boundaries was performed. For that purpose, the grain boundary lengths per unit area for both high angle (HAGB) and low angle (LAGB) grain boundaries were calculated. High and low angle grain boundary lengths per unit area are plotted for all of the

FCT and FCT-HT conditions in Figure 10. Regarding the low angle boundary length, no clear effect of fast heating was observed. This trend could explain the similar mean unit size measured for low angle boundaries before and after induction heat treatment (see Figure 9a). Huang et al. [20] also observed a negligible effect of recovery in bainitic microstructures during tempering. Conversely, for high angle boundaries, higher grain boundary lengths were estimated after HT, reflecting that a higher density of HAGB was obtained. The decomposition of MA islands during induction treatment can cause the formation of new fine grains, leading to an increment of high angle boundaries and a refinement of mean unit sizes (considering boundaries higher than 15°) after rapid heating (see Figure 9a).

From the EBSD analyses and using ϑ as the Kernel Average Misorientation (KAM) parameter, dislocation density was estimated for the different conditions. A deeper description of the followed methodology was shown in recent works [10,21]. For that purpose, in Figure 11a,c,e Kernel maps corresponding to FCT790, FCT720, and FCT650 (without heat treatment) are compared. Additionally, in Figure 11b,d,f, images corresponding to the HT samples are shown. Regarding the Fast Cooling Temperature, the lower FCT caused an increment in the dislocation density. The formation of more bainitic phases in the lowest FCT resulted in higher ϑ values. No considerable effect of induction treatment was observed when kernel maps obtained from the samples without and after heat treatment were compared. This trend was observed in both ferritic and bainitic microstructures. For example, ϑ values of 1.06 and 1.09° were quantified for FCT790 and FCT790-HT conditions, respectively (leading to dislocation density, ρ, values of $1.95 \cdot 10^{14}$ m^{-2} and $2 \cdot 10^{14}$ m^{-2}).

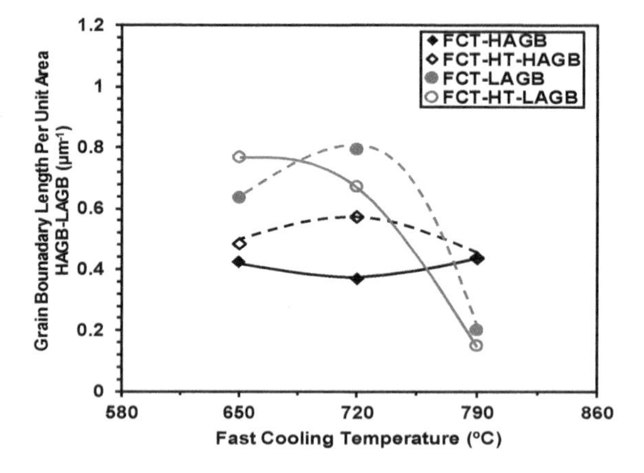

Figure 10. Grain boundary length per unit area as a function of the FCT. Results concerning high and low angle grain boundaries (HAGB and LAGB) are included.

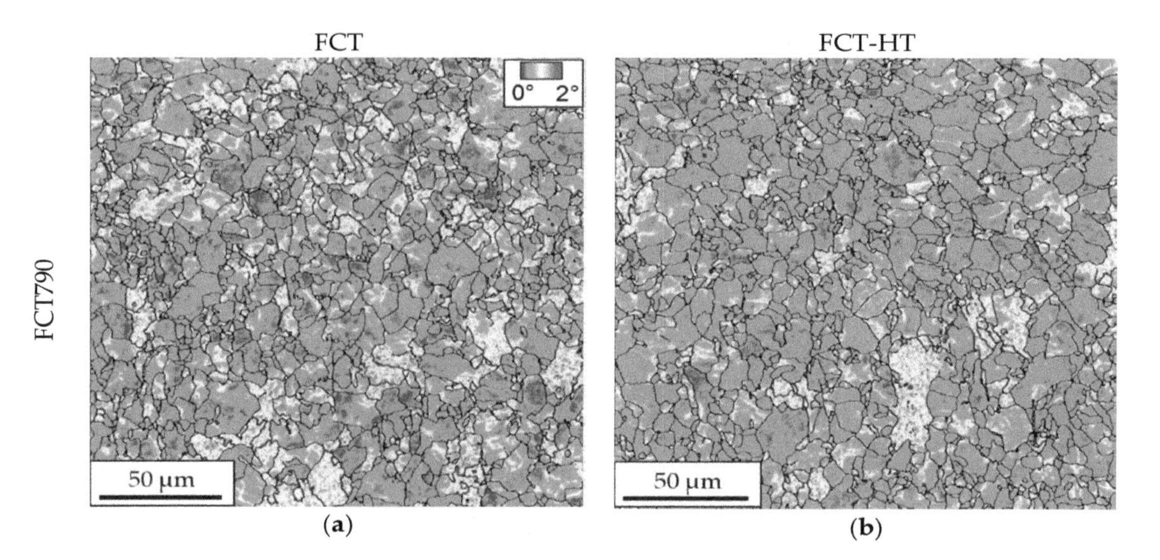

Figure 11. *Cont.*

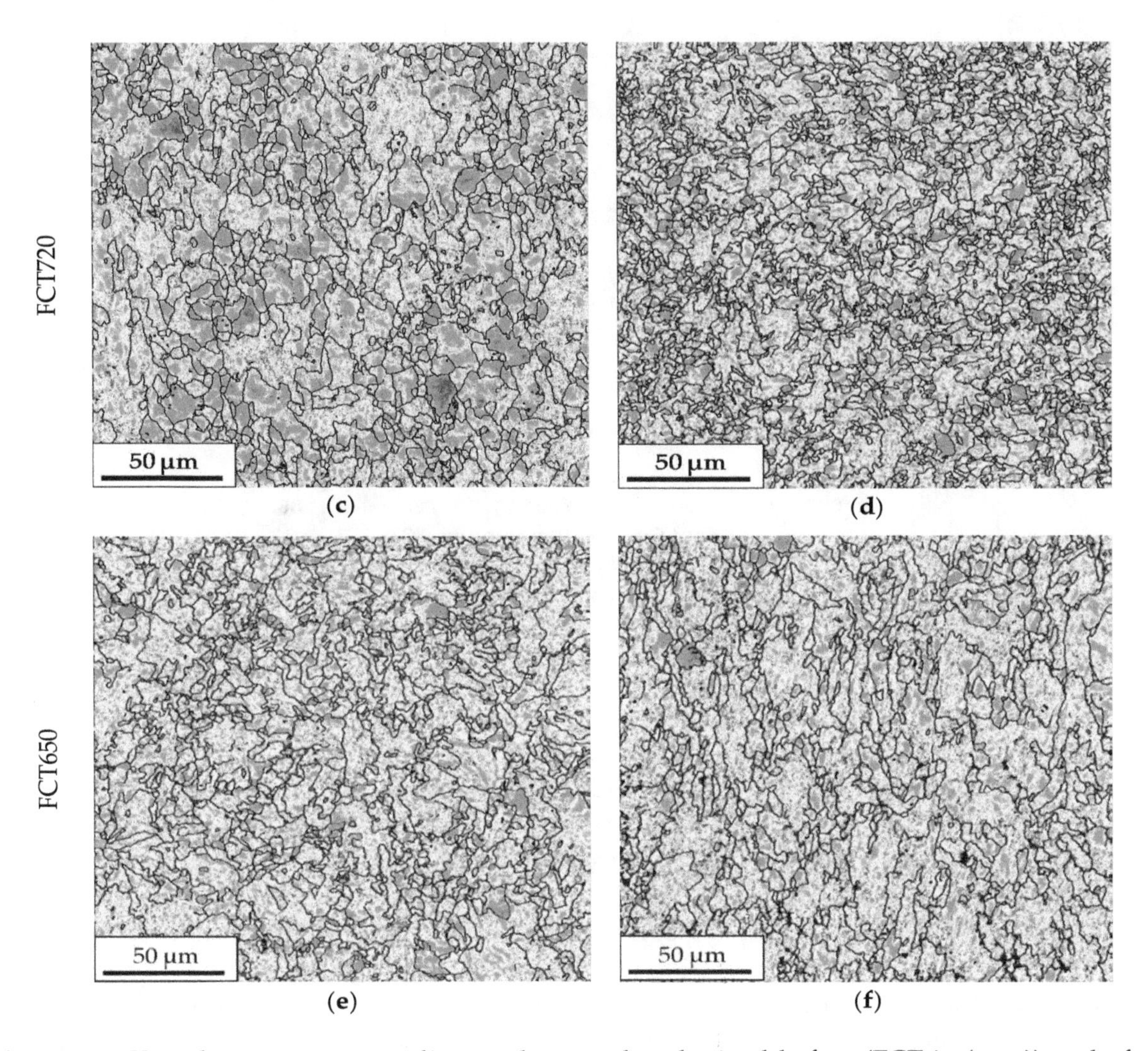

Figure 11. Kernel maps corresponding to the samples obtained before (FCT in (**a**,**c**,**e**)) and after induction treatment (FCT-HT in (**b**,**d**,**f**)) at different conditions: (**a**,**b**) FCT790 and (**c**,**d**) FCT720 and (**e**,**f**) FCT650.

In Figure 12, dislocation density measurements are represented as a function of the Fast Cooling Temperature for both FCT and FCT-HT conditions. Dislocation density values varied from $0.98{\cdot}10^{14}$ m^{-2} to $2{\cdot}10^{14}$ m^{-2} depending on the FCT and the applied cycle. When the FCT decreased from 790 to 720 °C, the formation of more bainitic microstructures with higher dislocation densities was shown (ρ increased from $1.03{\cdot}10^{14}$ m^{-2} to $1.82{\cdot}10^{14}$ m^{-2}). However, the dislocation density remained nearly constant for FCT720 and FCT650. As mentioned previously, induction heating does not considerably affect the dislocation density, and similar values were quantified for the samples obtained before and after the rapid heating step. Therefore, it can be concluded that there is a lack of microstructural variation in terms of dislocation density during induction heat treatment for the entire range of Fast Cooling Temperatures (for every microstructure). Similar behavior has already been reported in other studies [20], in which no effect of tempering on dislocation density was observed for mainly granular bainitic microstructures.

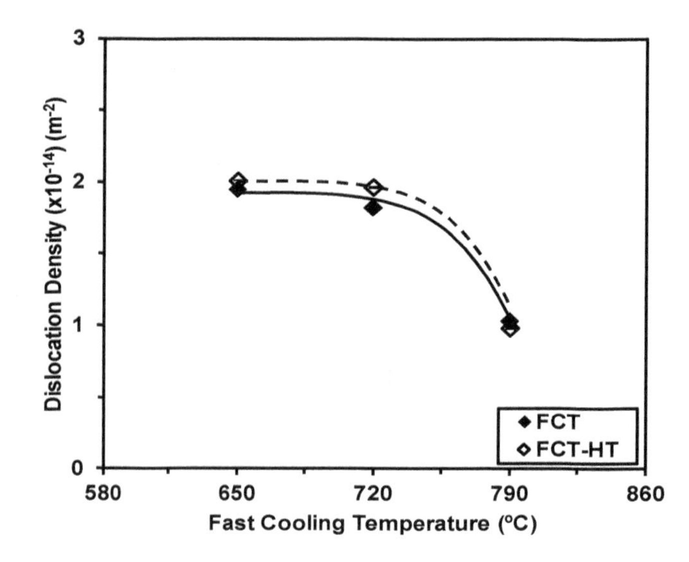

Figure 12. Influence of Fast Cooling Temperatures and induction heat treatment on dislocation density.

3.3. Precipitation Analysis

In order to evaluate the effect of FCT on precipitation as well as the influence of induction heat treatment, samples obtained before and after heat treatment were analyzed by TEM. Regarding the effect of the Fast Cooling Temperature and, therefore, the resulting microstructures, Figure 13a,b show the TEM micrographs corresponding to FCT790 and FCT650. From comparing both images, it is evident that FCT significantly modifies the precipitation taking place during continuous cooling after hot rolling [22]. When a high FCT of 790 °C is applied, interphase precipitation was clearly observed, while in the lowest FCT of 650 °C, low density of fine precipitates was detected. Therefore, the increment of FCT promotes the formation of a higher fraction of fine precipitates, ensuring a more pronounced strengthening contribution due to fine precipitation [23]. EDS analyses showed that the precipitates were Ti- and Mo-containing carbides [24]. It has been reported that the considerable strengthening effect observed in Ti–Mo microalloyed steels is attributed to the superior coarsening resistance of the (Ti, Mo)C carbide as compared to other carbides, such as TiC and (Ti, Nb)C [4]. In order to evaluate the hardening associated with the presence of fine particles, the precipitate size was measured for both FCT790 and FCT650 conditions, and the results are shown in Figure 13a,b. To quantify of precipitate size, particles smaller than 10 nm were taken into account, as they are supposed to be more efficient in terms of precipitation hardening. In Figure 13c, the precipitate size distributions (in terms of accumulated frequency) are presented for both FCTs. The results suggest that slightly finer precipitates formed as the FCT decreased. Mean precipitate sizes (Dppt) of 6.9 and 6.2 nm were quantified for FCT790 and FCT650 samples, respectively. Nevertheless, as mentioned previously, noticeably higher precipitate concentrations were distinguished for the FCT of 790 °C compared to 650 °C (see Figure 13a,b). In Ti–Mo microalloyed steels, the important contribution of fine precipitation has already been reported in several studies, and not just for plate hot rolling. In thin strip products, significant precipitation hardening can be achieved through an adequate coiling strategy [2,5,25].

Figure 14a–c show TEM micrographs at high magnifications in relation to the specimens obtained before heat treatment and the different Fast Cooling Temperatures. Moreover, in Figure 14d–f, the images corresponding to the heated samples are included for each FCT. By analyzing the micrographs corresponding to the samples without HT, as observed previously, the analysis using TEM suggests that the modification of FCT affects the size and the density of fine precipitates. As observed in Figure 13, for the highest FCTs (790 °C), the formation of aligned precipitates (interphase precipitation) was confirmed. However, at the intermediate FCT of 720 °C, random precipitation was clearly observed. When the lowest FCT of 650 °C was used, low precipitate fractions were shown. Therefore, lower precipitate densities were detected with a decreasing FCT. This could justify the yield strength enhancement that was previously observed when the FCT increased from 650 to 790 °C (see Figure 3).

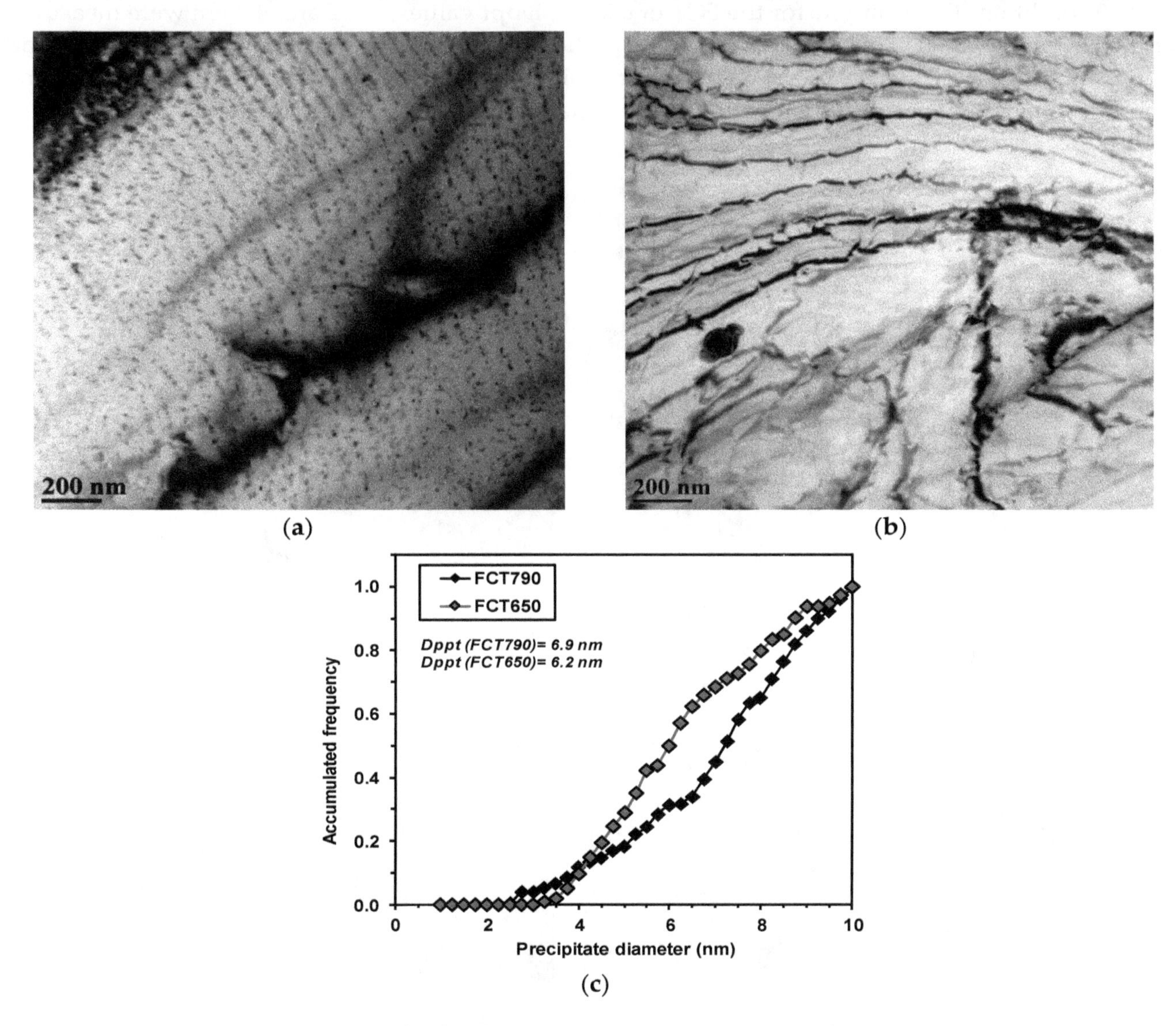

Figure 13. Thin foil TEM micrographs for the samples without induction treatment: (**a**) FCT790 and (**b**) FCT650. (**c**) Influence of the Fast Cooling Temperature on the precipitate size distribution (representing accumulated frequency).

Concerning the effect of induction heating on the strengthening due to precipitation, significantly higher precipitate densities were observed in the heated specimens than in the samples without heat treatment. In addition to interphase precipitates, random precipitates were observed in the ferritic microstructure obtained in the FCT790-HT condition. After an intermediate FCT of 720 °C, a higher density of fine precipitates was observed in the heated sample than in the sample obtained prior to heat treatment. Nevertheless, when a mainly bainitic microstructure formed at FCT650, the induction treatment promoted the formation of a more relevant precipitation. A higher precipitate concentration was observed in the FCT-HT sample compared to the FCT condition [9,15].

Quantification of the mean precipitate sizes was performed for all of the generated microstructures. In Figure 15, the evolution of the mean precipitate size as a function of the Fast Cooling Temperature was plotted for both FCT and FCT-HT conditions. In terms of the effect of FCT, as mentioned before, slightly fine mean precipitate sizes were achieved as the FCT decreased. Similar trends were observed for the samples obtained before and after induction treatment. The results shown in Figure 15 suggest that rapid heating led to the formation of finer precipitates, mainly in the bainitic matrix formed at the lowest FCT of 650 °C. Therefore, finer precipitates were formed during rapid heating mainly for the FCT650 condition. For example, for the FCT of 650 °C, Dppt values of 6.2 and 4.9 nm were measured for the FCT and FCT-HT conditions. In addition to the precipitate refinement, for both the ferritic and bainitic matrices, a higher concentration of fine precipitates was detected after heat treatment, ensuring a higher strengthening contribution owing to precipitation (see TEM micrographs shown in Figure 14).

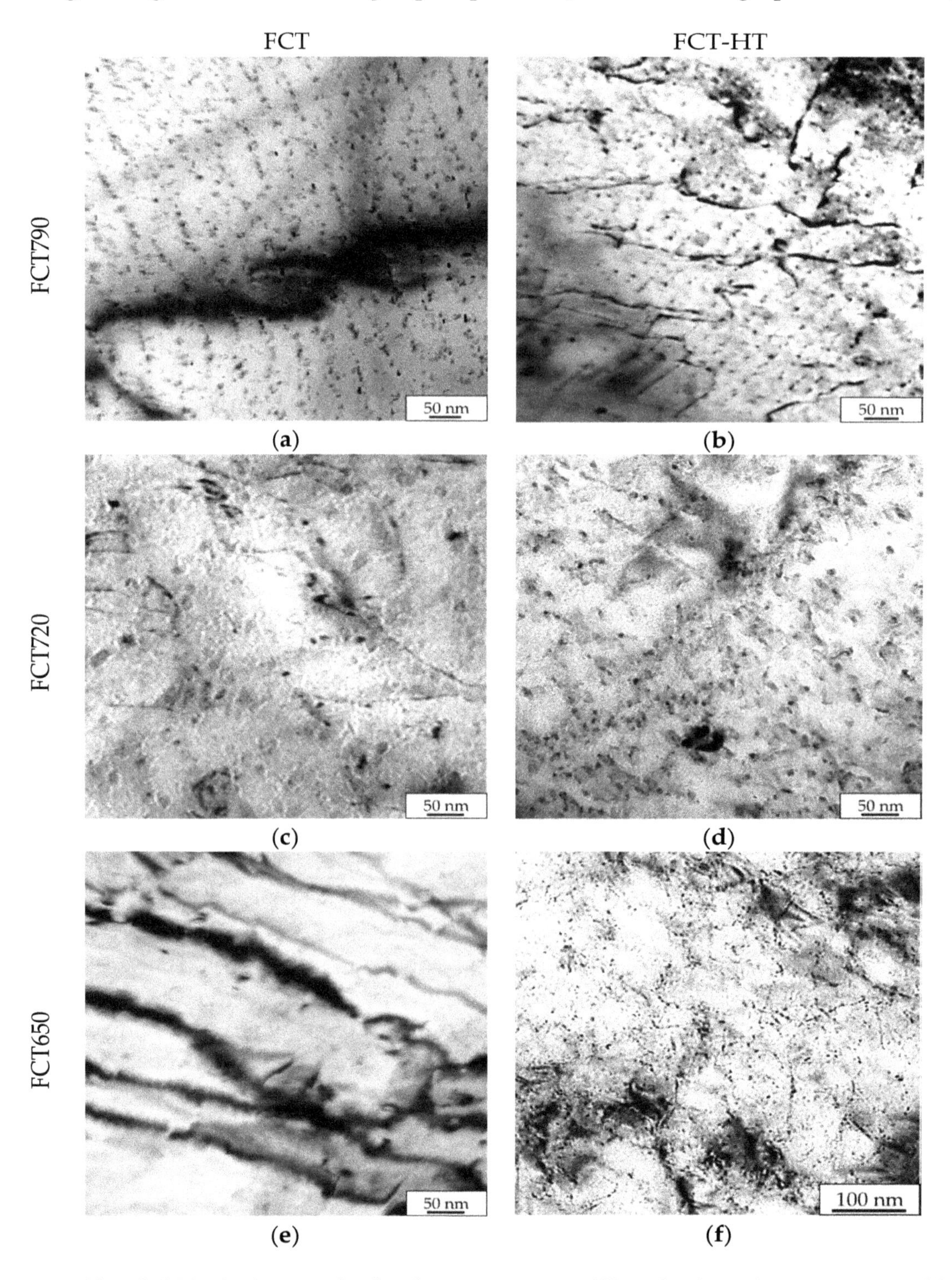

Figure 14. Thin foil TEM micrographs for the entire range of Fast Cooling Temperatures and both (**a**,**c**,**e**) FCT and (**b**,**d**,**f**) FCT-HT conditions: (**a**,**b**) FCT790, (**c**,**d**) FCT720 and (**e**,**f**) FCT650.

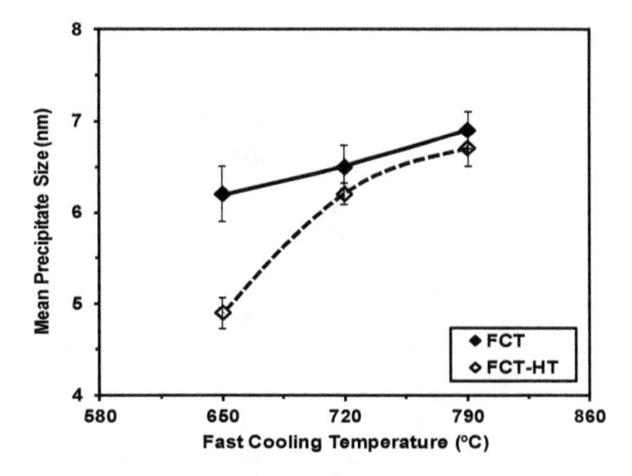

Figure 15. Mean precipitate sizes (Dppt) as a function of FCT. Measurements carried out in the FCT and FCT-HT samples have been included.

3.4. Interactions between Microstructure and Tensile Properties

In the current work, the yield strength was estimated according to Equation (1), in which a linear sum of the different strengthening contributions was considered (solid solution, grain size, dislocation density, presence of MA islands, and precipitation):

$$\sigma_y = \sigma_0 + \sigma_{ss} + \sigma_{gs} + \sigma_\rho + \sigma_{MA} + \sigma_{ppt}. \quad (1)$$

Given the limitations in the estimation of an accurate precipitate volume fraction, the term related to fine precipitation was estimated by subtracting the hardening due to the rest of the contributions from the experimental yield strength values. A more detailed explanation of the approaches as well as the methodology followed are be found in references [5,10].

In Figure 16, the contribution of the different hardening mechanisms is plotted for all the microstructures obtained from the different FCTs. Figure 16 indicates that the most important strengthening mechanism is the grain size. The analysis of the results concerning the samples without rapid heating showed that higher grain size contributions were measured as the FCT decreased. The formation of more bainitic matrices implies an increment in the term related to the dislocation density. Higher contributions for dislocation strengthening were achieved as the FCT decreased. The results plotted in Figure 16 also suggest that higher precipitate contributions were measured for the highest FCT of 790 °C. Precipitation hardening values of 113, 75, and 0 MPa was calculated for FCT790, FCT720, and FCT650, respectively. This is in line with the previously shown precipitation characterization, in which a higher precipitate density was detected in FCT790 compared to in the FCT720 and FCT650 conditions.

The data plotted in Figure 16 facilitates an evaluation of the influence of induction treatment on the strengthening mechanisms. Regarding the contribution related to grain size, slightly higher terms were estimated after rapid heating for every FCT and microstructure. This can be explained by the partial decomposition of the formed MA islands that takes place during heat treatment. However, no significant effect of heat treatment on dislocation density was detected. Similar dislocation density terms were estimated before and after HT. On the other hand, fast heating led to the reduction of the strengthening caused by MA islands, due to the partial dissolution of the secondary phase that occurs during treatment. In terms of the influence of heat treatment on precipitation, a higher hardening effect due to fine precipitates is estimated after induction treatment. When mainly ferritic phases are predominant (HT790), strengthening due to fine precipitation improved by about 95 MPa after induction treatment. Nevertheless, when more bainitic microstructures formed (HT650), an increment of yield strength of approximately 133 MPa was achieved through rapid heating. Therefore, it can be claimed that tensile property enhancement is more pronounced for bainitic microstructures than for ferritic microstructures. The high dislocation density that is typical of bainitic microstructures offered nucleation sites for precipitate to form during the fast heating.

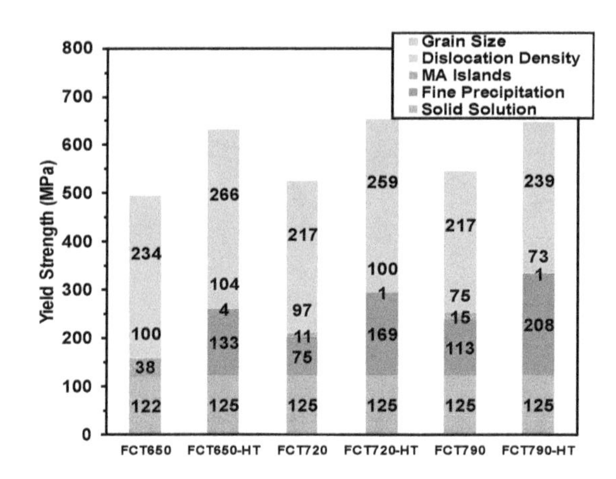

Figure 16. Individual strengthening contributions (grain size, dislocations, MA islands, precipitation and composition) for every microstructure (ferrite, ferrite/bainite and bainite, for FCT790, FCT720 and FCT650, respectively) and the samples obtained before (FCT) and after (FCT-HT) induction heat treatment.

3.5. Effect of Induction Strengthening on Toughness Properties

The improvement in tensile properties can impair toughness properties, mainly when precipitation hardening is the mechanism behind the strengthening. Therefore, an evaluation of the link between induction heating, microstructural parameters, and impact properties was carried out. Additionally, the effect of the initial microstructure morphology (ferritic or bainitic) on toughness property impairment was evaluated.

In the current work, Charpy tests were performed for the two extreme conditions of FCT790 and FCT650. In Figure 17, the Charpy curves obtained for the microstructures without treatment and treated samples are shown for both microstructure types (ferritic and bainitic in FCT790 and FCT650). In Figure 17a, the absorbed energy is plotted for the mentioned conditions, while Figure 17b shows the ductile fraction as a function of the test temperature.

Concerning the microstructures obtained before induction treatment (FCT samples), the results plotted in Figure 17a suggest that the initial microstructure strongly affects impact properties. Considerably lower impact transition temperatures were obtained when a low Fast Cooling Temperature of 650 °C was applied. Therefore, the formation of more bainitic phases in the FCT650 condition promotes the enhancement of toughness properties. The observed deterioration in toughness behavior in the FCT790 sample can be attributed mainly to the formation of coarser microstructures, the presence of MA islands, as well as the formation of a high density of fine precipitates (see the interphase precipitation shown in Figure 14a). This observation was also confirmed by the evolution of the ductile fraction presented in Figure 17b.

Regarding the influence of heat treatment on toughness properties, different trends were observed depending on the Fast Cooling Temperature and consequently, the initial microstructure type. For the mainly ferritic microstructure formed in FCT790, rapid heating did not significantly affect Charpy curves. Similarly, the absorbed energy evolution was clearly observed for both FCT790 and FCT790-HT conditions. Nevertheless, when bainite was the predominant phase in the microstructure prior to induction treatment (using a lower FCT of 650 °C), a considerable deterioration in toughness was observed. The impact transition curve (absorbed energy and ductile fraction, in Figure 17a,b, respectively) shifted to higher temperatures. This could be associated with the tensile property improvement mainly due to precipitation strengthening by the induction heat treatment in bainitic microstructures. Therefore, the fine precipitation that takes place during induction treatment could damage toughness properties. Therefore, the impact property impairment caused by rapid heating was more pronounced for the bainitic matrix compared to the mainly ferritic microstructures.

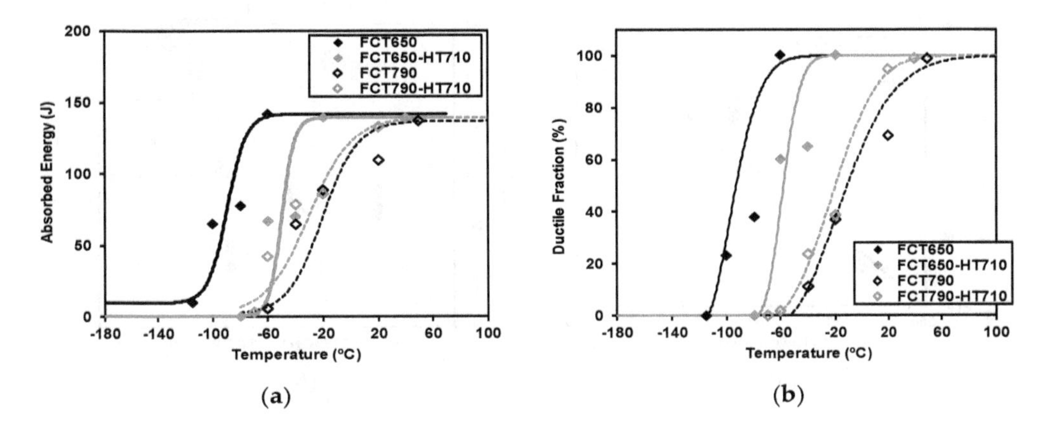

Figure 17. Comparison between Charpy curves corresponding to FCT650, FCT650-HT, FCT790 and FCT790-HT: (**a**) absorbed energy and (**b**) ductile fraction as a function of the test temperature.

In a recent study, the proposed empirical DBTT (50% ductile–brittle appearance transition temperature) equation (Equation (2)) was extended from Nb and Nb-Mo microalloyed steel to Ti-Mo microalloyed steel [5], where a considerably higher contribution of fine precipitation and dislocation density was observed.

$$\begin{aligned} \text{DBTT }(^\circ\text{C}) = & -11\text{Mn} + 42\text{Si} + 700\left(N_{free}\right)^{0.5} + 15(pct\ Pearlite + pct\ MA)^{\frac{1}{3}} + \\ & 0.26\Delta\sigma_\text{y} - 14(\text{D}15^\circ)^{-0.5} + 63\left(\tfrac{\text{Dc}20\%}{\text{D}15^\circ}\right)^{0.5} + 18(D_{MA})^{0.5} - 42. \end{aligned} \quad (2)$$

Hardening due to solid solution is included in the first two terms. Nitrogen is assumed to be zero, due to the hyperstoichiometric composition of Ti–Mo steel. The contribution of the secondary phase is considered through the fraction of pearlite and MA islands, as well as the size of the MA islands (D_{MA}). The evaluation of heterogeneity is introduced by including the Dc20%/D15° factor, reported in Figure 9b. Figure 18a shows the comparison between the experimental DBTT and the estimated DBTT values obtained from Equation (2) for all of the conditions shown in Figure 17. In addition, the results obtained in previously published works [5,10,17] have been included in Figure 18a. The comparison presented in Figure 18a suggests that a reasonable estimation of DBTT values can be reached using Equation (2) for all the conditions analyzed in the current work.

By using Equation (2), the estimation of individual contributions can be estimated. In Figure 18b, the contribution of each microstructural parameter is plotted for all of the conditions (FCT650, FCT650-HT, FCT790 and FCT790-HT). Besides the individual contributions to DBTT, the calculated DBTT value from Equation (2) is represented in the graph. The results showed that the term related to grain size together with the compositions are the only mechanisms that ensure an improvement in toughness properties. Moreover, the contribution related to grain size is the most relevant term of all the conditions. The terms associated with secondary phases varied significantly depending on the condition. Slightly higher secondary phase contributions were estimated for ferritic microstructures (FCT of 790 °C) than for bainitic microstructures. Concerning the effect of induction treatment, when a low Fast Cooling Temperature of 650 °C was used, a lowering of the secondary phase term was observed after rapid heating. This could be associated with the abovementioned decomposition of the secondary phase during induction treatment. Regarding $\Delta\sigma_y$ (fine precipitation + dislocation density), it is clear that the detrimental effect of $\Delta\sigma_y$ was more pronounced under FCT-HT conditions than under FCT conditions. In this case, the main mechanism that worsens transition temperatures is fine precipitation. For example, the contributions of $\Delta\sigma_y$ were about 27 and 62 °C, in the FCT650 and FCT650-HT conditions, respectively. Finally, the term due to heterogeneity is relevant under all conditions, varying from 107 to 138 °C. As observed in Figure 18b, slightly higher terms were estimated when bainitic phases were predominant (FCT of 650 °C) compared to mainly ferritic microstructures (FCT of 790 °C).

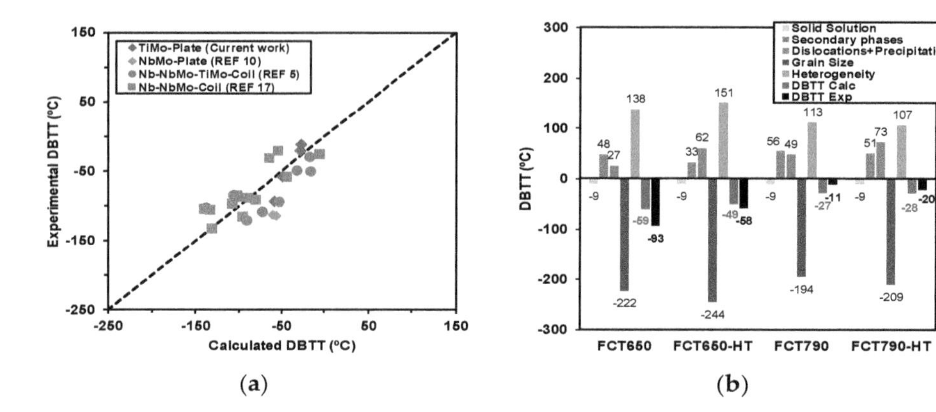

Figure 18. (**a**) Comparison between experimental and calculated DBTT values predicted by Equation (2). The data reported by Isasti et al. [17] and Larzabal et al. [5,10] have been included in the graph. (**b**) Estimated contributions (solid solution, secondary phases, dislocation + precipitation, grain size and heterogeneity) to the DBTT for every condition and comparison with experimental values.

To further evaluate the influence of the different metallurgical parameters, such as solid solution, secondary phase, dislocation, precipitation, grain size and heterogeneity, on yield strength and DBTT values, the vector diagram proposed by Gladman [26] was built. These diagrams summarize the different strengthening strategies that can be followed in the industry to achieve the required mechanical properties and to evaluate the balance between strength and toughness. The strengthening/embrittlement terms related to ach mechanism and the sum of all the contributions are represented in Figure 19. Figure 19a,b show the effect of induction treatment for both the ferritic (FCT790) and bainitic (FCT650) microstructures. The dotted lines are related to the microstructures without rapid heating, whereas continuous line vectors consider the contributions associated with treated microstructures. These diagrams summarize the different strategies and strengthening patterns that can be followed in industry to achieve the required mechanical properties and to evaluate the balance between strength and toughness. The differences between the samples before and after induction treatment that have already been highlighted are confirmed in the charts, such as the microstructural refinement effect, second phase dissolution, and/or heterogeneity. However, definitely, the effect of precipitation strengthening after rapid heating is the most relevant for both types of microstructures to achieve higher strengthening levels with a relatively low toughness impairment.

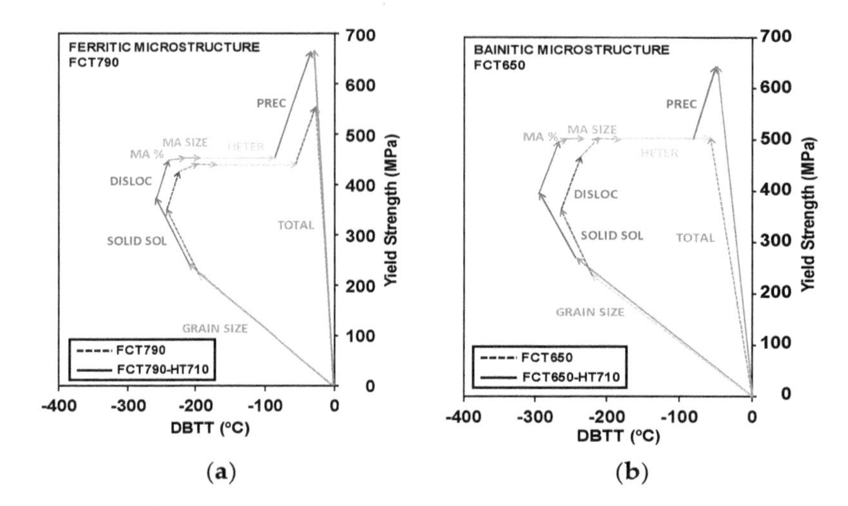

Figure 19. Relative contribution of each metallurgical parameter (grain size, solid solution, dislocations, martensite/austenite (MA) fraction, MA size, heterogeneity and precipitation) on both strength and toughness properties. FCT and FCT-HT conditions are compared for both types of microstructures: (**a**) ferritic in FCT790 and (**b**) bainitic in FCT650.

4. Conclusions

The suitability of titanium–molybdenum microalloyed steels for post-rolling induction strengthening was confirmed. This strengthening is relevant both for ferritic and bainitic microstructures. Additional precipitation strengthening was achieved through the formation of fine nanosized precipitates during induction heat treatment. The results suggest that tensile property improvement caused by induction treatment is more relevant in bainitic microstructures compared to ferritic microstructures. This fact can be attributed to the higher dislocation density of the bainitic microstructure that provides more nucleation sites for carbides to form during induction treatment. No significant microstructural changes were observed in terms of grain size and dislocation density after induction heating.

Author Contributions: G.L. carried out the experiments; N.I. analyzed the data and wrote the manuscript; J.M.R.-I. supervised the results and edited the manuscript; P.U. managed the project and edited the manuscript.

Nomenclature

σ_y	Yield Strength
σ_0	Lattice friction stress
σ_{ss}	Strengthening contribution due to solid solution
σ_{gs}	Strengthening contribution due to grain size
σ_ρ	Strengthening contribution due to dislocations
σ_{ppt}	Strengthening contribution due to precipitation
σ_{MA}	Strengthening contribution due to MA islands
D2°	Mean unit size using the 2° high angle boundary criterion
D15°	Mean unit size using the 15° high angle boundary criterion
Dc20%	20% critical grain size
ϑ	Kernel average misorientation
ρ	Dislocation density
Dppt	Mean precipitate size
Mn	Manganese content (wt%)
Si	Silicon content (wt%)
N_{free}	Free nitrogen (wt%)
pct pearlite	Volume fraction of pearlite
pct MA	Volume fraction of MA islands
D_{MA}	MA island mean size
$\Delta\sigma_y$	Yield strength increment due to dislocation and precipitation strengthening ($\sigma_\rho + \sigma_{ppt}$)

References

1. Isasti, N.; Jorge-Badiola, D.; Taheri, M.L.; Uranga, P. Microstructural Features Controlling Mechanical Properties in Nb-Mo Microalloyed Steels. Part I: Yield Strength. *Metall. Mater. Trans. A* **2014**, *45A*, 4960–4971. [CrossRef]
2. Lee, W.B.; Hong, S.G.; Park, C.G.; Park, S.H. Carbide precipitation and high-temperature strength of hot-rolled high-strength, low-alloy steels containing Nb and Mo. *Metall. Mater. Trans. A* **2002**, *33A*, 1689–1698. [CrossRef]
3. Funakawa, Y.; Shiozaki, T.; Tomita, K.; Yamamoto, T.; Maeda, E. Development of High Strength Hot-rolled Sheet Steel Consisting of Ferrite and Nanometer-sized Carbides. *ISIJ Int.* **2004**, *44*, 1945–1951. [CrossRef]
4. Chen, C.Y.; Yen, H.W.; Kao, F.H.; Li, W.C.; Huang, C.Y.; Yang, J.R.; Wang, S.H. Precipitation hardening of high-strength low-alloy steels by nanometer-sized carbides. *Mater. Sci. Eng. A* **2009**, *499*, 162–166. [CrossRef]
5. Larzabal, G.; Isasti, N.; Rodriguez-Ibabe, J.M.; Uranga, P. Evaluating Strengthening and Impact Toughness Mechanisms for Ferritic and Bainitic Microstructures in Nb, Nb-Mo and Ti-Mo Microalloyed Steels. *Metals* **2017**, *7*, 65. [CrossRef]
6. Huang, Y.; Zhao, A.; Wang, X.; Wang, X.; Yang, J.; Han, J.; Yang, F. A High-Strength High-Ductility Ti- and Mo-Bearing Ferritic Steel. *Metall. Mater. Trans. A* **2016**, *47A*, 450–460. [CrossRef]

7. Cheng, L.; Cai, Q.-W.; Xie, B.-S.; Ning, Z.; Zhou, X.-C.; Li, G.-S. Relationships among microstructure, precipitation and mechanical properties in different depths of Ti–Mo low carbon low alloy steel plate. *Mater. Sci. Eng. A* **2016**, *651*, 185–191. [CrossRef]
8. Xie, Z.J.; Fang, Y.P.; Han, G.; Guo, H.; Misra, R.D.K.; Shang, C.J. Structure–property relationship in a 960 MPa grade ultrahigh strength low carbon niobium–vanadium microalloyed steel: The significance of high frequency induction tempering. *Mater. Sci. Eng. A* **2014**, *618*, 112–117. [CrossRef]
9. Larzabal, G.; Isasti, N.; Pereda, B.; Rodriguez-Ibabe, J.M.; Uranga, P. Precipitation Strengthening by Induction Treatment in High Strength Low Carbon Microalloyed Hot Rolled Plates. In Proceedings of the Materials Science and Technology 2016 Conference, Salt Lake City, UT, USA, 23–27 October 2016; pp. 499–507.
10. Larzabal, G.; Isasti, N.; Rodriguez-Ibabe, J.M.; Uranga, P. Precipitation Strengthening by Induction Treatment, in High Strength Low Carbon Microalloyed Hot-Rolled Plates. *Metall. Mater. Trans. A* **2018**, *49A*, 946–961. [CrossRef]
11. Uranga, P.; Gutiérrez, I.; López, B. Determination of recrystallization kinetics from plane strain compression tests. *Mater. Sci. Eng. A* **2013**, *578*, 174–180. [CrossRef]
12. Wallin, K. *Modified Tank Fitting Algorithm for Charpy Impact Data*; Research Seminar on Economical and Safe Application of Modern Steels for Pressure Vessels: Aachen, Germany, 2003.
13. Kim, Y.M.; Kim, S.K.; Lim, Y.J.; Kim, N.J. Effect of Microstructure on the Yield Ratio and Low Temperature Toughness of Linepipe Steels. *ISIJ Int.* **2002**, *42*, 1571–1577. [CrossRef]
14. Araki, T.; Kozasu, I.; Tankechi, H.; Shibata, K.; Enomoto, M.; Tamehiro, H. *Atlas for Bainitic Microstructures*; ISIJ: Tokyo, Japan, 1992; Volume 1.
15. Xie, Z.J.; Ma, X.P.; Shang, C.J.; Wang, X.M.; Subramanian, S.V. Nano-sized precipitation and properties of a low carbon niobium micro-alloyed bainitic steel. *Mater. Sci. Eng. A* **2015**, *641*, 37–44. [CrossRef]
16. Iza-Mendia, A.; Gutiérrez, I. Generalization of the existing relations between microstructure and yield stress from ferrite-pearlite to high strength steels. *Mater. Sci. Eng. A* **2013**, *561*, 40–51. [CrossRef]
17. Isasti, N.; Jorge-Badiola, D.; Taheri, M.L.; Uranga, P. Microstructural Features Controlling Mechanical Properties in Nb-Mo Microalloyed Steels. Part II: Impact Toughness. *Metall. Mater. Trans. A* **2014**, *45*, 4972–4982. [CrossRef]
18. Olasolo, M.; Uranga, P.; Rodriguez-Ibabe, J.M.; López, B. Effect of austenite microstructure and cooling rate on transformation characteristics in a low carbon Nb–V microalloyed steel. *Mater. Sci. Eng. A* **2011**, *528*, 2559–2569. [CrossRef]
19. Isasti, N.; Jorge-Badiola, D.; Taheri, M.L.; López, B.; Uranga, P. Effect of Composition and Deformation on Coarse-Grained Austenite Transformation in Nb-Mo Microalloyed Steels. *Metall. Mater. Trans. A* **2011**, *42A*, 3729–3742. [CrossRef]
20. Huang, B.M.; Yang, J.R.; Yen, H.W.; Hsu, C.H.; Huang, C.Y.; Mohrbacher, H. Secondary hardened bainite. *Mater. Sci. Technol.* **2014**, *30*, 1014–1023. [CrossRef]
21. Calcagnotto, M.; Ponge, D.; Demir, E.; Raabe, D. Orientation gradients and geometrically necessary dislocations in ultrafine grained dual-phase steels studied by 2D and 3D EBSD. *Mater. Sci. Eng. A* **2010**, *257*, 2738–2746. [CrossRef]
22. Chen, C.Y.; Yang, J.R.; Chen, C.C.; Chen, S.F. Microstructural characterization and strengthening behavior of nanometer sized carbides in Ti–Mo microalloyed steels during continuous cooling process. *Mater. Charact.* **2016**, *114*, 18–29. [CrossRef]
23. Bu, F.Z.; Wang, X.M.; Yang, S.W.; Shang, C.J.; Misra, R.D.K. Contribution of interphase precipitation on yield strength in thermomechanically simulated Ti–Nb and Ti–Nb–Mo microalloyed steels. *Mater. Sci. Eng. A* **2015**, *620*, 22–29. [CrossRef]
24. Larzabal, G. Efecto de los Parámetros de Laminación y Post-Tratamiento Térmico por Inducción en la Mejora de Propiedades Mecánicas de Aceros Microaleados. Ph.D. Thesis, Universidad de Navarra, San Sebastian, Spain, 2017.
25. Sanz, L.; Pereda, B.; López, B. Effect of thermomechanical treatment and coiling temperature on the strengthening mechanisms of low carbon steels microalloyed with Nb. *Mater. Sci. Eng. A* **2017**, *685*, 377–390. [CrossRef]
26. Gladman, T. *The Physical Metallurgy of Microalloyed Steels*, 2nd ed.; The Institute of Materials: London, UK, 1997; pp. 62–68.

10

Physically-Based Modeling and Characterization of Hot Flow Behavior in an Interphase-Precipitated Ti-Mo Microalloyed Steel

Chuanfeng Wu [1], Minghui Cai [1,*], Peiru Yang [1], Junhua Su [1] and Xiaopeng Guo [2]

1 School of Materials Science and Engineering, Northeastern University, Shenyang 110819, China; chf_wu@163.com (C.W.); yangpeiru1995@126.com (P.Y.); juh_su@163.com (J.S.)
2 School of Mechanical Engineering and Automation, Northeastern University, Shenyang 110819, China; m15804050383@163.com
* Correspondence: cmhing@126.com or caimh@smm.neu.edu.cn

Abstract: In this contribution, a series of hot compression tests was conducted on a typical interphase-precipitated Ti-Mo steel at relatively higher strain rates of 0.1~10 s^{-1} and temperatures of 900~1150 °C using a Gleeble-2000 thermo-mechanical simulator. A combination of Bergstrom and Kolmogorov–Johnson–Mehl–Avrami models was first used to accurately predict the whole flow behaviors of Ti-Mo steel involving dynamic recrystallization, under various hot deformation conditions. By comparing the characteristic stresses and material parameters, especially at the higher strain rates studied, the dependence of hot flow behavior on strain rate and deformation temperature was further clarified. The hardening parameter U and peak density ρ_p exhibited an approximately positive linear relationship with the Zener–Hollomon (Z) parameter, while the softening parameter Ω dropped with increasing Z value. The Avrami exponent n_A varied between 1.2 and 2.1 with lnZ, implying two diverse nucleation mechanisms of dynamic recrystallization. The experimental verification was performed as well based on the microstructural evolution and mechanism analysis upon straining. The proposed constitutive models may provide a powerful tool for optimizing the hot working processes of high performance Ti-Mo microalloyed steels with interphase precipitation.

Keywords: Ti-Mo steel; hot deformation; constitutive model; microstructural evolution

1. Introduction

A new generation of advanced high strength steels (AHSSs) with both high/ultrahigh strength and good cold formability is becoming more commonly used in commercial vehicles to meet the continuous demand for better fuel economy and passenger safety [1–3]. However, many of the conventional AHSSs, e.g., dual-phase (DP) steel, have limited the stretch-flangeability during cold stamping in view of the notable difference in hardness between the soft ferrite and the hard martensite [2,3]. As a solution, a ferritic matrix with interphase precipitation has been proposed to improve the stretch-flangeability of AHSSs through optimizing alloy design and thermomechanical processes [4,5].

Funakawa et al. [6] first reported a Ti-Mo ferritic steel strengthened by period arrangement of nanoscale interphase precipitates, which exhibited an ultimate tensile strength (UTS) of ~780 MPa and a hole expansion ratio (HER) of 120%. The basic principle designed for this type of steel was to facilitate the formation of a large fraction of nanoscale precipitates by controlling the coarsening kinetics of (Ti, Mo)C particles [1,4,5,7–10], thus giving rise to a large contribution to the strength of steels. More recently, some research has been carried out with respect to the influence of Mo on the formation of interphase precipitates. It has been demonstrated by atom probe tomography (APT) that the size of (Ti, Mo)C precipitates could be reduced to as fine as 1–3 nm [1].

Additionally, recent research on the hot-compressed Ti-Mo steel [8] revealed that hot deformation promoted segregation of C to grain boundaries and simultaneously reduced the size of precipitates at grain boundaries during the isothermal holding process. Kim et al. [9,10] also pointed out that precipitation hardening exhibited a stronger relation to rolling temperature rather than coiling temperature. However, very few studies have considered the hot flow behaviors of Ti-Mo microalloyed steels, especially at relatively higher strain rates of 0.1~10 s^{-1}, which was relevant to the industrial hot rolling processes [11]. A full list of nomenclature for all physical parameters is presented in Table 1.

Table 1. A full list of nomenclature.

Symbols	Parameters
A	Material constant (s^{-1})
AARE	Average absolute relative error
k	Material constant
M	Dislocation strengthening constant
n	Stress exponent
n_A	Avrami's exponent
Q	Activation energy for hot deformation ($kJ \cdot mol^{-1}$)
R	Universal gas constant ($J \cdot K^{-1} \cdot mol^{-1}$)
T	Absolute deformation temperature (K)
U	Hardening parameter (m^{-2})
X	Recrystallized volume fraction
Z	Zener–Hollomon parameter (s^{-1})
α	Stress multiplier (MPa^{-1})
γ	Austenite phase
ε	True strain
$\dot{\varepsilon}$	Strain rate (s^{-1})
μ	Shear modulus (GPa)
ρ	Dislocation density (m^{-2})
σ	True stress (MPa)
σ_0	Lattice friction stress (MPa)
σ_p	Peak stress (MPa)
σ_{Rex}	Steady state stress after dynamic recrystallization (MPa)
Ω	Softening parameter

Therefore, the main aim of the present work was to study the hot flow behavior of an interphase-precipitated Ti-Mo microalloyed steel with a typical composition of Fe-0.04C-0.2Si-1.5Mn-0.08Ti-0.22Mo at strain rates of 0.1~10 s^{-1} and a series of deformation temperatures ranging from 900–1150 °C. The activation energy for hot deformation and peak stress of both steels were determined, in conjunction with physically-based modeling of hot flow behavior. The experimental verification was performed through microstructural evolution and mechanism analysis during hot deformation.

2. Experimental Procedures

A 30-kg ingot with a typical composition of Fe-0.04C-0.2Si-1.5Mn-0.08Ti-0.22Mo was prepared, followed by heating to 1200 °C for 2 h and hot rolling to ~20 mm in thickness. The calculated austenite to ferrite transformation start temperature Ar_3 by Thermo-calc software is approximately 858.1 °C. The as hot-rolled microstructure consisted of ferrite with an average grain size of ~40 μm, along with a small amount of pearlite (~7%), as presented in Figure 1. Cylindrical specimens of Φ 8 × 15 mm were machined from the hot-rolled plates, with the axis parallel to the rolling direction.

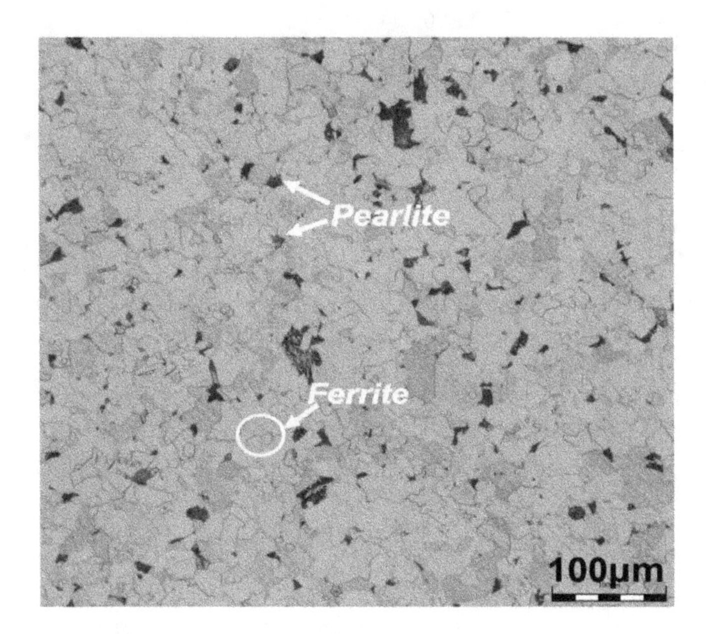

Figure 1. The initial hot rolled microstructure of Ti-Mo steel.

A series of hot compression tests was conducted on a Gleeble-2000 thermo-mechanical simulator. Specimens were reheated at a rate of 10 °C/s to 1200 °C for 3 min for homogenization purposes, followed by cooling at 10 °C/s to diverse deformation temperatures. After soaking for 10 s, they were compressed in a single hit to $\varepsilon = 0.8$ at temperatures of 900~1150 °C and strain rates of 0.1~10 s^{-1}, followed by immediate quenching in water. The selection of temperatures considered is based on the following two facts: (1) the austenite was deformed in the single austenitic region; (2) all three common deformation mechanisms appeared on the flow curves in this study.

Microstructural characterization after hot deformation was performed using an optical microscope (OM, OLYMPUS DSX500, Tokyo, Japan) and electron backscattered diffraction (EBSD) in a JEOL JSM-7100F (Jeol, Tokyo, Japan) field emission gun scanning electron microscope. The scanning step size was approximately 0.5 μm, and data post-processing was performed using a Tex-SEM Laboratories orientation-imaging microscope (OIM) system. Specimens for EBSD were prepared by standard mechanical grinding and electro-polishing using a perchloric acid-alcohol solution (1:15) at 30 V.

3. Results and Discussion

3.1. Hot Flow Behaviors

Figure 2 shows the representative true stress-strain curves obtained during hot deformation of Fe-0.04C-0.2Si-1.5Mn-0.08Ti-0.22Mo steel at temperatures of 900~1150 °C and strain rates of 0.1~10 s^{-1}. As commonly observed, the flow stress level of experimental steel was elevated with increasing strain rate or dropping deformation temperature. In the relatively low strain rate (0.1 s^{-1}), the flow curves were characterized by dynamic recovery (DRV) and dynamic recrystallization (DRX), except at 900 °C. Increasing strain rate was found to retard the progress of DRV or DRX, and no DRX phenomenon was observed at the stain rate of 10 s^{-1}, irrespective of deformation temperature. The change in flow stress level with deformation temperature or strain rate was probably due to an increase in the rate of restoration processes and a decrease in the strain hardening rate [12]. The lower strain rate can provide enough time for energy accumulation, and the higher deformation temperature can accelerate the dissolution of precipitates, dislocation movement and grain boundary migration for the formation of DRX nuclei, which in turn lowered the stress level [12,13]. It is worth mentioning that the flow curves exhibited a prominent oscillation at all temperatures for the lowest strain rate of 0.1 s^{-1}, while the oscillations were much slower with increasing strain rate. The oscillation phenomenon could be mainly attributed to the reduced accuracy and stability of sensors at the lowest strain rate of 0.1 s^{-1}, as well as the competing effect between various deformation mechanisms [13].

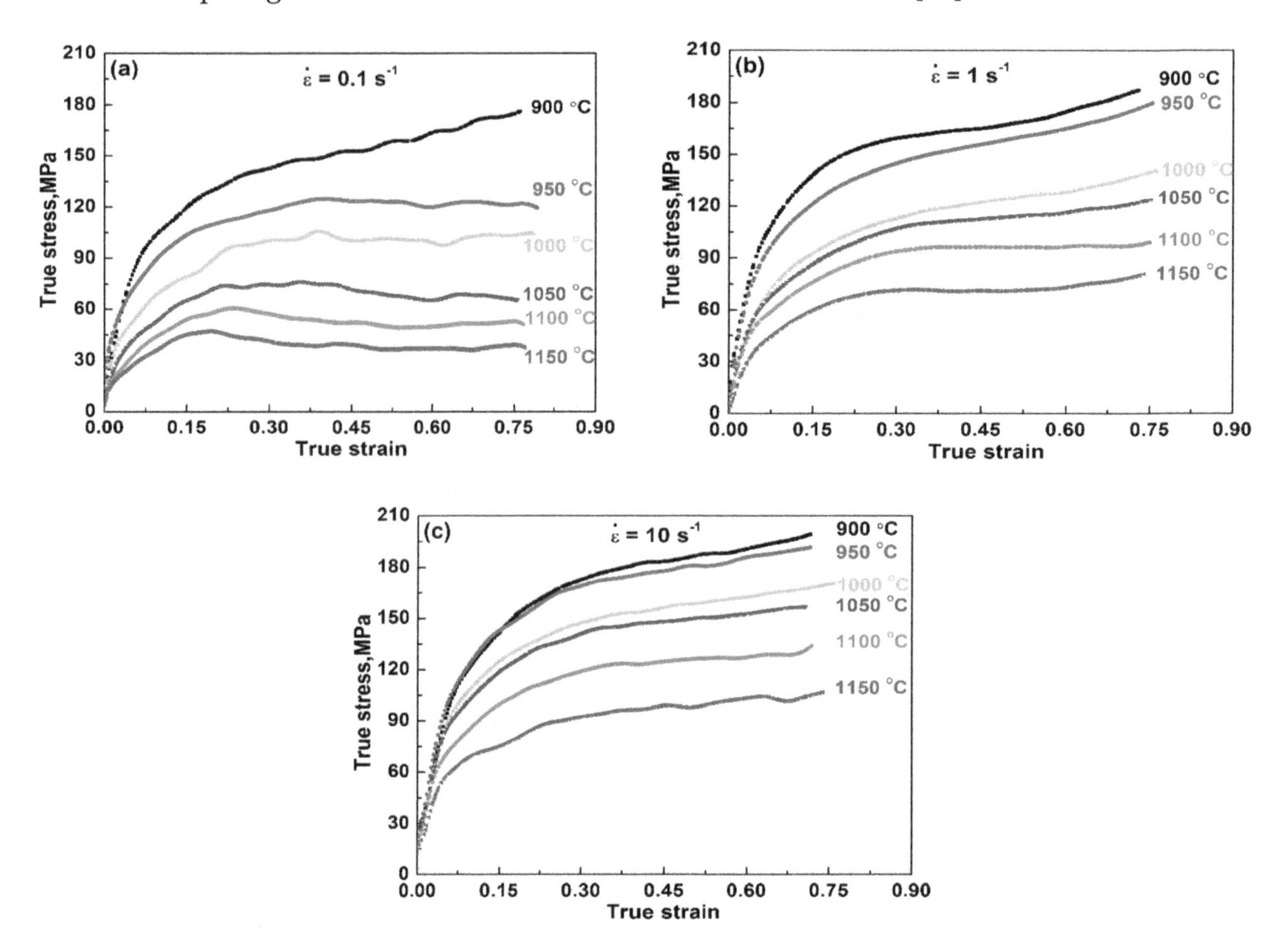

Figure 2. Typical flow curves of Ti-Mo steel obtained under various temperatures in the range of 900–1150 °C and strain rates of (**a**) 0.1 s^{-1}, (**b**) 1 s^{-1} and (**c**) 10 s^{-1}.

3.2. Peak Stress Analysis

In general, the peak stress was one of the most widely-accepted parameter used to find the hot working constants, especially for DRV and DRX types [14–16]. According to the analysis of flow curves, both DRV and DRX behaviors were found to happen in the temperature range from 1000–1150 °C for all strain rates. To elucidate the influence of deformation temperature and strain rate on the peak stress of experimental steel, a peak stress model, which was suitable for a wide range of strain rates and deformation temperatures, was expressed in the hyperbolic sine form [17].

$$\dot{\varepsilon} = A[\sinh(\alpha\sigma_p)]^n \exp(-Q/RT) \tag{1}$$

where σ_p is the peak stress, $\dot{\varepsilon}$ is the strain rate, n is the stress exponent, A is a material constant, α is the stress multiplier, Q is the activation energy for hot deformation, R is the universal gas constant and T is the absolute deformation temperature.

In terms of various stress levels, Equation (1) can be simplified in the power law and exponential law:

$$\dot{\varepsilon} = A_1 {\sigma_p}^{n_1} \exp(-Q/RT) \qquad (\alpha\sigma_p < 0.8) \tag{2}$$

$$\dot{\varepsilon} = A_2 exp(\beta\sigma_p) \exp(-Q/RT) \qquad (\alpha\sigma_p > 1.2) \tag{3}$$

where A_1, A_2, n_1, β are material constants. According to the linear regression of experimental data, the values of n_1 and β were determined using plots of $\ln\dot{\varepsilon}$-$\ln\sigma_p$ and $\ln\dot{\varepsilon}$-σ_p at each deformation temperature [18], as displayed in Figure 3. Thus, the stress multiplier is determined as $\alpha = \beta/n_1 = 0.06653/6.736 = 0.009876$.

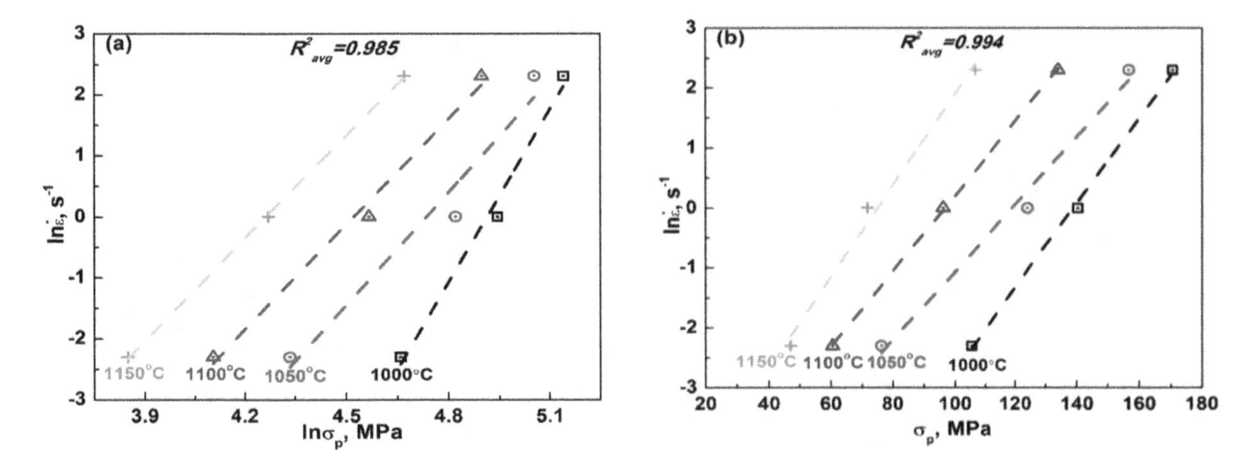

Figure 3. Plots of (**a**) $\ln\dot{\varepsilon}$ vs. $\ln\sigma_p$ and (**b**) $\ln\dot{\varepsilon}$ vs. σ_p at each deformation temperature of 1000–1150 °C for determining material constants, n_1 and β.

To derive the value of n in Equation (1), plots of $\ln\dot{\varepsilon}$-$\ln[\sinh(\alpha\sigma_p)]$ at each deformation temperature are shown in Figure 4a. By linear fitting, the average value of n was approximately 5.0. Thus, the value of Q was obtained by taking the partial derivative of Equation (1) in the form of the natural logarithm.

$$Q = Rn\left[\frac{\partial \ln(\sinh(\alpha\sigma_p))}{\partial(1/T)}\right]_{\dot{\varepsilon}} = RnS \tag{4}$$

where the value of S was determined as 10.31, based on the average slope of plots of $\ln[\sinh(\alpha\sigma_p)]$ − 1000/T at each strain rate (Figure 4b). The Q value of ~428.5 kJ/mol obtained here was much higher than the activation energy for self-diffusion in γ-iron (~280 kJ/mol) [19], indicating that the deformation mechanism at high temperature was not controlled by atomic diffusion.

A widespread parameter, the Zener–Hollomon (Z) parameter [20], was introduced to determine the material constant A, as expressed in Equation (5).

$$Z = \dot{\varepsilon}\exp\left(\frac{Q}{RT}\right) \tag{5}$$

By combining Equations (1) and (5), the following natural logarithm form was obtained:

$$\ln Z = \ln A + n\ln[\sinh(\alpha\sigma_p)] \tag{6}$$

According to the plot of $\ln Z$-$\ln[\sinh(\alpha\sigma_p)]$ (Figure 4c), the value of A was determined as 1.69×10^{16} s^{-1}. The detailed fit parameters associated with the peak stress model for both steels are summarized in Table 2. The correlation coefficients (R^2) of all fit parameters were beyond 0.984, implying a desirable fit to experimental data.

Table 2. The fit parameters associated with the peak stress model for Ti-Mo steel.

Sample	Q, kJ/mol	n	A, s^{-1}	α, MPa^{-1}
Ti-Mo	428.5	5.00	1.69×10^{16}	0.009876

Note: Q is the activation energy for hot deformation; n is the stress exponent; A is a material constant; α is the stress multiplier.

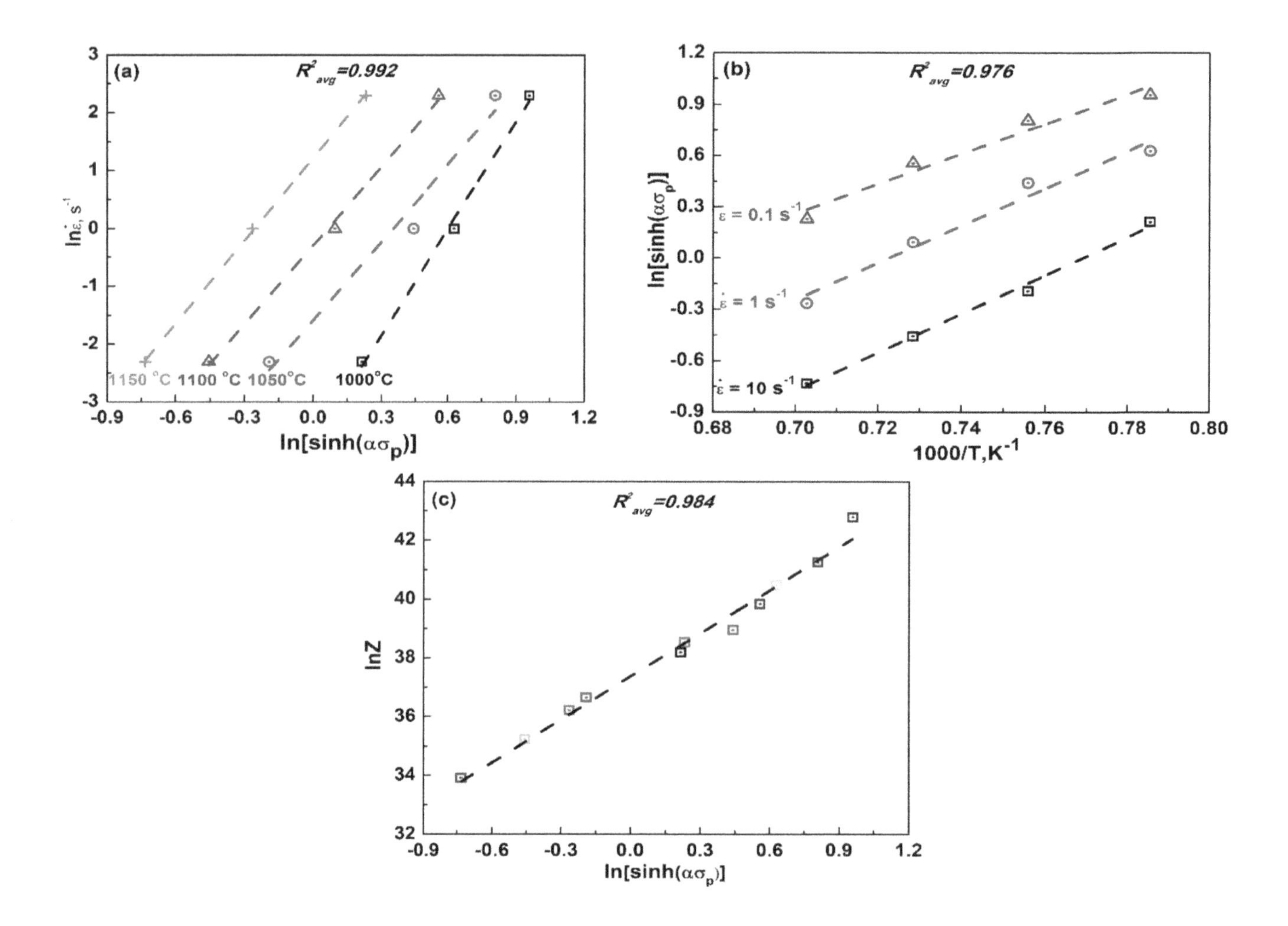

Figure 4. Plots of (**a**) ln[sinh($\alpha\sigma_p$)] vs. 1000/T, (**b**) $\ln\dot{\varepsilon}$ vs. ln[sinh($\alpha\sigma_p$)] and (**c**) lnZ vs. ln[sinh($\alpha\sigma_p$)] under different experimental conditions for the values of n, S and A.

By substituting the fit parameters into Equation (1), the σ_p as a function of deformation temperature and strain rate can be expressed as:

$$\dot{\varepsilon} = 1.69 \times 10^{16}(\sinh 0.009876\sigma_p)^5 \exp(-428500/RT) \tag{7}$$

When introducing the Z parameter into Equation (7), the σ_p can be also obtained in the following form.

$$\begin{cases} Z = \dot{\varepsilon}\exp(\frac{428500}{RT}) \\ \sigma_p = \frac{1}{0.009876}\ln\left\{(\frac{Z}{1.69\times10^{16}})^{1/5} + \left[(\frac{Z}{1.69\times10^{16}})^{2/5}+1\right]^{1/2}\right\} \end{cases} \tag{8}$$

The experimental peak stresses were compared with those predicted using the constitutive Equation (8) under different deformation conditions, see Figure 5. The validity of the σ_p model was further examined by an average absolute relative error (AARE):

$$\text{AARE} = \frac{1}{N}\sum_{i=1}^{N}\left|\frac{E_i - P_i}{E_i}\right| \times 100\% \tag{9}$$

where E_i is experimental data, P_i is the predicted value and N is the number of fit data used in this study. The AARE value was about 3.69%, reflecting the accurate estimation of peak stress. The satisfactory agreement between the predicted and experimental data suggests that the constitutive equations we proposed herein gave an accurate prediction of flow stress during hot deformation.

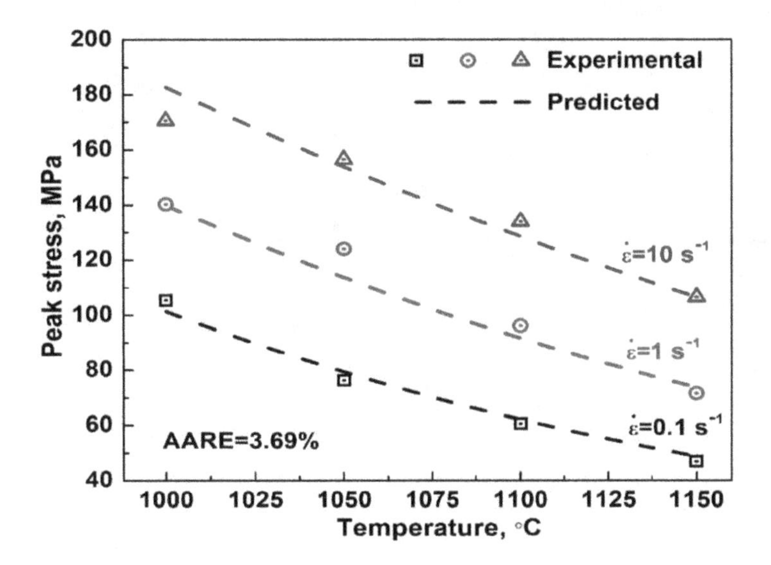

Figure 5. Comparison of the peak stress model and experiment data for Ti-Mo microalloyed steel at different temperatures of 1000–1150 °C and strain rates of 0.1, 1 and 10 s^{-1}. The average absolute relative error was about 3.69%.

3.3. *Physically-Based Constitutive Analysis*

In general, the dislocation density increased with strain during hot deformation, leading to work hardening (WH) or the DRV phenomenon. Regarding the flow curves of WH and DRV, the dislocation density (ρ)-based flow stress model proposed by Bergstrom [21] was used in this study as follows:

$$\begin{aligned} \sigma &= \sigma_0 + M\mu b\sqrt{\rho} \\ \frac{d\rho}{d\varepsilon} &= U - \Omega\rho \end{aligned} \tag{10}$$

where σ_0 is the lattice friction stress, M is a dislocation strengthening constant, μ is the shear modulus, b is the burgers vector, U is the hardening parameter and Ω is the softening parameter. The value of M ranged from 0.88–0.9 [22], and the b value was taken as 2.5×10^{-10} m [6]. The value of μ varied from 43.5–48.4 GPa at temperatures of 900–1150 °C by extrapolating from the experimental data from the literature [23]. Thus, the values of σ_0, U, Ω could be obtained based on the present experiment data.

After the onset of DRX, the hot flow behavior of Ti-Mo steel was described, based on the Kolmogorov–Johnson–Mehl–Avrami (KJMA) softening model as follows [24,25]:

$$\begin{aligned} \sigma &= \sigma_p - X(\sigma_p - \sigma_{Rex}) \\ X &= 1 - \exp(-kt^{n_A}) \end{aligned} \tag{11}$$

where σ_{Rex} is the steady state stress after DRX, X is the recrystallized volume fraction, k is a material constant and n_A is Avrami's exponent. Thus, the flow behavior of Ti-Mo steel involving DRX was modeled by the coupling of the Bergstrom and KJMA's models. The former was used to model the flow stress before reaching peak stress; whereas the later for that after the onset of DRX. All undetermined parameters in Equations (10) and (11) can be obtained based on the MATLAB programing.

To verify the accuracy of the developed models, the predicted and experimental flow curves under different deformation conditions were compared, as plotted in Figure 6. According to Equation (9), the accurate estimation of flow stress using the Bergstrom and KJMA models was about 1.68%, which reflected a satisfactory agreement between the predicted and experimental data. In contrast, the combined Bergstrom and KJMA models could precisely predict the whole hot deformation behaviors of Ti-Mo steel involving DRX (Figure 6a).

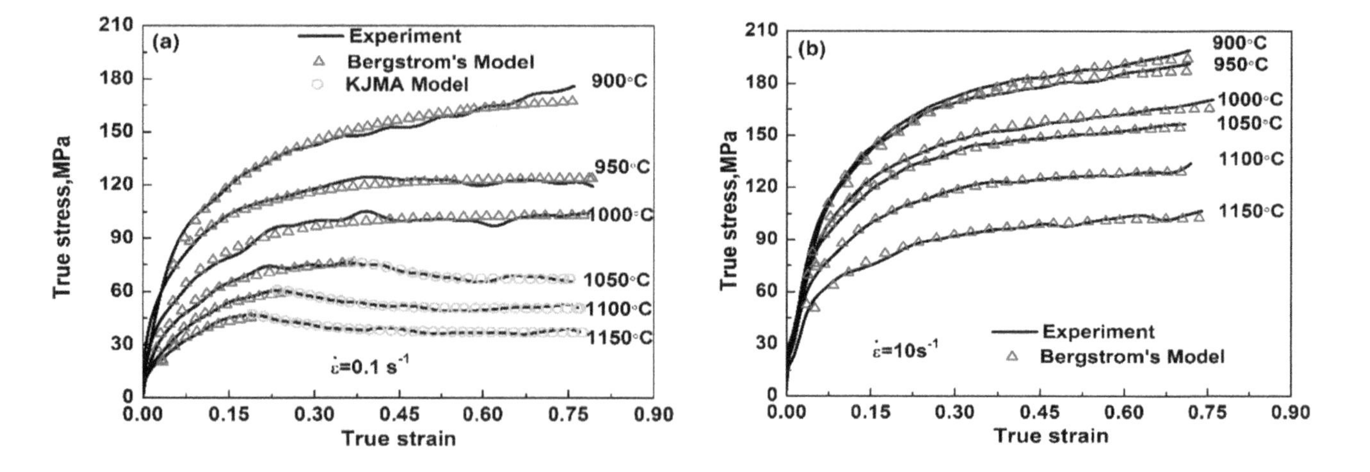

Figure 6. The typical predicted and experimental flow curves of Ti-Mo microalloyed steel at temperatures of 900–1150 °C and different strain rates: (**a**) $\dot{\varepsilon} = 0.1\ s^{-1}$; (**b**) $\dot{\varepsilon} = 10\ s^{-1}$.

Meanwhile, several important parameters such as U, Ω and n_A were utilized to compare the variations in hot deformation behavior of Ti-Mo steel as a function of the Z parameter. From Figure 7, the U value was found to exhibit an approximately positive linear relationship with ln$Z$ (Figure 7a). An increase in the Z value corresponded to the larger strain rate or the lower deformation temperature, implying more effective work hardening. On the contrary, the Ω value dropped with increasing Z value (Figure 7b). Furthermore, the peak dislocation density (ρ_p) exhibited a similar tendency with the U (Figure 7c), due to the proportional relationship between these two parameters.

Additionally, the values of the Avrami exponent, n_A, were found to vary between 1.2 and 2.1 with ln$Z$, as presented in Figure 7d, which was comparable with those values reported in previous studies [11,26]. The variation in n_A indicated different mechanisms of DRX. In the lower Z region, the n_A value was close to one, suggesting a larger probability of nucleation sites on the interfacial surface of grain and twin boundaries. In the larger Z region, however, the n_A value reached ~2, implying that nucleation of DRX happened on the grain and twin edges [27].

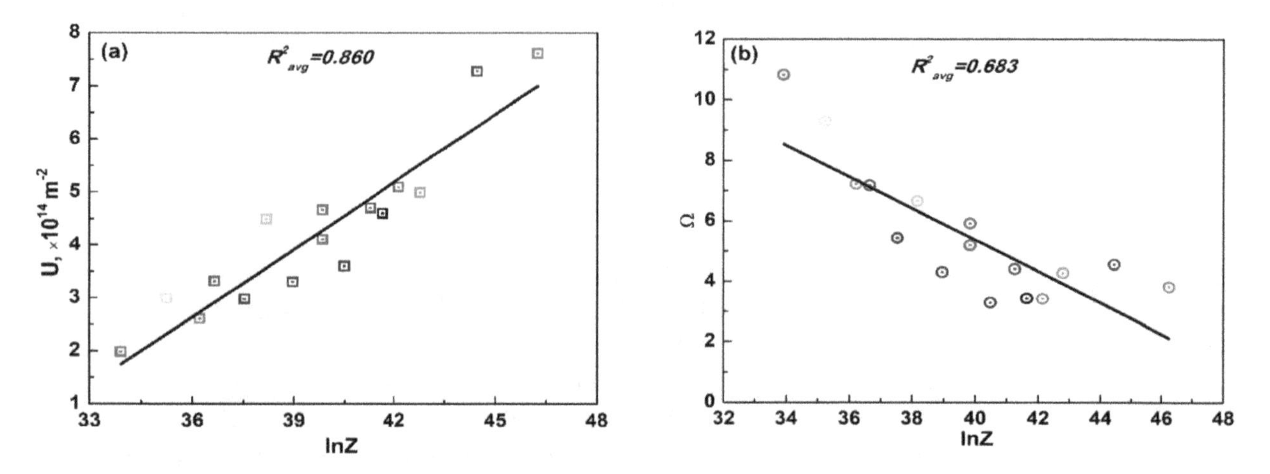

Figure 7. *Cont.*

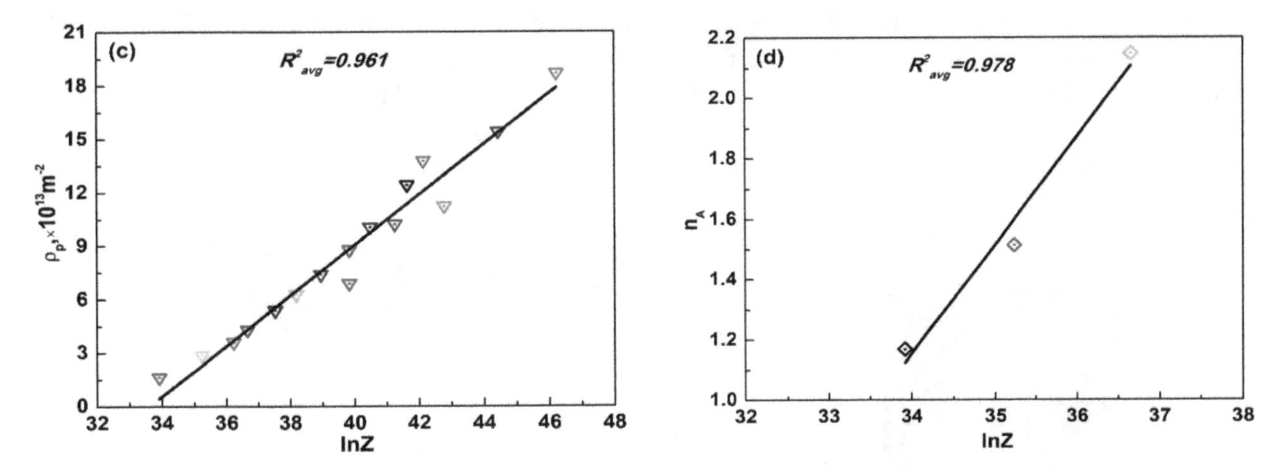

Figure 7. The relationship between typical parameters for physical models and the Z parameter. (**a**) Hardening parameter, U; (**b**) softening parameter, Ω; (**c**) Avrami's parameter, n_A; (**d**) peak dislocation density, ρ_p.

3.4. Experimental Verification and Mechanism Analysis

To analyze the influence of hot deformation on the microstructural features of Ti-Mo steel, both OM and EBSD were employed to study the deformed specimens. When the specimen was deformed at 900 °C to a strain of 0.8 at 0.1 s^{-1}, the severely pancaked prior austenitic grains were observed (Figure 8a), implying that WH was the dominant deformation mechanism. Furthermore, the fresh ferrite formed at the prior grain boundaries (inset of Figure 8a), which suggests that the strain-induced ferrite (SIF) occurred due to the lower deformation temperature, just above the Ar_3 temperature (~858.1 °C). With increasing deformation temperature to 1150 °C at 0.1 s^{-1}, the fully-equiaxed grains were observed (see the black lines in Figure 8b), indicating that the hot deformation of Ti-Mo steel was mainly controlled by DRX. This was in agreement with the continuous softening mechanism observed on the flow curves in Figure 2a. Accordingly, the formation of new equiaxed grains was mainly associated with the rearrangement or annihilation of dislocations [28,29]. Meanwhile, further straining was accompanied by an acceleration of the transformation of low to high angle grain boundaries resulting from the recovery-assisted absorption of dislocations in the existing sub-boundaries. The whole microstructural evolution upon straining at 1150 °C and 0.1 s^{-1} is schematically illustrated in Figure 8d as well.

However, at the higher stain rate of 10 s^{-1} and deformation temperature of 1150 °C, some deformation bands and substructures were still observed in the large pancaked prior austenite grains (Figure 8c). This suggests that DRV played an important role during hot compression in the high strain rate domain (e.g., 10 s^{-1}) at higher temperatures, which was also in accordance with the steady flow stress with strain in Figure 2c. As schematically displayed in Figure 8d, the new grain evolution involved the development of strain-induced HAGBs resulting from an increase in the misorientations of individual LAGBs and progressive subgrain rotation [30,31] upon further straining.

The change in boundary misorientation angle as a function of deformation temperature and strain rate is displayed in Figure 9. In EBSD analysis, the number of high angle austenitic/martensitic boundaries was distinguished, in view of the variation between their crystal orientations. When deformation was performed at 900 °C and 0.1 s^{-1}, a large number of low-angle subgrain boundaries of 2–5° (>40%) was observed inside the deformed grains (Figure 9a). As the deformation temperature increased to 1150 °C at the same strain rate of 0.1 s^{-1}, the fraction of LAGBs of 2–5° dramatically dropped by ~33.6% with a number of high angle austenitic grain boundaries (~46.2%), implying the occurrence of a pronounced DRX [32]. Consequently, the DRX was accelerated, and the flow stress was greatly reduced due to the decreased density dislocation, which corresponded to Figures 2a and 8b. Having deformed at higher strain rates (e.g., 10 s^{-1}) and 1150 °C, the fraction of high angle austenitic grain boundaries was reduced to ~29.7% with a higher fraction of LAGBs of 2–5° (~37.5%), implying that the DRX process was impeded and that the flow stress increased, owing to the accumulating strain and hardening substructure, which correspond to the flow curves in Figures 2c and 8c.

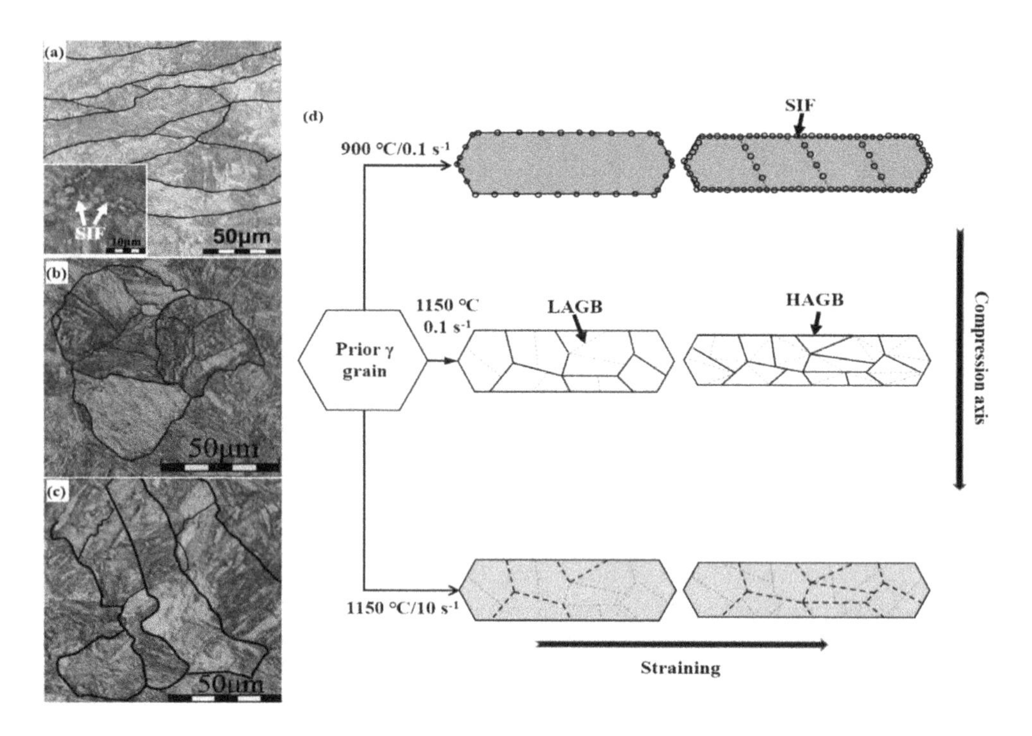

Figure 8. Microstructural evolution of a Ti-Mo microalloyed steel during hot compression at different temperatures and strain rates: (**a**) 900 °C/0.1 s^{-1}, (**b**) 1150 °C/0.1 s^{-1} and (**c**) 1150 °C/10 s^{-1}; as well as the corresponding schematic diagrams (**d**). LAGBs and HAGBs indicate the low/high angle grain boundaries. The background colors represent the low/high density of dislocations, respectively. SIF, strain-induced ferrite.

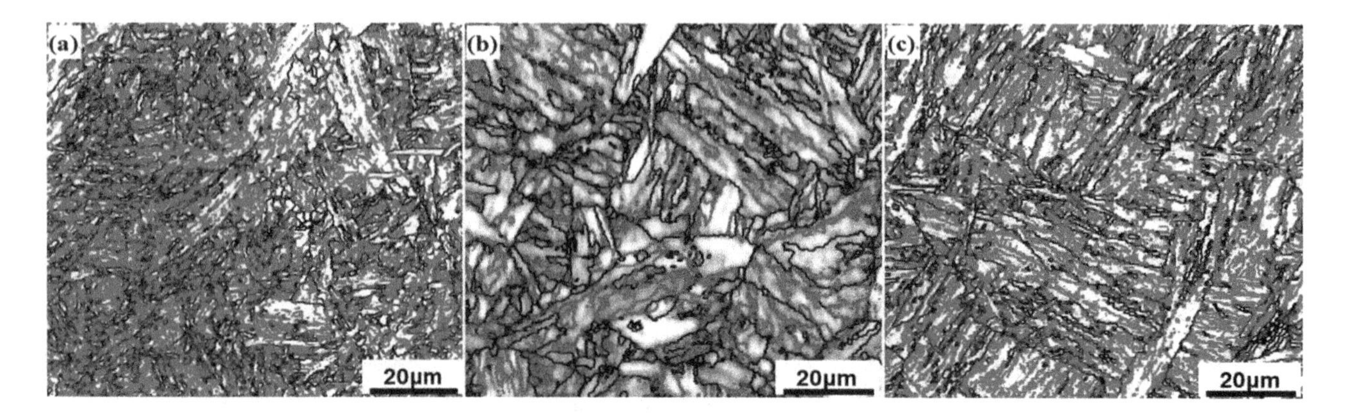

Figure 9. Electron backscattered diffraction (EBSD) maps of Ti-Mo microalloyed steel under different deformation conditions: (**a**) 900 °C/0.1 s^{-1}, (**b**) 1150 °C/0.1 s^{-1} and (**c**) 1150 °C/10 s^{-1}. The red and black solid lines represent the low-angle grain boundaries (LAGBs) of 2–15° and high-angle grain boundaries (HAGBs) of above 15°. The plane of EBSD observation was parallel to the compression axis.

The present study demonstrates that a combination of Bergstrom and KJMA models, which we proposed herein, could be used to predict the whole flow behaviors of the interphase-precipitated microalloyed steels under various hot deformation conditions, involving DRX. By comparing the characteristic stresses and material parameters over all deformation conditions, especially at the higher strain rates studied, the dependence of hot flow behavior on strain rate and deformation temperature was further clarified. Therefore, the constitutive model may provide a powerful tool for optimizing the hot working processes of high performance Ti-Mo steels with interphase precipitation.

4. Conclusions

The hot flow behaviors of an interphase-precipitated Ti-Mo steel were investigated by conducting a series of hot compression tests at higher strain rates of 0.1~10 s^{-1} and temperatures of 900~1150 °C on a Gleeble-2000 thermo-mechanical simulator. The main conclusions involved:

(1) The peak stress as a function of deformation temperature and strain rate was determined as:

$$\dot{\varepsilon} = 1.69 \times 10^{16}(\sinh 0.009876\sigma_p)^5 \exp(-428500/RT) \quad (12)$$

(2) Under conditions where DRV or WH was dominant, the hot flow behavior could be modeled using the Bergstrom equation; once DRX was initiated, the coupling of the Bergstrom and KJMA's models was capable of predicting the flow behaviors in the whole hot deformation.

(3) The values of U and ρ_p exhibited an approximately positive linear relationship with the lnZ parameter, while the Ω dropped with increasing Z value. Meanwhile, the n_A value varied between 1.2 and 2.1 with lnZ, indicating the variation in the mechanisms of DRX with deformation conditions.

(4) After the microstructure deformed at 900 °C and 0.1 s^{-1}, the severely pancaked prior austenitic grains remained; with increasing deformation temperature above 950 °C and 0.1 s^{-1}, the fully-equiaxed DRX grains formed through rearrangement or annihilation of dislocations. However, as the strain rate increased to 10 s^{-1} for different deformation temperatures, WH and DRV were dominant, resulting in some deformation bands and substructures in the large pancaked prior austenite grains.

(5) With increasing the temperature from 900–1150 °C at 0.1 s^{-1}, the fraction of LAGBs of 2–5° dropped by ~33.6%, with greatly increased high angle austenitic grain boundaries (~46.2%) due to a pronounced DRX phenomenon. With increasing strain rate to 10 s^{-1} at 1150 °C, the fraction of LAGBs of 2–5° increased to ~37.5%, which was associated with inhibition of DRX.

Acknowledgments: The present work was financially supported by the 3rd Key Scientific Research Project for Undergraduates at Northeastern University, China (Grant No. ZD1709), the Fundamental Research Funding of the Central Universities, China (Grant No. N160204001), and National Natural Science Foundation, China (Grant Nos. 51400150 and 51671149).

Author Contributions: Chuanfeng Wu and Peiru Yang performed the physically-based modeling and microstructural characterization. Junhua Su and Xiaopeng Guo designed the experimental procedures and carried out the hot compression tests. Minghui Cai was responsible for the discussion, plotting figures and paper writing.

References

1. Mukherjee, S.; Timokhina, I.B.; Zhu, C.; Ringer, S.P.; Hodgson, P.D. Three-dimensional atom probe microscopy study of interphase precipitation and nanoclusters in thermomechanically treated titanium–molybdenum steels. *Acta Mater.* **2013**, *61*, 2521–2530. [CrossRef]
2. Cai, M.H.; Ding, H.; Lee, Y.K.; Tang, Z.Y.; Zhang, J.S. Effects of Si on Microstructural Evolution and Mechanical Properties of Hot-rolled Ferrite and Bainite Dual-phase Steels. *ISIJ Int.* **2011**, *51*, 476–481. [CrossRef]
3. Lee, J.; Lee, S.J.; De Cooman, B.C. Effect of micro-alloying elements on the stretch-flangeability of dual phase steel. *Mater. Sci. Eng. A* **2012**, *536*, 231–238. [CrossRef]
4. Kamikawa, N.; Abe, Y.; Miyamoto, G.; Funakawa, Y.; Furuhara, T. Tensile behavior of Ti,Mo-added low carbon steels with interphase precipitation. *ISIJ Int.* **2014**, *54*, 212–221. [CrossRef]

5. Rahnama, A.; Clark, S.; Janik, V.; Sridhar, S. A phase-field model investigating the role of elastic strain energy during the growth of closely spaced neighbouring interphase precipitates. *Comput. Mater. Sci.* **2018**, *142*, 437–443. [CrossRef]
6. Funakawa, Y.; Shiozaki, T.; Tomita, K.; Yamamoto, T.; Maeda, E. Development of high strength hot-rolled sheet steel consisting of ferrite and nanometer-sized carbides. *ISIJ Int.* **2004**, *44*, 1945–1951. [CrossRef]
7. Yen, H.W.; Huang, C.Y.; Yang, J.R. Characterization of interphase-precipitated nanometer-sized carbides in a Ti–Mo-bearing steel. *Scr. Mater.* **2009**, *61*, 616–619. [CrossRef]
8. Wang, J.; Hodgson, P.D.; Bikmukhametov, I.; Miller, M.K.; Timokhina, I. Effects of hot-deformation on grain boundary precipitation and segregation in Ti-Mo microalloyed steels. *Mater. Des.* **2018**, *141*, 48–56. [CrossRef]
9. Kim, Y.W.; Kim, J.H.; Hong, S.G.; Lee, C.S. Effects of rolling temperature on the microstructure and mechanical properties of Ti–Mo microalloyed hot-rolled high strength steel. *Mater. Sci. Eng. A* **2014**, *605*, 244–252. [CrossRef]
10. Yong, W.K.; Hong, S.G.; Huh, Y.H.; Chong, S.L. Role of rolling temperature in the precipitation hardening characteristics of Ti–Mo microalloyed hot-rolled high strength steel. *Mater. Sci. Eng. A* **2014**, *615*, 255–261. [CrossRef]
11. Hamada, A.; Khosravifard, A.; Porter, D.; Pentti Karjalainen, L. Physically based modeling and characterization of hot deformation behavior of twinning-induced plasticity steels bearing vanadium and niobium. *Mater. Sci. Eng. A* **2017**, *703*, 85–96. [CrossRef]
12. Mirzadeh, H.; Cabrera, J.M.; Najafizadeh, A. Constitutive relationships for hot deformation of austenite. *Acta Mater.* **2011**, *59*, 6441–6448. [CrossRef]
13. Feng, H.; Jiang, Z.H.; Li, H.B.; Jiao, W.C.; Li, X.X.; Zhu, H.C.; Zhang, S.C.; Zhang, B.B.; Cai, M.H. Hot deformation behavior and microstructural evolution of high nitrogen martensitic stainless steel 30Cr15Mo1N. *Steel Res. Int.* **2017**, *88*, 1700149. [CrossRef]
14. Cram, D.G.; Fang, X.Y.; Zurob, H.S.; Bréchet, Y.J.M.; Hutchinson, C.R. The effect of solute on discontinuous dynamic recrystallization. *Acta Mater.* **2012**, *60*, 6390–6404. [CrossRef]
15. Marandi, A.; Zarei-Hanzaki, R.; Zarei-Hanzaki, A.; Abedi, H.R. Dynamic recrystallization behavior of new transformation–twinning induced plasticity steel. *Mater. Sci. Eng. A* **2014**, *607*, 397–408. [CrossRef]
16. Zangeneh Najafi, S.; Momeni, A.; Jafarian, H.R.; Ghadar, S. Recrystallization, precipitation and flow behavior of D3 tool steel under hot working condition. *Mater. Charact.* **2017**, *132*, 437–447. [CrossRef]
17. Sellars, C.M.; McTegart, W.J. On the mechanism of hot deformation. *Acta Metall.* **1966**, *14*, 1136–1138. [CrossRef]
18. McQueen, H.J.; Ryan, N.D. Constitutive analysis in hot working. *Mater. Sci. Eng. A* **2002**, *322*, 43–63. [CrossRef]
19. McQueen, H.J.; Yue, S.; Ryan, N.D.; Fry, E. Hot working characteristics of steels in austenitic state. *J. Mater. Process. Technol.* **1995**, *53*, 293–310. [CrossRef]
20. Zener, C.; Hollomon, J.H. Effect of strain rate upon plastic flow of steel. *J. Appl. Phys.* **1944**, *15*, 22–32. [CrossRef]
21. Bergstrom, Y. A Dislocation Model for the Stress-strain behaviour of polycrystalline a-fe with special emphasis on the variation of the densities of mobile and immobile dislocations. *Mater. Sci. Eng.* **1969**, *5*, 193–200. [CrossRef]
22. Bergitrom, Y. The Plastic Deformation of Metals—A dislocation model and its applicability. *Rev. Powder Metall. Phys. Ceram.* **1983**, *2*, 79–265.
23. Hamada, A.S.; Khosravifard, A.; Kisko, A.P.; Ahmed, E.; Porter, D.A. High temperature deformation behavior of a stainless steel fiber-reinforced copper matrix composite. *Mater. Sci. Eng. A* **2016**, *669*, 469–479. [CrossRef]
24. Jonas, J.J.; Quelennec, X.; Jiang, L.; Martin, É. The Avrami kinetics of dynamic recrystallization. *Acta Mater.* **2009**, *57*, 2748–2756. [CrossRef]
25. Zhang, C.; Zhang, L.; Xu, Q.; Xia, Y.; Shen, W. The kinetics and cellular automaton modeling of dynamic recrystallization behavior of a medium carbon Cr-Ni-Mo alloyed steel in hot working process. *Mater. Sci. Eng. A* **2016**, *678*, 33–43. [CrossRef]
26. Haghdadi, N.; Martin, D.; Hodgson, P. Physically-based constitutive modelling of hot deformation behavior in a LDX 2101 duplex stainless steel. *Mater. Des.* **2016**, *106*, 420–427. [CrossRef]

27. El Wahabi, M.; Cabrera, J.M.; Prado, J.M. Hot working of two AISI 304 steels: A comparative study. *Mater. Sci. Eng. A* **2003**, *343*, 116–125. [CrossRef]
28. Dong, J.; Li, C.; Liu, C.; Huang, Y.; Yu, L.; Li, H.; Liu, Y. Hot deformation behavior and microstructural evolution of Nb–V–Ti microalloyed ultra-high strength steel. *J. Mater. Res.* **2017**, *32*, 3777–3787. [CrossRef]
29. Qin, F.; Zhu, H.; Wang, Z.; Zhao, X.; He, W.; Chen, H. Dislocation and twinning mechanisms for dynamic recrystallization of as-cast Mn18Cr18N steel. *Mater. Sci. Eng. A* **2017**, *684*, 634–644. [CrossRef]
30. Sun, H.; Sun, Y.; Zhang, R.; Wang, M.; Tang, R.; Zhou, Z. Hot deformation behavior and microstructural evolution of a modified 310 austenitic steel. *Mater. Des.* **2014**, *64*, 374–380. [CrossRef]
31. Wang, S.; Zhang, M.; Wu, H.; Yang, B. Study on the dynamic recrystallization model and mechanism of nuclear grade 316LN austenitic stainless steel. *Mater. Charact.* **2016**, *118*, 92–101. [CrossRef]
32. Parthiban, R.; Ghosh Chowdhury, S.; Harikumar, K.C.; Sankaran, S. Evolution of microstructure and its influence on tensile properties in thermo-mechanically controlled processed (TMCP) quench and partition (Q&P) steel. *Mater. Sci. Eng. A* **2017**, *705*, 376–384. [CrossRef]

11

Effect of Cold-Deformation on Austenite Grain Growth Behavior in Solution-Treated Low Alloy Steel

Xianguang Zhang [1,*], Kiyotaka Matsuura [2] and Munekazu Ohno [2]

[1] School of Metallurgical and Ecological Engineering, University of Science and Technology Beijing (USTB), Beijing 100083, China

[2] Division of Materials Science and Engineering, Faculty of Engineering, Hokkaido University, Kita 13 Nishi 8, Kita-ku, Sapporo 060-8628, Hokkaido, Japan; matsuura@eng.hokudai.ac.jp (K.M.); mohno@eng.hokudai.ac.jp (M.O.)

* Correspondence: xgzhang@ustb.edu.cn

Abstract: The occurrence of abnormal grain growth (AGG) of austenite during annealing is a serious problem in steels with carbide and/or nitride particles, which should be avoided from a viewpoint of mechanical properties. The effects of cold deformation prior to annealing on the occurrence of AGG have been investigated. It was found that the temperature range of the occurrence of AGG is shifted toward a low temperature region by cold deformation, and that the shift increases with the increase of the reduction ratio. The lowered AGG occurrence temperature is attributed to the fine and near-equilibrium AlN particles that are precipitated in the cold-deformed steel, which is readily dissolved during annealing. In contrast, coarse and non-equilibrium AlN particles precipitated in the undeformed steel, which is resistant to dissolution during annealing.

Keywords: austenite; abnormal grain growth; cold-deformation; precipitate

1. Introduction

Microalloying has been used as an important austenite grain refinement technology in steels. Austenite grain growth can be inhibited by microalloying, which causes the precipitation of second-phase particles in austenite [1–5]. Abnormal grain growth (AGG) frequently occurs in micro-alloyed steels during annealing. The austenite grain growth of the micro-alloyed steels usually experiences "normal → abnormal → normal" growth modes in sequence, with an increase of the annealing temperature [5]. AGG occurs within a certain temperature range during annealing, which is referred to as an AGG temperature range. The AGG temperature range of steels is an important factor in designing the heat treatment to achieve fine-grained products.

The austenite grain growth behavior in low-alloy steels is largely dependent on the stability of second-phase particles. At low temperatures, the second-phase particles are stable, and the austenite grain boundaries can be effectively pinned by the particles. The grain growth is thus greatly retarded. However, at relatively high temperatures, the second-phase particles become unstable, and they readily dissolve or coarsen [5]. Then, some of the austenite grains may become unpinned, and they may grow much faster than the matrix grains. This leads to the occurrence of AGG [6]. There is a transition temperature at which the austenite grain growth switches from normal grain growth (NGG) to AGG mode. This transition temperature is referred to as a grain coarsening temperature, T_c [7]. In addition, at temperatures that are much higher than the T_c, NGG resumes, due to the almost complete dissolution of the pinning particles, and the grain growth becomes free [5]. Hence, there is another transition temperature, the AGG finishing temperature (T_f), at which the grain growth switches from AGG to NGG mode. T_c and T_f are important parameters for understanding the grain growth behavior of austenite.

Cold-deformation, such as cold-rolling, hammering, or cold-forging, has been widely used in the steel manufacturing process. It was found that hot-deformation could accelerate the nucleation and coarsening of the second-phase particles due to the large amount of dislocation that is introduced into the steels by deformation [8–10]. In addition, previous studies [11–13] have revealed that cold-deformation also has a strong influence on the precipitation and growth of the precipitates during subsequent annealing. It was found that the second-phase particles were finer and distributed more uniformly by cold-deformation [11]. Meanwhile, cold-deformation makes the precipitates resistant to coarsening [11,12].

The grain growth behavior of austenite relies on the presence of second-phase particles. Therefore, it is expected that the cold-deformation should have a strong influence on the AGG temperature range of the low-alloy steels. However, previous works have mainly focused on the effect of cold-deformation on the precipitation or coarsening behaviors of second-phase particles. Rare attention has been paid to its influence on austenite grain growth behavior. Recently, it was reported by the present authors [13,14] that an inhomogeneous distribution of AlN particles in ferrite/pearlite banded steel became homogeneous, due to cold-deformation prior to austenization, which increased the low T_c of the banded steel significantly. However, to the best of the authors' knowledge, no systematic work has been reported regarding the effects of cold deformation on the AGG temperature range of austenite. The present study has been undertaken to clarify the effects of cold-deformation on austenite grain growth behavior in solution-treated low-alloy steel. It was found that the AGG temperature range of austenite is shifted toward the lower temperature region by applying prior cold-deformation.

2. Materials and Methods

The composition of the low-alloy steel used in the present study is displayed in Table 1. The solubility of AlN in austenite can be written as [15]:

$$\text{Log}[\text{Al}][\text{N}] = 1.03 - 6770/T \quad (1)$$

Where [Al] and [N] represent the solubilities of aluminum and nitrogen at the absolute temperature T. According to Equation (1), the complete dissolution temperature of AlN in the current steel was calculated to be 1453 K. To completely dissolve the AlN, and to eliminate the banded segregation of Mn, the solution treatment temperature and time was designed to be 1573 K for 2 h. After the solution treatment, the specimen was air-cooled to room temperature.

Table 1. Composition of the steel used in this study (wt %).

C	Mn	Si	P	Al	N	Fe
0.2	0.8	0.2	0.015	0.04	0.006	Bal.

The cold-deformation was carried out by cold-rolling, with reduction ratios ranging from 10 to 70%. Both the solution-treated (ST), and solution-treated and deformed (STD) specimens were annealed at temperatures between 1173–1473 K for various periods, as schematically shown in Figure 1. The heating rate of the specimens was about 5 K/s. After annealing, the specimens were quenched in ice water to freeze the microstructure. The specimens were cut and mechanically polished to obtain a mirror-like surface. To reveal the prior austenite grain boundaries, the specimens were etched in a supersaturated picric acid solution at 333 K. The microstructure was characterized by using an optical microscope (OM). The well-known linear intercept method was used to evaluate the mean austenite grain size. About 450 to 550 intercepts were counted for each sample.

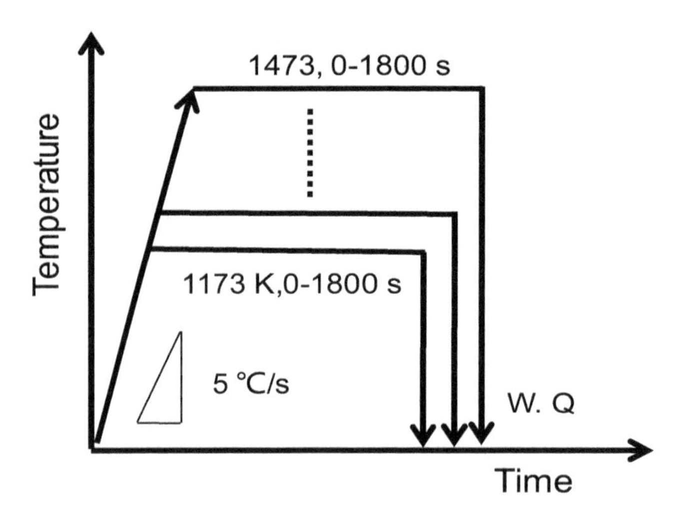

Figure 1. Schematic illustrations of the annealing treatment.

The AlN particles were characterized by using transmission electron microscopy (TEM, JEOL JEM-2010, JEOL Ltd., Tokyo, Japan) equipped with energy dispersive spectroscopy (EDS) operating at 200 kV on carbon extraction replicas eroded off the specimens. At least a thousand AlN precipitates were measured for each case to evaluate the volume fraction and particle sizes.

The volume fraction of the AlN precipitates from the carbon extraction replica can be calculated based on the McCall-Boyed method [16] as follows:

$$f = \left(\frac{1.4\pi}{6}\right)\left(\frac{ND^3}{V}\right), \tag{2}$$

where N is the number of the precipitates, D is the mean diameter of the particles, and V is the volume of the matrix from which the precipitates were extracted. The thickness of the matrix can be assumed as the mean diameter of the particles [17]. Thus, Equation (2) can be written as:

$$f = \left(\frac{1.4\pi}{6}\right)\left(\frac{ND^2}{S}\right), \tag{3}$$

where S is the area of the corresponding matrix.

3. Results and Discussion

3.1. Austenite Grain Growth Behavior

Figure 2 shows the prior austenite grain structures (now martensite after quenching) of the solution-treated (ST) specimens after annealing at different temperatures for various periods of time. Microstructure observation revealed that the austenite grain growth during annealing at 1298 K is very slow and the prior austenite grain size almost has no change. Hence, only the microstructure after annealing at 1298 K for 1800 s is shown Figure 2a. It is clear that fine and uniform prior austenite grain structures were formed, which is attributed to the occurrence of NGG. However, mixed prior austenite grain structures were formed after the specimen was annealed at 1323 K for 60 s, as shown in Figure 2b, where abnormal coarse grain (as indicated by the red arrow) mixed with fine matrix grain structures (as indicated by the black arrow) were observed. After a prolonged holding time of up to 1800 s, extremely coarse grain structures were developed (Figure 2c). Hence, typical AGG occurred at this temperature. After the sample was annealed at 1423 K for 1800 s, slightly coarse but uniform grain structures were formed (Figure 2d). This indicates that the grain growth switched to NGG mode again.

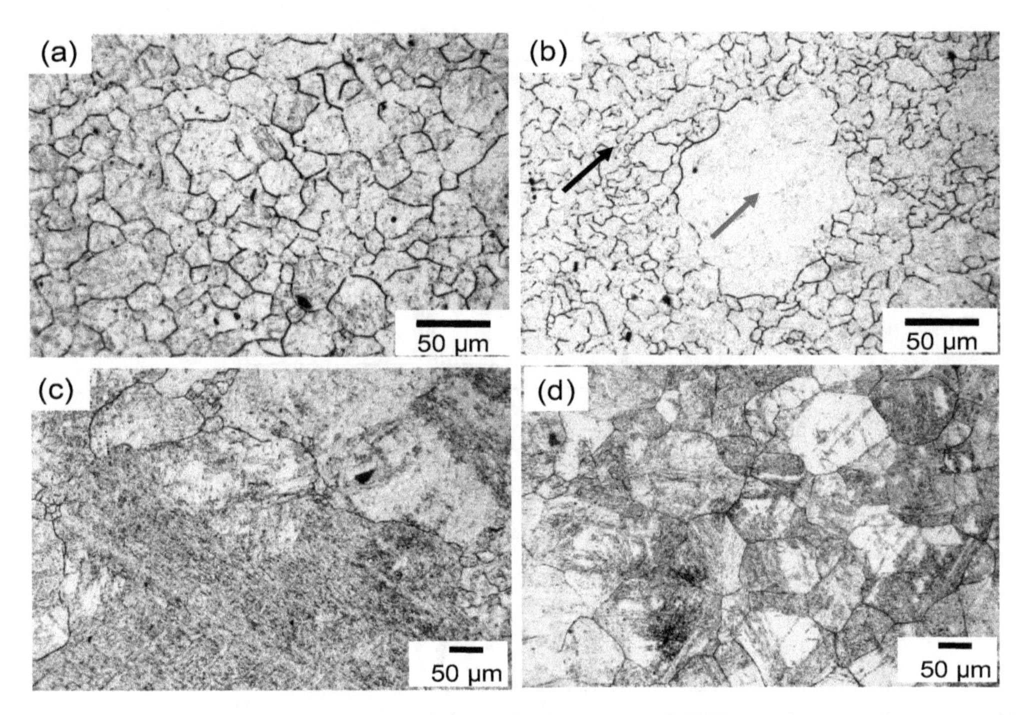

Figure 2. Prior austenite grain structures of the solution-treated (ST) specimens after annealing at (**a**) 1298 K for 1800 s, (**b**) 1323 K for 60 s, (**c**) 1323 K for 1800 s, and (**d**) 1423 K for 1800 s. The red and black arrows in (**b**) to indicate the abnormal coarse and fine matrix grains, respectively.

According to the above results, the occurrence AGG led to extremely coarse grain structures, while the NGG led to relatively fine grain structures after annealing. Therefore, the mean austenite grain size of the ST specimens after annealed at various temperatures for 1800 s were plotted in Figure 3a to show the austenite grain growth behavior during annealing. The average austenite grain sizes after annealing at 1298 K (Figure 2a), 1323 K (Figure 2c), and 1423 K (Figure 2d) are indicated by a black solid, red solid, and open arrows, correspondingly. From Figure 3a, it can be seen that the prior austenite grain size is fine, and that it has almost no change below 1298 K. This indicates that the austenite grain growth was greatly retarded, and it grew under the NGG mode below 1298 K. The coarse grains that were formed by AGG appeared at intermediate temperatures. Slightly coarse grains that were formed by NGG appeared at high temperatures. Therefore, the ST specimens experienced "normal → abnormal → normal" grain growth modes that were usually observed in the micro-alloyed steels [5]. The grain growth modes and the transition temperatures of T_c and T_f are indicated in Figure 3a. It is necessary to point out that the grain growth behaviors were examined carefully by changing the annealing temperatures within 25 K intervals, and that only partial results are shown in Figure 3a.

To understand the effects of cold-deformation on the austenite grain growth behavior, the average austenite grain sizes at various temperatures in the solution-treated and cold-deformed by 50% (STD-50%) specimens were examined. The measured austenite grain size as a function of annealing temperature is summarized in Figure 3b. A similar grain structure changing tendency was observed as in that of the ST specimens; viz., a fine grain structure at low temperatures, a coarse grain structure at intermediate temperatures, and slightly coarse and uniform grain structures at high temperatures. Therefore, the STD-50% specimen experienced "normal → abnormal → normal" grain growth modes as well. The prior austenite grain structure evolution process was similar to that of the ST specimens, but at different temperatures; therefore, the microstructures are not shown here again. Importantly, comparing Figure 3a,b shows that the AGG temperature range was shifted toward a low temperature by cold-deformation.

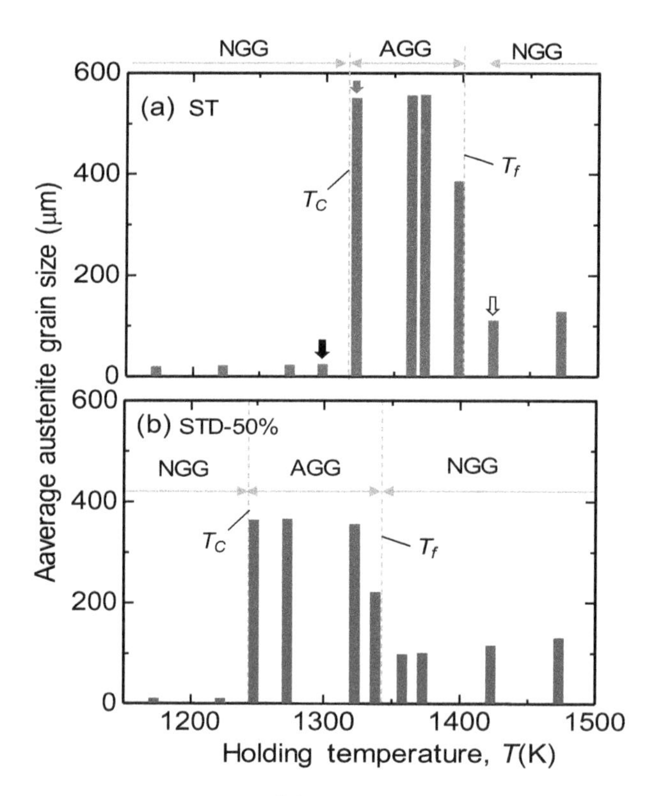

Figure 3. Average prior austenite grain sizes of the (**a**) ST and (**b**) solution-treated and cold-deformed by 50% (STD-50%) specimens after annealing at various temperatures for 1800 s.

To gain an overall image of the effect of cold-deformation on austenite grain growth behavior, the austenite grain growth modes against annealing temperatures in the specimens deformed with various reduction ratios are summarized in Figure 4. The "normal → abnormal → normal" grain growth modes occurred in all the specimens. Importantly, the AGG temperature range was gradually depressed toward low temperatures with the increase in the reduction ratio, which was almost saturated up to 70%. The decreased T_c and T_f temperatures, induced by the cold-deformation, can be explained by its influences on the evolution of the AlN precipitates, which is discussed in the following section.

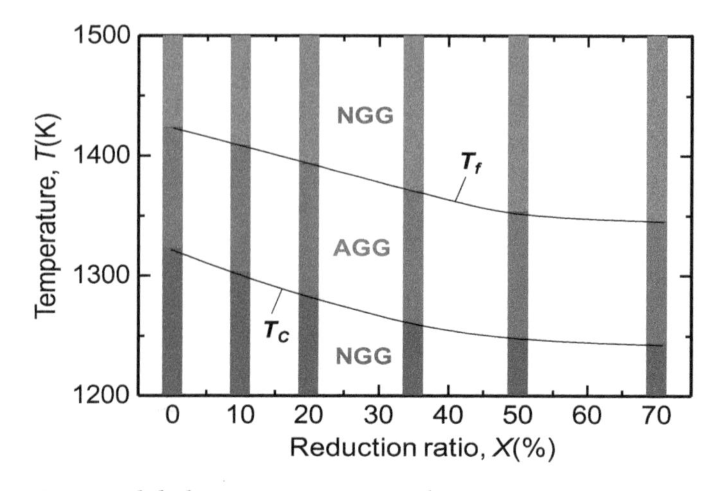

Figure 4. Austenite grain growth behavior against annealing temperatures in the specimens deformed at various reduction ratios.

3.2. The Evolution of AlN Precipitates

As reviewed in the Introduction, the stability of the second-phase particles during annealing plays a key role in the austenite grain growth behavior. To understand the reasons for the suppression of the

AGG temperature range by the cold-deformation, the evolution of the AlN particles were carefully studied by TEM observations. The volume fractions and sizes of the AlN particles in both the ST and STD-50% specimens quenched from various temperatures from 1073 to 1353 K are summarized in Figure 5a,b, respectively. For comparison, the equilibrium volume fraction of the AlN particles calculated from Equation (1) is also plotted in Figure 5a, as shown by the blue broken line.

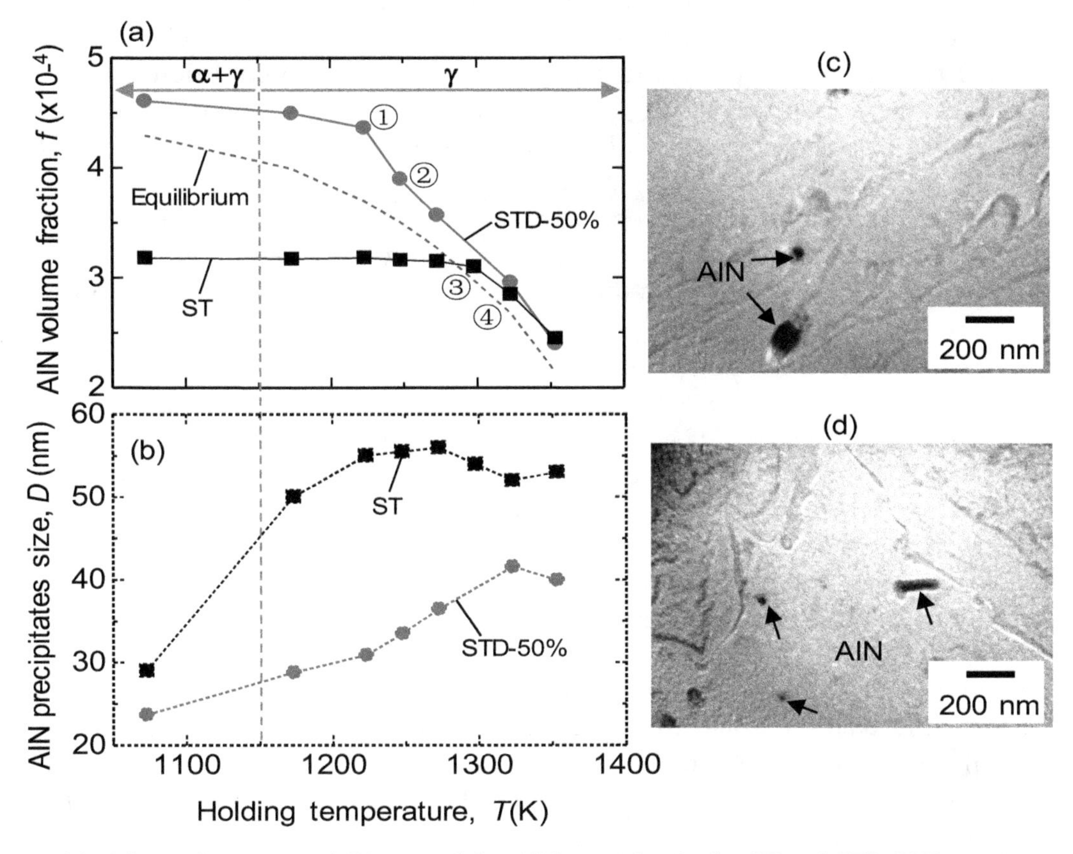

Figure 5. (**a**) Volume fraction and (**b**) size of the AlN particles in the ST and STD-50% specimens after annealing at various temperatures for 1800 s. TEM images of the AlN particles (indicated by the black arrows) in (**c**) ST and (**d**) STD-50% specimens after annealing at 1173 K for 1800 s.

It should be pointed out that after the solution treatment, the AlN precipitates should have been fully dissolved into the austenite. It was reported that the precipitation of AlN is difficult during the cooling process, and that it is sensitive to the cooling rate [18–20]. The precipitation of AlN can be entirely suppressed at a cooling rate larger than 1 K/s [21], which is approximately equivalent to air cooling a 50 mm diameter steel bar [22]. Therefore, the AlN precipitation during the air-cooling process is negligible in this study. The AlN precipitates should mainly be re-precipitated during the reheating process in both the ST and STD specimens.

According to the precipitation-time-temperature (PTT) diagram of AlN [23], the AlN precipitation should have been completed below 1073 K in the current steel during the annealing process. Therefore, the AlN particles within the specimens quenched from 1073 K (ferrite + austenite two phase region) were examined. According to Figure 5a,b, fine- and near-equilibrium AlN particles precipitated in the STD-50% specimen at 1073 K. Meanwhile, a much lower volume fraction of the AlN particles than the equilibrium was precipitated in the ST specimen. These indicate that the AlN particles have been almost fully precipitated in the STD-50% specimen, while the AlN particles were only partially been precipitated in the ST specimen. The near-equilibrium precipitation in the STD-50% specimen should result from the dislocation-induced precipitation of AlN [11]. It needs to point out that the slightly overestimated AlN volume fraction in the STD-50% specimen compared to the equilibrium prediction should be caused by the differences between the metallographic observation and theoretical calculation.

It is necessary to point out that as the annealing temperature was raised up to 1173 K (just after the ferrite-to-austenite transformation), the AlN particles in the STD-50% specimen were less coarsened than that of the ST specimen (Figure 5b). This could be more clearly seen from the TEM images of the AlN particles in both the undeformed and cold-deformed specimens after annealing at 1173 K for 1800 s, as shown in Figure 5c,d, respectively. Smaller AlN particles were formed in the STD-50% specimen, than that of the ST specimen. Therefore, the AlN particles in the prior cold-deformed specimen had sluggish growth kinetics. This may strongly influence the dissolution behavior of the precipitates at high temperatures. Similar phenomena have also been reported by Furubayashi et al. [11] and Kesternich [12], while the mechanism(s) for this are still not very clear.

Regarding the pinning effect of the second-phase particles, there is a maximum radius of the precipitates, r_{crit}, for pinning the grain growth [6]. That is, only the fine second phase particles with a radius smaller than r_{crit}, can effectively pin the grain boundaries. The r_{crit} can be expressed as [6]:

$$r_{crit} = 3\frac{R_0 f}{\pi}(3 - 4/Z)^{-1}, \tag{4}$$

where R_0 is the average grain size, Z is a parameter that is used to represents the inhomogeneity of the initial grains, and f is the volume fraction of second-phase particles. It is obvious that for a given uniform initial grain structure, r_{crit} linearly decreases with the decrease of f [6].

According to Figure 5a, the fraction of the AlN particles is stable in the STD-50% specimen below 1223 K (as indicated by symbol ① in Figure 5a). Therefore, both the r_{crit} and the pinning force, which is proportional to f/r [6] (here, r is the average radius of the pinning particles), may have no obvious change, and the grain boundaries can be well-pinned. This results in the sluggish austenite grain growth kinetics in the STD-50% specimens below 1223 K, as shown in Figure 3b. However, when the annealing temperature was increased up to 1248 K (position ② in Figure 5a), the fraction of the AlN particles in the STD-50% specimen decreased obviously. This may be caused by the accelerated diffusion rates of the elements at elevated temperatures, the increased solubility of the AlN in austenite, and the easy dissolution of the fine AlN particles in the STD-50% specimen. Hence, the r_{crit} may decrease, according to Equation (4). Then, the volume faction of the AlN precipitates available to pin the grain boundaries may decrease steeply at 1248 K. This may cause the occurrence of AGG at this temperature.

As for the ST specimen, the fraction of AlN particles were almost constant until 1298 K (position ③ in Figure 5a). Therefore, both r_{crit} and the pinning force were stable, and thus the grain growth could be effectively retarded. This agrees with the observed sluggish grain growth kinetics below 1298 K in Figure 3a. However, as the annealing temperature was raised up to 1323 K (position ④ in Figure 5a), the fractions of the AlN particles were decreased, and the r_{crit} may decrease accordingly. This may result in the occurrence of AGG in the ST specimen, as shown in Figure 2b,c.

According to the discussion above, it can be concluded that the change in the volume fraction of the AlN particles during annealing plays a key role in the austenite grain growth behavior. The rapid decrease in the AlN volume fraction may greatly change the r_{crit}, and then the effective volume fraction of the second phase particles to pin the gain boundaries may decrease dramatically, which may result in the occurrence of AGG.

As reviewed in the Introduction, the transition from AGG to NGG mode occurs at high temperatures when the AlN precipitates are nearly completely dissolved, and the grain growth becomes free. According to Figure 5a, the fractions of the AlN particles in both the deformed and undeformed specimens were low at 1353 K. Experimental results revealed that AGG occurred within 60 s when the ST specimen was annealed at 1353 K, while NGG occurred in the STD-50% specimen, even after a longer holding time (Figure 3b). The average diameters of the AlN particles in ST and STD-50% specimens after annealing at 1353 K for 0 s were evaluated to be around 60 and 30 nm, correspondingly. It was reported that the dissolution time and the size of the precipitates had a parabolic relationship [23]. Accordingly, the dissolution time of the AlN particles in the ST specimen was around four times that in the STD-50% specimen. Therefore, it was possible that the fine AlN precipitates in the STD-50% specimen dissolved very quickly during annealing, which made the grain growth become free and change to the NGG mode. However, since the dissolution of the coarse precipitates in the ST specimen took a longer time, AGG could occur. To quickly dissolve the relatively large AlN precipitates in the ST specimen, this required higher temperatures. This may promote the AGG-to-NGG transition temperature to the higher temperature in the ST specimen.

According to the above discussion, cold-deformation results in the fine and near-equilibrium precipitation of the AlN particles, which are readily dissolved during the heating process, due to the increase in the solubilities of Al and N in austenite. However, non-equilibrium and coarse AlN precipitates formed in the undeformed specimen, which were resistant to dissolution during the heating process. These should be the reasons for the suppression of the AGG temperature range by the prior cold-deformation.

4. Summary

In summary, the study of austenite grain growth behavior in a solution-treated low-alloy steel has shown that cold-deformation prior to annealing shifts the abnormal grain growth (AGG) temperature range of austenite toward a lower temperature region. The amount of shift increases with the increase of the reduction ratio. The lowered AGG occurrence temperature is attributed to the effects of cold deformation on the evolution of the AlN precipitates during annealing. TEM observations confirmed that a much lower volume fraction than the equilibrium and coarse AlN precipitates formed in the undeformed steel, which is resistant to dissolution during annealing. However, prior cold-deformation results in the fine and near-equilibrium precipitation of AlN, which is readily dissolved at elevated temperatures. This results in the shift of the AGG temperature range to lower temperatures from the application of prior cold-deformation.

Author Contributions: X.Z., K.M. and M.O. conceived and designed the experiments; X.Z. performed the experiments and collected the data; X.Z., and K.M. analyzed the data; X.Z. wrote the paper; K.M. and M.O. discussed with the results and revised the manuscript.

References

1. Morrison, W.B. Influence of small niobium additions on properties of carbon-manganese steels. *J. Iron Steel Inst.* **1963**, *201*, 317–325.
2. Webster, D.; Woodhead, J.H. Effect of 0.03 percent niobium on ferrite grain size of mild steel. *J. Iron Steel Inst.* **1964**, *202*, 987–994.
3. Irvine, K.J.; Pickering, F.B. Impact properties of low carbon bainitic steels. *J. Iron Steel Inst.* **1963**, *201*, 944–960.
4. Irvine, K.J.; Pickering, F.B.; Gladman, T. Grain-refined C-Mn steels. *J. Iron Steel Inst.* **1967**, *205*, 161–182.
5. Doğan, Ö.N.; Michal, G.M.; Kwon, H.W. Pinning of austenite grain boundaries by AlN precipitates and abnormal grain growth. *Metall. Trans. A* **1992**, *23*, 2121–2129. [CrossRef]
6. Gladman, T. On the theory of the effect of precipitate particles on grain growth in metals. *Proc. R. Soc.* **1966**, *294*, 298–309. [CrossRef]
7. Cuddy, L.J.; Raley, J.C. Austenite grain coarsening in microalloyed steels. *Metall. Trans. A* **1983**, *14*, 1989–1995. [CrossRef]
8. Dutta, B.; Palmiere, E.J.; Sellars, C.M. Modelling the kinetics of strain induced precipitation in Nb microalloyed steels. *Acta Mater.* **2001**, *49*, 785–794. [CrossRef]
9. Palmiere, E.J.; Garcia, C.I.; DeArdo, A.J. Compositional and microstructural changes which attend reheating and grain coarsening in steels containing niobium. *Metall. Mater. Trans. A* **1994**, *25*, 277–286. [CrossRef]
10. Liu, W.J. A new theory and kinetic modeling of strain-induced precipitation of Nb (CN) in microalloyed austenite. *Metall. Mater. Trans. A* **1995**, *26*, 1641–1657. [CrossRef]
11. Furubayashi, E.; Endo, H.; Yoshida, H. Effects of prior plastic deformation on the distribution and morphology of AlN precipitates in α iron. *Mater. Sci. Eng.* **1974**, *14*, 123–130. [CrossRef]

12. Kesternich, W. Dislocation-controlled precipitation of TiC particles and their resistance to coarsening. *Philos. Mag. A* **1985**, *52*, 533–548. [CrossRef]
13. Zhang, X.; Matsuura, K.; Ohno, M. Increase of austenite grain coarsening temperature in banded ferrite/pearlite steel by cold deformation. *Metall. Mater. Trans. A* **2015**, *46*, 32–36. [CrossRef]
14. Zhang, X.; Matsuura, K.; Ohno, M. Abnormal grain growth in austenite structure reversely transformed from ferrite/pearlite-banded structure. *Metall. Mater. Trans. A* **2014**, *45*, 4623–4634. [CrossRef]
15. Leslie, W.C.; Rickett, R.L.; Dotson, C.L.; Walton, C.S. Solution and precipitation of aluminum nitride in relation to the structure of low carbon steels. *Trans. Am. Soc. Met.* **1954**, *46*, 1470–1499.
16. McCall, J.L.; Boyd, J.E. Quantitative metallography of dispersion-strengthened alloys from extraction replicas-volume fraction and interparticle spacing. In Proceedings of the First Annual Technical Meeting of the International Metallographic Society, Denver, CO, USA, 8 September 1969; p. 153.
17. Ma, H.; Li, Y. Measure of size distribution and volume fraction of precipitates in silicon steel. *Mater. Sci. Eng.* **2002**, *20*, 328–330. (In Chinese)
18. Wilson, F.G.; Gladman, T. Aluminium nitride in steel. *Int. Mater. Rev.* **1988**, *33*, 221–286. [CrossRef]
19. Gladman, T.; Pickering, F.B. Grain-coarsening of austenite. *J. Iron Steel Inst.* **1967**, *205*, 653–664.
20. Hannerz, N.E. Influence of cooling rate and composition on intergranular fracture of cast steel. *Met. Sci. J.* **1968**, *2*, 148–152. [CrossRef]
21. Honer, K.E.; Baliktay, S. Susceptibility of steel castings to aluminium nitride-induced primary grain boundary fracture. In Proceedings of the 44th International Foundry Congress, Florence, Italy, 11–14 September 1977; pp. 125–140.
22. Atkins, M. *Atlas of Continuous Cooling Transformation Diagrams for Engineering Steels*; ASM International: Novelty, OH, USA, 1980.
23. Chen, Y.L.; Wang, Y.; Zhao, A.M. Precipitation of AlN and MnS in low carbon aluminium-killed steel. *J. Iron Steel Res. Int.* **2012**, *19*, 51–56. [CrossRef]

12

Local Characterization of Precipitation and Correlation with the Prior Austenitic Microstructure in Nb-Ti-Microalloyed Steel by SEM and AFM Methods

Lena Eisenhut *, Jonas Fell and Christian Motz

Department for Materials Science and Engineering, Saarland University, 66123 Saarbrücken, Germany; j.fell@matsci.uni-saarland.de (J.F.); motz@matsci.uni-sb.de (C.M.)

* Correspondence: l.eisenhut@matsci.uni-sb.de

Abstract: Precipitation is one of the most important influences on microstructural evolution during thermomechanical processing (TMCP) of micro-alloyed steels. Due to precipitation, pinning of prior austenite grain (PAG) boundaries can occur. To understand the mechanisms in detail and in relation to the thermomechanical treatment, a local characterization of the precipitation state depending on the microstructure is essential. Commonly used methods for the characterization, such as transmission electron microscopy (TEM) or matrix dissolution techniques, only have the advantage of local or statistically secured characterization. By using scanning electron microscopy (SEM) and atomic force microscopy (AFM) techniques, both advantages could be combined. In addition, in the present work a correlation of the precipitation conditions with the prior austenite grain structure for different austenitization states could be realized by Electron Backscatter Diffraction (EBSD) measurement and reconstruction methods using the reconstruction software Merengue 2.

Keywords: micro-alloyed steels; precipitations; Zener pinning; atomic force microscopy (AFM); precipitation-microstructure correlation; EBSD; reconstruction methods

---

1. Introduction

The microstructural evolution during thermomechanical processing (TMCP) is not only determined by temperature and deformation, but also by the steel composition. If precipitation and grain boundary pinning occurs during TMCP because of the addition of micro-alloying elements such as titanium and niobium, the driving force for grain boundary motion is reduced by the Zener pinning force [1–4]. Furthermore, during hot-deformation processing strain-induced precipitation can also occur. This precipitation process results in a retardation of the recrystallization [5–7]. In addition to the thermomechanical processing, precipitation formation can occur during welding processes in the heat-affected zone. This leads to an improvement in the fracture toughness and precipitation strengthening, but the precipitates can also act as cleavage initiators [8–11]. To describe all the mechanisms due to precipitation, many investigations were made using different characterization methods [3,7,12–16].

Sha et al. [3] used TEM to locally characterize larger precipitations in the range of 20–150 nm in a coarse-grained austenitic microstructure in Nb-V-Ti microalloyed steel. A characterization of strain-induced precipitations due to torsion tests was made for Nb-microalloyed steel by Badiola et al. using TEM on carbon extraction replicas. They found Nb(C, N) precipitations less than 10 nm in size and Ti-Nb-rich precipitations between 20 and 60 nm. The smaller ones were determined to be the ones with the higher influence on the recrystallization behavior [12]. TEM methods were also used to characterize strain-induced precipitation behavior in hot-deformed steel by many others [7,13].

Using the matrix dissolution technique proposed by Lu et al. [14] and Hegetschweiler et al. [15] provides the opportunity to analyze a large amount of particles, extracted from the steel matrix. Lu et al. characterized more than 2000 precipitates by TEM measurements. Based on this technique, there is the potential that extracted particles could not only be analyzed and characterized by TEM measurements but also by colloidal methods like field-flow fractionation methods. This offers new chances for a statistically satisfying analysis of the size distribution of a very high amount (>2000) of precipitates.

Rentería-Borja et al. [16] developed a method for the quantification of nanometer-sized precipitates in microalloyed steels using the AFM and an etching with 0.5%-Nital for the selective preparation of the particles. Precipitate size distribution was obtained for three different steels. By chemical etching, the microstructure could also be etched selectively and therefore, the detection possibility of precipitates by AFM measurements could be influenced.

As precipitates influence the recrystallization behavior and the grain growth, their local distribution in the austenitic grain structure is important. Due to the low amount of residual elements as sulfur or phosphorus in low-carbon steel grades, the reconstruction of the PAG structure by chemical etching is not a reliable method [17]. Therefore, the correlation of the prior austenite grain structure with the detected precipitations was made by reconstructing the PAG by EBSD and reconstruction method using the software Merengue 2 [18].

2. Materials and Methods

The investigated steel grade is a low-carbon steel, containing a small amount of the micro-alloying elements niobium (Nb) and titanium (Ti). The chemical composition is presented in Table 1. During reheating of the samples (as-rolled condition), carbides, nitrides and carbo-nitrides will form due to the micro-alloying contents and pre-existing precipitations will grow.

Table 1. Chemical composition of the low carbon steel (wt %).

C	Mn	Si	Nb	Ti
0.09	1.6	0.2	0.038	0.015

For the investigations, samples with a diameter of 8 mm and a height of 5 mm were used. The heat treatment for the setup of two different prior austenite grain structures and corresponding precipitation conditions consists of three main segments. First, the samples were heated to an austenitization temperature $T_a = 1000$ °C and then held for a specific time t_a (min). After the holding time, the austenitic samples were water-quenched to produce a fully martensitic microstructure with very thin laths. This is important for the reconstruction of the PAGs out of orientation measurement data (EBSD) because it results in more variants for the reconstruction and therefore increases the reliability of the reconstruction process with Merengue 2. The different applied heat treatment parameters are summarized in Table 2. The expected mean particle diameter from MatCalc 6.01 (MatCalc Engineering GmbH, Vienna, Austria) simulations for both heat treatments is also listed in Table 2, where significant differences can be seen.

Table 2. Heat treatments of the sample fabrication.

Sample	T_a (°C)	t_a (min)	Expected Mean Particle Diameter (MatCalc Simulations) (nm)
1	1000	10	25
2	1000	240	45.4

After the sample fabrication, the samples were cut in the middle, to avoid influences of decarburization on the edge of the samples. Then the metallographic preparation was applied to the samples, as listed in Table 3. The last step of polishing with the Masterprep® Polishing Suspension (Buehler, ITW Test & Measurement GmbH European Headquarters, Esslingen am Neckar, Germany) is essential for the separation of the precipitations from the steel matrix by AFM analysis and the reduction of remaining surface artifacts due to the previous preparation steps. Subsequently, an area of about $100 \times 100\ \mu m^2$ was marked on the sample surface using the micro-indentation method. This marking is essential for the correlation of the AFM measurements with the reconstructed austenitic microstructure.

Table 3. Metallographic sample preparation.

Step	Characteristics	Time
Grinding	800, 1200, 2500	-
Polishing	6, 3, 1, 0.25 μm	5 min
Final Polishing	Masterprep® (Buehler), 0.05 μm	180 s

If the samples are prepared as described before, the precipitation detection with AFM is possible. The AFM measurements were made using a Dimension Fast Scan AFM (Bruker Corporation, Billerica, MA, USA) with Scan Asyst. One part of this work was to optimize the AFM detection parameters and to find the mode of operation which gives the best results. The parameters used are listed in Table 4. Using the Bruker Peak-Force Tapping Mode with the Scan Asyst-Air Tip gave the best results.

Table 4. AFM detection parameters.

Parameter	Value
Scan Size	$100 \times 100\ \mu m^2$
Scan Rate	0.0494 Hz
Samples/Line	10,240
Peak Force Amplitude	150 nm
Peak Force Frequency	2 kH
Lift Height	47.9 nm
Spring Constant	0.4 N/m

Before correlating the PAG structure with the precipitation state, it is necessary to be sure that the detected particles really are precipitates and that there are no artifacts of the microstructure or preparation method. Therefore, all the particles of an area of $50 \times 25\ \mu m^2$ detected by AFM were chemically analyzed in a Zeiss Sigma SEM (Carl Zeiss Microscopy GmbH, Jena, Germany) using energy dispersive X-ray spectroscopy (EDX, X-Max, Oxford Instruments, Witney, Oxon, UK) measurements.

Afterwards an area of $100 \times 100\ \mu m^2$ was investigated for both samples using the final preparation and AFM detection parameters. Then the sample was transferred to the SEM to perform the EBSD scan using an Oxford EBSD system (Nord-Lys, Oxford Instruments, Witney, Oxon, UK) and the AZtecHKL software (3.1 SP1, Oxford Instruments, Witney, Oxon, UK). The step size was adapted mainly to the martensitic microstructure to a value of 0.075 μm but also to get a high resolution of the prior austenitic grain boundaries after the reconstruction step. By using the CHANNEL5 data processing software (5.12.61.0, Oxford Instruments, Witney, Oxon, UK), the EBSD maps were slightly smoothed and zero-solutions were excluded.

In the following step, the austenite microstructures were reconstructed from the martensitic EBSD measurements. The reconstruction of the parent austenite phase from martensitic EBSD orientation maps is based on two assumptions. First, there must be an orientation relationship (OR) between the parent and the daughter phase, which is strictly respected. In addition, the parents must have a unique local crystallographic orientation. Different ORs between parent (face-centered cubic) and daughter (body-centered cubic) phase were figured out depending on the investigated material and its microstructural condition [19]. Besides the Kurdjumov-Sachs and Nishiyama-Wassermann ORs, there is the OR of Greninger and Troiano (GT), which was applied in this work. The reconstruction method used in this work was also successfully applied in other works to investigate other mechanisms such as grain growth behavior or recrystallization mechanisms in microalloyed steels [20–22].

As other studies have shown, using the Merengue 2 reconstruction software with assuming the GT Orientation Relationship as the starting OR gives very good results for the used steel grade [21]. The chosen OR was refined during the reconstruction using local misorientations [23]. The reconstruction method with Merengue 2 is explained in detail in [18].

Basically, it involves two main steps:

1. Identification of clusters of at least three neighboring domains related to a unique parent orientation within a tolerance of 3°, called fragments.
2. All child domains were browsed recursively starting from a fragment within a tolerance of 3°, from neighbor to neighbor. If they share the parent orientation of the fragment, they were assigned to the according parent.

Then the austenite grains could be detected in the CHANNEL5 data processing software by defining a misorientation angle for the separation of two grains. The detected grain boundaries were segmented and aligned to a SE-image to correct the drift that may occur during the long duration of the EBSD measurement. The overall reconstruction process is schematically shown in Figure 1. Figure 1a,b show the unreconstructed and corresponding reconstructed EBSD map. Figure 1c only contains the segmented grain boundaries that were finally aligned to the corresponding SE-image to correct possible errors due to thermal drift.

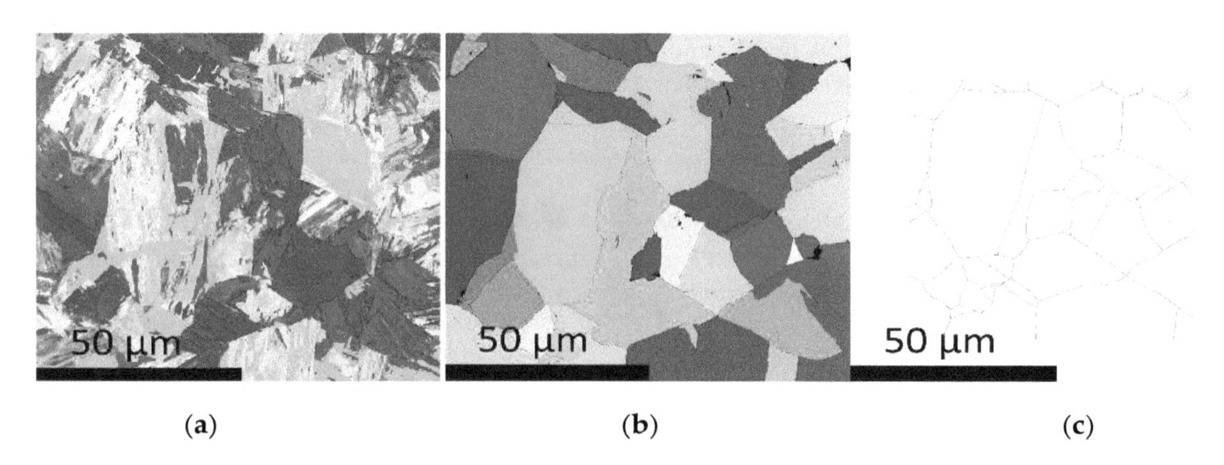

Figure 1. Maps for the steps of the PAG (prior austenite grain) boundary segmentation: (**a**) EBSD map of the martensitic microstructure; (**b**) reconstructed map of the prior austenitic grain structure; (**c**) segmented PAG boundaries.

In addition to the segmentation of the grain boundaries, the precipitations have to be extracted from the AFM measurements. This was made with the Axio Vision 4.8 analyzing software (Carl Zeiss Microscopy GmbH, Jena, Germany) by a segmentation of the particles from the matrix. Figure 2 shows an original AFM image (a) and the corresponding image with the segmented particles in red (b). For the segmentation of the particles a shading correction or a high-pass filter is applied to the AFM images. In a second step the contrast, brightness and gamma values have to be optimized. Then the segmentation takes places by defining a threshold to the grayscale values that excludes the background (matrix). All grayscale values that are higher than this threshold are defined as particles and statistically evaluated.

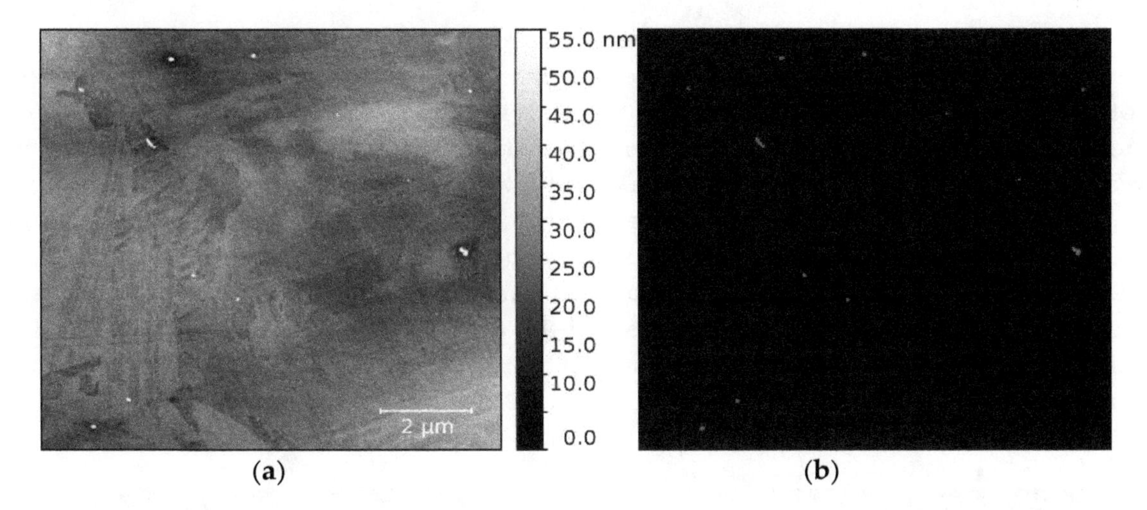

Figure 2. Maps of the two steps of the particle segmentation: (**a**) original AFM map; (**b**) segmented particles.

Now it was possible to correlate the grain boundaries and the particle distribution by adjusting both resulting maps based on the micro-indents. After the alignment, a statistical evaluation was carried out including the total number of precipitates, the number of precipitates on the grain boundary, and the size distribution of the particles depending on the austenitization state.

3. Results

3.1. Precipitation Detection with AFM and Chemical Verification by EDX

Figure 3a shows a typical result of the AFM measurements. Different particles in different size ranges could be detected within the grains as well as on grain boundaries. The line profile of the height for three different particles exemplifies the difference between matrix and particles in height due to the selective erosion by preparation (Figure 3b). This difference increases with the particle diameter, but is also big enough to be used for smaller particles (~10 nm) as a good indicator for the differentiation from the matrix.

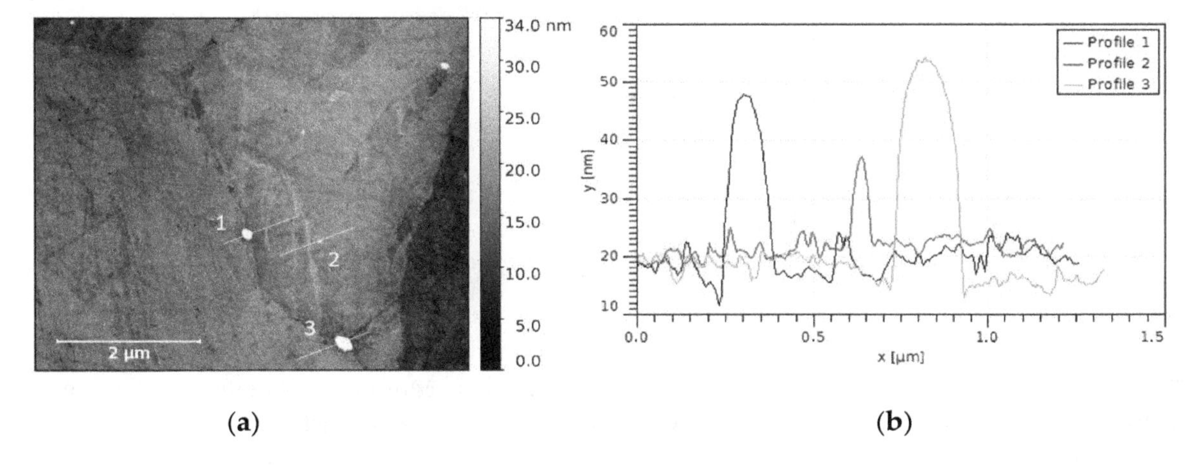

Figure 3. AFM (atomic force microscopy) measurement of particles in a martensitic microstructure with particles on grain boundaries and within the grains (**a**) and the corresponding line profiles of three selected particles of different sizes (**b**).

The precipitates detected by AFM were verified by chemical analysis with EDX, as shown in Figure 4. The comparison of one AFM and SE image is shown where the particles and matrix can be easily distinguished because of their differences in contrast and brightness. The EDX verification measurement is also displayed for one particle and one point in the steel matrix. The occurrence and height of the titanium and niobium peaks in the EDX spectrum of the particle is the most obvious difference compared to the spectrum of the matrix. This peak is the basis for the definition of the measured particles as precipitates. Although a quantitative determination of the titanium or niobium content of one particle is not possible, there are differences between titanium and titanium-free precipitates. In total, this verification was made for an area of $50 \times 25\ \mu m^2$ with at least 200 particles. This verification showed that 97.5% of all analyzed particles were precipitates containing titanium and/or niobium. Hence, the chemical analysis of the precipitations confirms the assumption that all detected particles are indeed precipitations and there are no artifacts from the microstructure or preparation method.

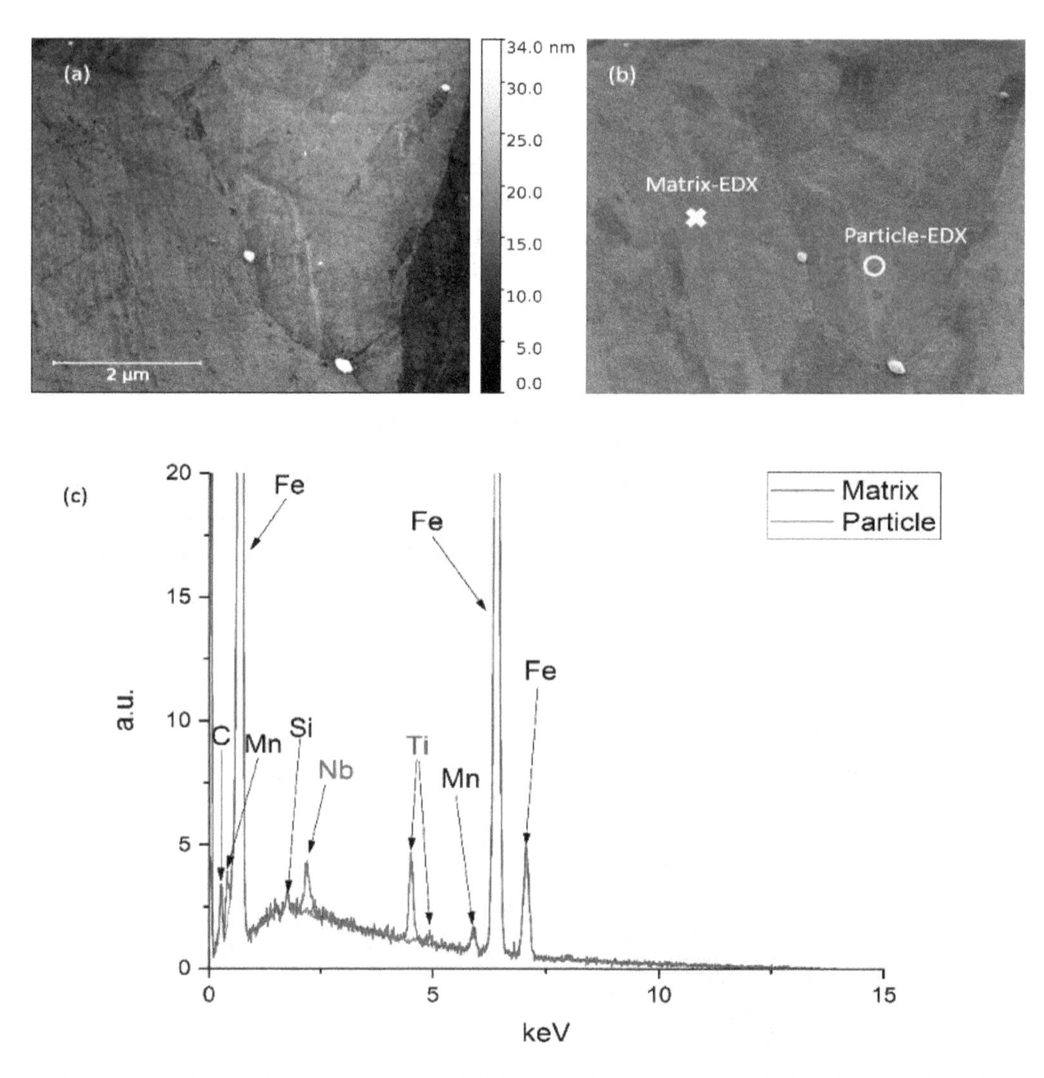

Figure 4. AFM map (**a**) and corresponding SE image (**b**); spectrum of the chemical analysis of the matrix (**c**) and one precipitate, which is shown for all other measured particles.

Figure 5 shows the result of an AFM measurement at a very high resolution of 1280 lines per µm. At this resolution, particles in the size range of around 10 nm in diameter could also be detected. The diameter of the four biggest particles of this map was measured as 9.1 to 13.3 nm, but the detection of even smaller particles should be possible by modifying the AFM parameters and the preparation method.

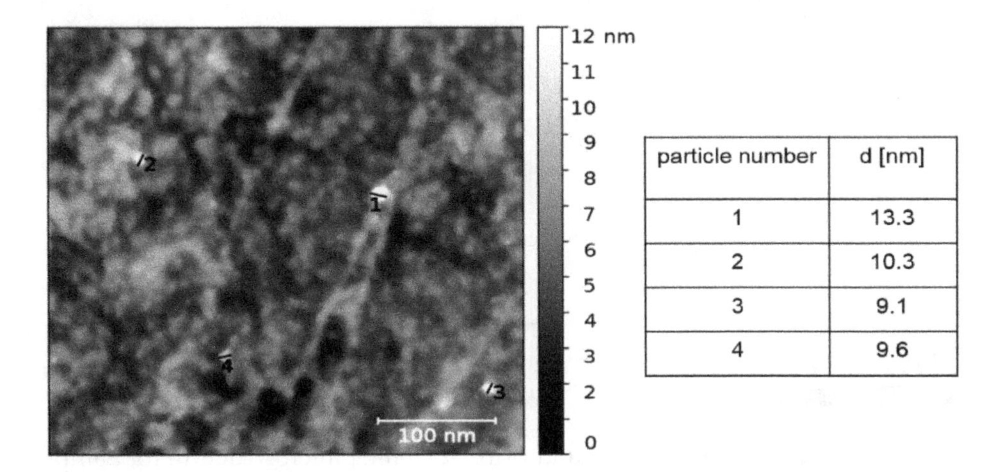

particle number	d [nm]
1	13.3
2	10.3
3	9.1
4	9.6

Figure 5. AFM map with very high resolution and four particles with their corresponding diameter.

3.2. Correlation of Precipitates with the PAG Structure and Comparison of Two Austenitization States

Figure 6 shows the correlation of the segmented PAG with the segmented precipitates for the austenitization state of 10 min austenitization time at 1000 °C austenitization temperature. The size distribution of the mean particle diameter for both samples is compared in Figure 7. One aim of the present study was to analyze differences between these two microstructures, depending on their precipitation state. Table 5 summarizes the results of the statistical analysis for both samples.

For both samples, a comparable number of particles could be detected by AFM measurements. Comparing the particle size distribution of both samples (Figure 7), the mean particle diameter of the sample with 10-min austenitization time is lower than the particle diameter for long austenitization times. This confirms the presumption that the particle diameter will increase with a longer austenitization time and also confirms our expectations after the pre-simulations with MatCalc. The simulations with MatCalc were only slightly optimized for the used steel grade and only made for a simple pre-estimation of the particle diameters depending on different heat treatments. The difference in the mean diameter in the experimental results is much smaller than in the simulations. One reasonable explanation is that in the simulations only two particle classes of pure TiN and pure NbC particles are assumed. In the real material complex particles, containing both Ti and Nb, may also occur. The size range of the investigated and simulated mean particle sizes is comparable.

Comparing the number of particles on the PAG boundaries by the correlation of the microstructure and the precipitation states shows that for Sample 1 the number of particles on the grain boundary is higher than for Sample 2. In both cases, around 5% of all detected particles are lying on the PAG. Differences arise when comparing the number of particles on the grain boundaries relative to the whole grain boundary lengths, which could be measured after reconstruction. Therefore, the number of particles on the PAG for Sample 1 is about 11 per 100 µm grain boundary length and for Sample 2 only five. Because of holding Sample 2 at high temperatures for a longer period of time, the total number of precipitates is reduced. During holding the bigger particles grow whereas the smaller ones will dissolve.

Table 5. Statistical evaluation and comparison of both austenitization states.

Sample	Mean Prior Austenite Grain Size (µm)	# of Measured Particles	Mean Particle Diameter (nm)	# of Particles on the PAG
1	30	2090	50	111
2	32.6	1670	65.3	79

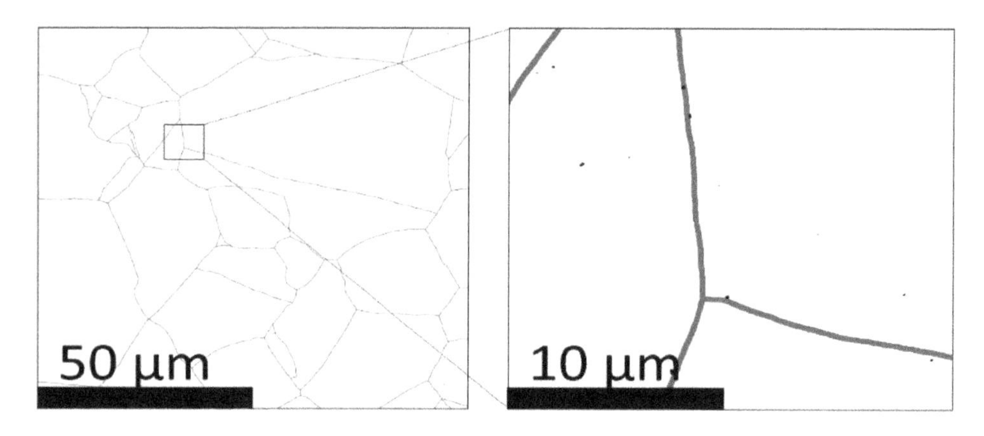

Figure 6. Correlation of precipitates and prior austenitic microstructure for Sample 1.

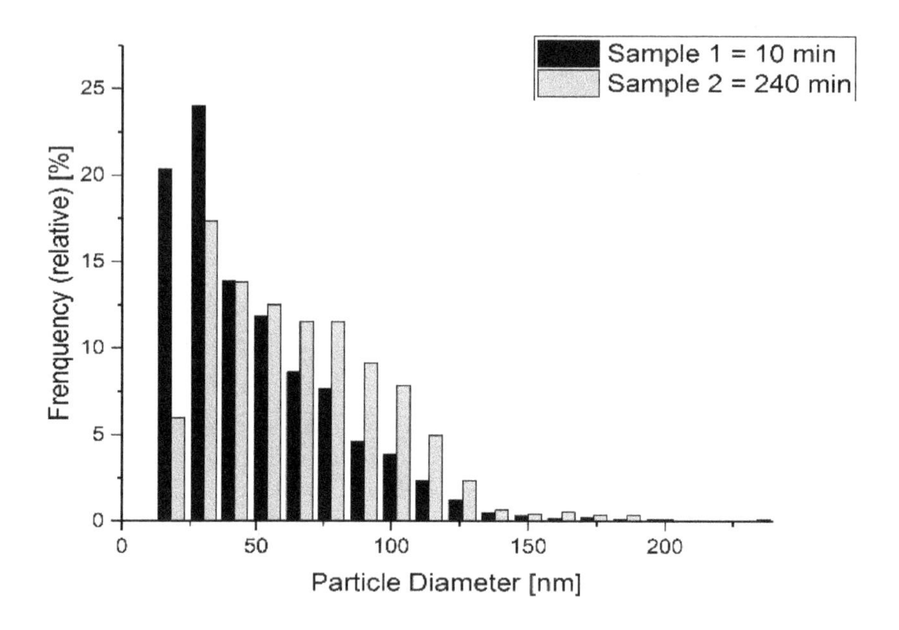

Figure 7. Precipitation size distribution compared for both austenitization states.

4. Discussion

With this preparation and characterization method, precipitates on the boundaries of different shapes and sizes could have been detected. The size of precipitates that could be detected with this method is 10 nm or even smaller.

The comparison of AFM and SE images and the chemical verification by EDX measurements ensures that nearly all particles detected by AFM are indeed precipitates containing niobium and/or titanium. The advantage of the method developed in this work is the detection of a high number of particles, comparable to other methods such as the matrix dissolution technique, with a comparatively low effort of sample preparation and implementation.

By correlating the PAG structure with the measured precipitation state, not only statistical assumptions related to the total number of particles could be made. This way to investigate the precipitation state retains its local resolution and enables a clear statement about how many of the detected particles are placed on the PAG boundaries and within the grains.

To avoid errors due to the PAG reconstruction, the quality of EBSD measurements was verified by the MAD (mean angular deviation) value, which is the quality parameter of the Oxford EBSD system and also the indexing rate of the EBSD depending on this MAD value (MAD > 1: no indexing, zero solution). The maximum value of the MAD was 0.42 and the indexing rate 97.57% at the minimum. In addition, small grains (below 3 pixels) were removed before reconstruction and single-pixel grains were removed prior to the reconstruction step. Because the reconstruction step is very sensitive to the microstructure, very high cooling rates were realised by water quenching. This leads to a fully martensitic microstructure with more variants available for the reconstruction step [24]. The best OR was found by avoiding local minima, which is described in detail in [23]. The angular tolerance for the reconstruction was set to 3°, which is smaller than other authors have reported [25].

However, due to possible errors in the EBSD and reconstruction step, the statistical evaluation could be influenced. There can be particles that are lying on the PAG boundary but do not occur in the corresponding statistical values due to possible reconstruction and correlation errors. Especially for Sample 2, the reconstruction of the PAG structure from the orientation data was not reliable in every part of the map. Therefore, a manual correction under consideration of the real microstructure measured in AFM was made.

Possible errors because of the thermal drift of the samples in the AFM and SEM were corrected by adjusting the different measurements on the micro-indents before the statistical evaluation of the correlation was made. Even this method only delivers insights in two dimensions; the results are statistically secured by investigating a minimal area of $100 \times 100\ \mu m^2$ with a relatively high number of more than 1500 particles for each sample.

5. Conclusions

The presented AFM method in synergy with EBSD/SEM methods may be used to analyze and understand the precipitation states in different microstructures depending on the processing parameters. Moreover, a statistical evaluation of the mean particle size and the number of particles on the grain boundaries is possible. By refining the detection parameters and increasing resolution of the AFM scan, particles smaller than 10 nm in diameter could also be detected, which also makes it possible to characterize other precipitation states in future projects as strain induced precipitation. By using EBSD and PAG reconstruction methods, recrystallized and unrecrystallized grains can be distinguished and then correlated with very small, strain-induced precipitations detected by the AFM. In future investigations, the three-dimensional precipitation state including the correlation with the prior austenitic microstructure could also be investigated using tomographic methods.

Author Contributions: Conceptualization, L.E. and J.F.; Methodology, J.F.; Validation, J.F., L.E.; Formal Analysis, J.F., L.E.; Investigation, J.F.; Writing-Original Draft Preparation, L.E.; Writing-Review & Editing, L.E.; Visualization, L.E. and J.F.; Supervision, L.E.; Project Idea: C.M.; Project Administration, C.M.; Funding Acquisition, C.M.

Acknowledgments: The authors would like to thank Dillinger for providing the sample material and Lionel Germain for providing the reconstruction software Merengue 2. They also acknowledge the Deutsche Forschungsgemeinschaft (German Research Foundation, DFG) and the state of Saarland for the financial support for the atomic force microscope (INST 256/455-1) and the scanning electron microscope (INST 256/340-1 FUGG) used in this work.

References

1. Chapa, M.; Medina, S.F.; Lopez, V.; Fernandez, B. Influence of Al and Nb on Optimum Ti/N Ratio in Controlling Austenite Grain Growth at Reheating Temperatures. *ISIJ Int.* **2002**, *42*, 1288–1296. [CrossRef]
2. Manohar, P.A.; Ferry, M.; Chandra, T. Five Decades of the Zener Equation. *ISIJ Int.* **1998**, *38*, 913–924. [CrossRef]
3. Sha, Q.Y.; Sun, Z.Q. Grain growth behavior of coarse-grained austenite in a Nb-V-Ti microalloyed steel. *Mater. Sci. Eng. A* **2009**, *523*, 77–84. [CrossRef]
4. Medina, S.F.; Chapa, M.; Valles, P.; Quispe, A.; Vega, M.I. Influence of Ti and N contents on austenite grain control and precipitate size in structural steels. *ISIJ Int.* **1999**, *39*, 930–936. [CrossRef]
5. Dutta, B.; Valdes, E.; Sellars, C.M. Mechanism and kinetics of strain induced precipitation of Nb(C, N) in austenite. *Acta Metall. Mater.* **1992** *40*, 653–662. [CrossRef]
6. Le Bon, A.; Rofes-Vernis, J.; Rossard, C. Recrystallization and Precipitation during Hot Working of a Nb-Bearing HSLA Steel. *Met. Sci.* **1975**, *9*, 36–40. [CrossRef]
7. Pereloma, E.V.; Crawford, B.R.; Hodgson, P.D. Strain-induced precipitation behaviour in hot rolled strip

steel. *Mater. Sci. Eng. A* **2001**, *299*, 27–37. [CrossRef]

8. Shanmugam, S.; Ramisetti, N.K.; Misra, R.D.K.; Hartmann, J.; Jansto, S.G. Microstructure and high strength-toughness combination of a new 700 MPa Nb-microalloyed pipeline steel. *Mater. Sci. Eng. A* **2008**, *478*, 26–37. [CrossRef]
9. Shanmugam, S.; Misra, R.D.K.; Hartmann, J.; Jansto, S.G. Microstructure of high strength niobium-containing pipeline steel. *Mater. Sci. Eng. A* **2006**, *441*, 215–229. [CrossRef]
10. Charleux, M.; Poole, W.J.; Militzer, M.; Deschamps, A. Precipitation behavior and its effect on strengthening of an HSLA-Nb/Ti steel. *Metall. Mater. Trans. A* **2001**, *32*, 1635–1647. [CrossRef]
11. Zhang, L.P.; Davis, C.L.; Strangwood, M. Effect of TiN particles and microstructure on fracture toughness in simulated heat-affected zones of a structural steel. *Metall. Mater. Trans. A* **1999**, *30*, 2089–2096. [CrossRef]
12. Jorge-Badiola, D.; Gutiérrez, I. Study of the strain reversal effect on the recrystallization and strain-induced precipitation in a Nb-microalloyed steel. *Acta Mater.* **2004**, *52*, 333–341. [CrossRef]
13. Jia, Z.; Misra, R.D.K.; O'Malley, R.; Jansto, S.J. Fine-scale precipitation and mechanical properties of thin slab processed titanium-niobium bearing high strength steels. *Mater. Sci. Eng. A* **2011**, *528*, 7077–7083. [CrossRef]
14. Lu, J.F.; Wiskel, J.B.; Omotoso, O.; Henein, H.; Ivey, D.G. Matrix dissolution techniques applied to extract and quantify precipitates from a microalloyed steel. Metall. *Mater. Trans. A* **2011**, *42*, 1767–1784. [CrossRef]
15. Hegetschweiler, A.; Kraus, T.; Staudt, T. Colloidal analysis of particles extracted from microalloyed steel. *Metall. Ital.* **2017**, *3*, 23–28.
16. Rentería-Borja, L.; Hurtado-Delgado, E.; Garnica-González, P.; Domínguez-López, I.; García-García, A.L. Atomic force microscopy applied to the quantification of nano-precipitates in thermo-mechanically treated microalloyed steels. *Mater. Charact.* **2012**, *69*, 9–15. [CrossRef]
17. Weyand, S.; Britz, D.; Rupp, D.; Mücklich, F. Investigation of Austenite Evolution in Low-Carbon Steel by Combining Thermo-Mechanical Simulation and EBSD Data. *Mater. Perform. Charact.* **2015**, *4*, 322–340. [CrossRef]
18. Germain, L.; Gey, N.; Mercier, R.; Blaineau, P.; Humbert, M. An advanced approach to reconstructing parent orientation maps in the case of approximate orientation relations: Application to steels. *Acta Mater.* **2012**, *60*, 4551–4562. [CrossRef]
19. Nolze, G. Improved determination of FCC/BCC orientation relationships by use of high-indexed pole figures. *Cryst. Res. Technol.* **2006**, *41*, 72–74. [CrossRef]
20. Humbert, M.; Germain, L.; Gey, N.; Boucard, E. Evaluation of the orientation relations from misorientation between inherited variants: Application to ausformed martensite. *Acta Mater.* **2015**, *82*, 137–144. [CrossRef]
21. Krämer, T.; Eisenhut, L.; Germain, L.; Rupp, D.; Detemple, E.; Motz, C. Assessment of EBSD Analysis and Reconstruction Methods as a Tool for the Determination of Recrystallized Fractions in Hot-Deformed Austenitic Microstructures. *Metall. Mater. Trans. A* **2018**, *49*, 2795–2802. [CrossRef]
22. Eisenhut, L.; Rupp, D.; Motz, C. Evolution of the austenitic grain growth of micro-alloyed steels was studied by metallography and EBSD analysis and compared to a grain growth model. In Proceedings of the 5th International Conference on Thermomechanical Processing (TMP), Milan, Italy, 26–28 October 2016; p. 5026.
23. Humbert, M.; Blaineau, P.; Germain, L.; Gey, N. Refinement of orientation relations occurring in phase transformation based on considering only the orientations of the variants. *Scr. Mater.* **2011**, *64*, 114–117. [CrossRef]
24. Blaineau, P.; Germain, L.; Humbert, M.; Gey, N. A New Approach to Calculate the γ Orientation Maps in Steels. *Solid State Phenom.* **2010**, *160*, 203–210. [CrossRef]
25. Bernier, N.; Bracke, L.; Malet, L.; Godet, S. An alternative to the crystallographic reconstruction of austenite in steels. *Mater. Charact.* **2014**, *89*, 23–32. [CrossRef]

13

Comparative Effect of Mo and Cr on Microstructure and Mechanical Properties in NbV-Microalloyed Bainitic Steels

Andrii Kostryzhev [1,*], Navjeet Singh [1], Liang Chen [2], Chris Killmore [3] and Elena Pereloma [2,4]

[1] Australian Steel Research Hub, University of Wollongong, Wollongong, NSW 2500, Australia; ns106@uowmail.edu.au
[2] School of Mechanical, Materials, Mechatronic and Biomedical Engineering, University of Wollongong, Wollongong, NSW 2500, Australia; lchen@uow.edu.au (L.C.); elenap@uow.edu.au (E.P.)
[3] BlueScope Steel Limited, Five Islands Road, Port Kembla, NSW 2505, Australia; chris.killmore@bluescopesteel.com
[4] UOW Electron Microscopy Centre, University of Wollongong, Wollongong, NSW 2519, Australia
* Correspondence: andrii@uow.edu.au

Abstract: Steel product markets require the rolled stock with further increasing mechanical properties and simultaneously decreasing price. The steel cost can be reduced via decreasing the microalloying elements contents, although this decrease may undermine the mechanical properties. Multi-element microalloying with minor additions is the route to optimise steel composition and keep the properties high. However, this requires deep understanding of mutual effects of elements on each other's performance with respect to the development of microstructure and mechanical properties. This knowledge is insufficient at the moment. In the present work we investigate the microstructure and mechanical properties of bainitic steels microalloyed with Cr, Mo, Nb and V. Comparison of 0.2 wt. % Mo and Cr additions has shown a more pronounced effect of Mo on precipitation than on phase balance. Superior strength of the MoNbV-steel originated from the strong solid solution strengthening effect. Superior ductility of the CrNbV-steel corresponded to the more pronounced precipitation in this steel. Nature of these mechanisms is discussed.

Keywords: steel; thermomechanical processing; microstructure characterisation; mechanical properties; molybdenum

---

1. Introduction

In high strength low alloyed steels Mo is well known to provide phase balance strengthening, via facilitating the bainite transformation [1–5], and solid solution strengthening [6–9]. It can decrease the rate of dynamic recrystallization of austenite [10–12], which may lead to grain refinement. Sometimes Mo can contribute to precipitation strengthening through formation of Mo-rich carbides [13–16]. Although its main effect on precipitation is via the increase in solubility of Ti [17,18] and Nb [10] in austenite, resulting in decreased sizes and increased number densities of Ti- and Nb-rich particles [19–23], which are essential for the precipitation strengthening from Ti- and Nb-rich particles.

Similar to Mo, Cr facilitates the bainite transformation [24–26], may precipitate in complex Cr-rich carbides [27–29], and increases solubility of Ti [30,31] and Nb [30,32,33]. In particular, Cr was observed delaying Fe_3C precipitation in low carbon steel [34]. However, the solid solution strengthening effect of Cr is ~6 times weaker than this of Mo [35], and, therefore, Cr is less affective in retarding recrystallization [36,37].

Amongst the published data, effects of Mo and Cr in multi-microalloyed steels, in particular containing V, are rarely reported. In 0.042C-0.3Mo-1.0Cr-0.08V steel coiled in the temperature range of 180–530 °C, Hutchinson et al. [38] observed bainitic microstructures with average ferrite grain size of 3 μm. Increased to 640–770 MPa proof stress was suggested to originate mainly from high dislocation density in bainite and, in particular, dislocation pinning by V(C,N) precipitates. In another work, Kong et al. [39] investigated mechanical properties of 0.064C-0.22Mo-0.21Cr-0.031Nb-0.031V steel thermomechanically processed in a temperature range of 1150–800 °C and cooled at the rate of 20–30 °C/s to 430–550 °C finish cooling temperature. The yield stress in the range of 530–710 MPa was attributed to the narrow width of bainitic ferrite lath (about 0.52 μm), although precipitation of TiNbV-rich particles was also observed. Abbasi and Rainforth [40] studied the microstructure and mechanical properties in MoNbV microalloyed ferritic steel. Simultaneous additions of 0.08 wt. % Mo and 0.04 wt. % Nb to 0.12C-0.16V ferritic steel resulted in precipitation of MoNbVC and decreased size of VC particles, which was explained by the improved temperature stability and reduced coarsening rate of multi-element precipitates. Increased steel hardness with Nb and Mo microalloying was related to finer ferrite grain size and higher number density of VC particles in the NbMoV-microalloyed steel. In this work we advance the knowledge of multi-microalloyed steels in the following aspects: (i) compare the effects of minor Mo and Cr additions on phase transformation and particle precipitation in low carbon NbV-microalloyed bainitic steels; (ii) analyse the microstructure-property relationship in the newly developed steels with 700–850 MPa of yield stress; and (iii) investigate the effect of high temperature strain (in the recrystallization temperature region) on room temperature microstructure and mechanical properties. The effect of high temperature (>1000 °C) strain is important to study because increased strain values may enhance recrystallization of austenite (refine grain size) and accelerate precipitation of MoNbV-rich particles [10,23,28,41] (reduce Mo solid solute concentrations). Consequently, the grain size, precipitate number density and solute atom concentrations will affect the ambient temperature mechanical properties.

2. Materials and Methods

Two steels containing 0.08C, 1.5Mn, 0.3Si, 0.2Ni, 0.03Al, 0.003S, 0.015P, 0.01N, 0.06Nb, 0.12V and either 0.3Cr-0.2Mo or 0.5Cr-0Mo (wt. %), denoted below as MoNbV-steel and CrNbV-steel respectively, were melted in a 60 kg induction furnace and cast as 75 × 100 × 150 mm^3 blocks by Hycast Metals Pty, Sydney, Australia. The blocks were homogenised at 1250 °C for 30 h, to equalise chemical composition, then forged in the temperature range of 1250–900 °C along the 100 mm side to 28 mm plate thickness, to assure 3.5 times reduction of the as-cast microstructure. The forged plates were cut into standard 20 × 15 × 10 mm^3 Gleeble samples. Thermomechanical processing in Gleeble (manufactured by Dynamic Systems Inc., Poestenkill, NY, USA) was conducted using two schedules:

- Austenitising at 1250 °C for 180 s, followed by cooling to 1175 °C at a cooling rate of 1 $°C·s^{-1}$;
- First deformation at 1175 °C to 0.3 (low strain schedule) or 0.35 (high strain schedule) strain at 5 s^{-1} strain rate, followed by cooling to 1100 °C at a cooling rate of 2 $°C·s^{-1}$;
- Second deformation at 1100 °C to 0.35 (low strain schedule) or 0.50 (high strain schedule) strain at 5 s^{-1} strain rate, followed by cooling to 1000 °C at a cooling rate of 25 $°C·s^{-1}$;
- Third deformation at 1000 °C to 0.25 strain at 5 s^{-1} strain rate, followed by cooling to 900 °C at a cooling rate of 30 $°C·s^{-1}$;
- Fourth deformation at 900 °C to 0.25 strain at 5 s^{-1} strain rate, followed by holding at this temperature for 10 s and cooling to 500 °C at a cooling rate of 30 $°C·s^{-1}$ to assure bainite transformation;
- Holding at 500 °C for 900 s to simulate coiling, followed by air cooling to room temperature.

The processing schedule parameters (deformation temperature range, total strain and strain per pass, strain rate, and cooling rate between passes) have been defined to model the industrial rolling process within reasonable limits of the Gleeble simulator.

Microstructure characterisation for the four studied conditions was carried out using optical, scanning (SEM) and transmission (TEM) electron microscopy. For optical and SEM microscopy, the Gleeble samples were cut parallel to the normal direction (ND)–rolling direction (RD) plane, where ND is the compression direction and RD represents the rolling direction in Gleeble simulation. For TEM and tensile properties testing the samples were cut parallel to the normal direction (ND)–transverse direction (RD) plane. Optical and SEM sample preparation included polishing with SiC papers and diamond suspensions followed by etching with 5% Nital. Foils for TEM were prepared by hand polishing with a number of SiC papers, pre-thinning on a dimple grinder, and ion milling on a Gatan PIPS machine (manufactured by Gatan, Pleasanton, CA, USA). Optical microscopy was conducted on a Leica DM6000M microscope (manufactured by Leica Microsystems, Wetzlar, Germany) equipped with Leica Application Suite (LAS) 4.0.0 image processing software (developed by Leica Microsystems). Scanning electron microscopy was carried out using a JEOL 7001F FEG scanning electron microscope (manufactured by JEOL, Tokyo, Japan) operating at 15 kV for imaging and 7 kV for energy dispersive X-ray spectroscopy (EDS) of precipitates. For the determination of size of bainitic ferrite (the shortest distance between the martensitic grains) more than 400 randomly located areas were manually measured in SEM images for each of four studied conditions. In the SEM visible size range precipitation was scarce. Thus, only a limited number of 50 particles was analysed for the determination of precipitate sizes, number density and area fraction values for each of four studied conditions. The EDS semi-quantitative point analysis was carried out for 20 particles for each studied condition using an AZtec 2.0 Oxford SEM EDS system (manufactured by Oxford Instruments, Abingdon, UK). Transmission electron microscopy was conducted on a JEOL JEM2010 TEM microscope (manufactured by JEOL, Tokyo, Japan). For the analysis of <15 nm particle parameters, 200–500 precipitates were imaged for each of four studied conditions. The precipitates type was analyzed using selected area diffraction. The foil thickness was measured to be ~80 nm; a convergent beam diffraction technique was applied for this measurement [42]. Imaging of dislocation structure was performed for the beam direction being close to [011] grain zone axis. Tensile testing for the four studied conditions was carried out on a Kammrath and Weiss GmbH tensile stage. Testing was performed using 3 mm wide, 1 mm thick and 7 mm gauge length flat specimens. The constant crosshead speed of 7 $\mu m \cdot s^{-1}$ was applied and resulted in 1×10^{-3} s^{-1} strain rate. Two specimens were tested per condition.

3. Results

3.1. Grain Structure and Phase Balance

Optical, SEM and TEM microscopy showed in both steels a microstructure of mixed granular bainite and bainitic ferrite (Figure 1). Blocky or elongated martensite was present as the second phase. In both steels, the martensite crystals comprise a number of sub-grains with low angle (~10°) boundaries between them in accordance with possible intervariant misorientation of 10.53° [43] for Kurdjumov–Sachs relationship between the parent austenite (face centred cubic (fcc) crystal structure) and product martensite (body centred cubic (bcc) crystal structure, observe the rotation of diffraction patterns in neighbouring sub-grains in Figures 2 and 3). The diffraction analysis (points A, B and C in Figure 2b, and points A and B in Figure 3b) confirmed the bcc type of crystal structure of martensite. Retained austenite was not observed. The average size of bainitic ferrite (the shortest distance between bainitic ferrite-martensite boundaries across the bainitic ferrite area) was measured to be below 1 μm (Table 1). The variation in the average sizes and shape of size distributions (Figure 4b) of bainitic ferrite with steel composition and processing schedule was insignificant and could result from the measurement error. However, a noticeable variation in the size of martensite grains was observed (Figure 4c,d). Thus, for low strain processing, the maximum size of blocky grains was 16% smaller and the maximum length of elongated grains was 40% shorter in the CrNbV-steel. For high strain processing an opposite trend was observed: the average and maximum sizes of blocky grains were 64% and 3.7 times, respectively, larger in the CrNbV-steel; and average and maximum length of elongated grains were 20% and 30%, respectively, larger in the CrNbV-steel. In addition for high strain processing, the total fraction of martensite was 1.5 times higher in the CrNbV-steel.

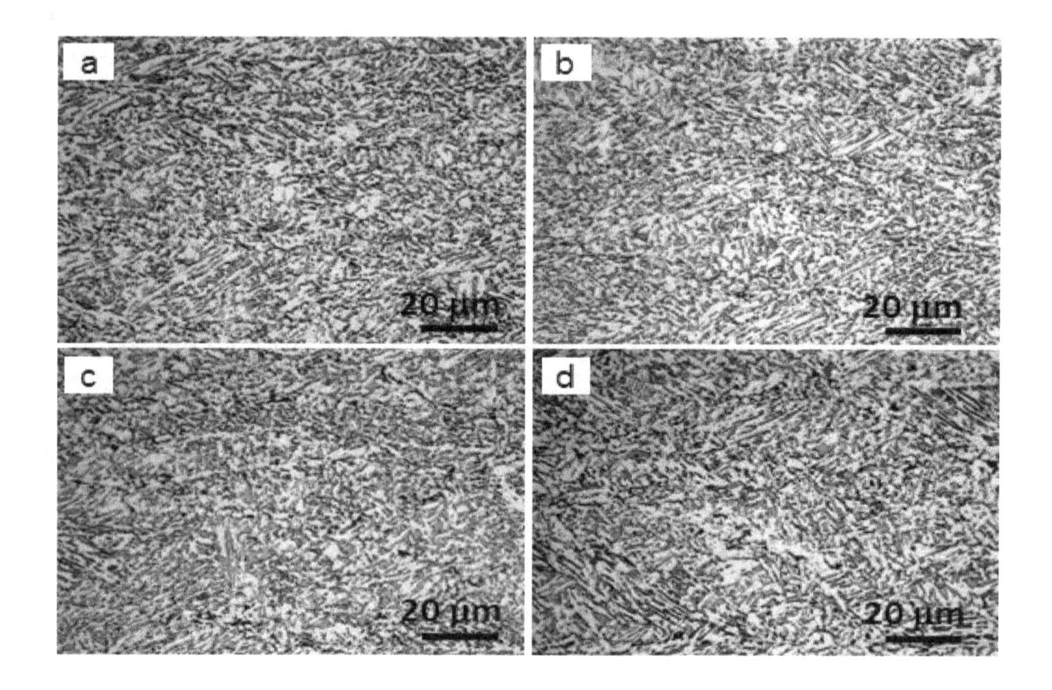

Figure 1. Optical images of microstructures in (**a**,**b**) MoNbV-steel and (**c**,**d**) CrNbV-steel after (**a**,**c**) low and (**b**,**d**) high strain processing.

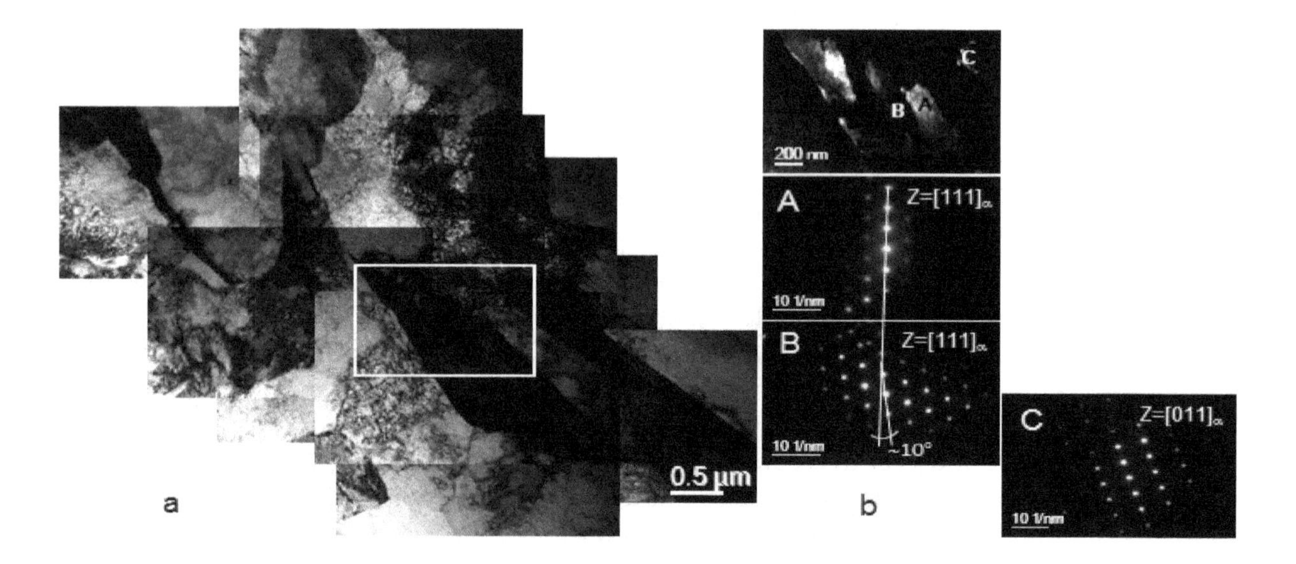

Figure 2. TEM (**a**) bright field image of microstructure and (**b**) dark field image of martensite in MoNbV-steel after low strain processing; (**b**) is from the white frame in (**a**); A, B and C diffraction patterns were taken from the corresponding points in the dark field image.

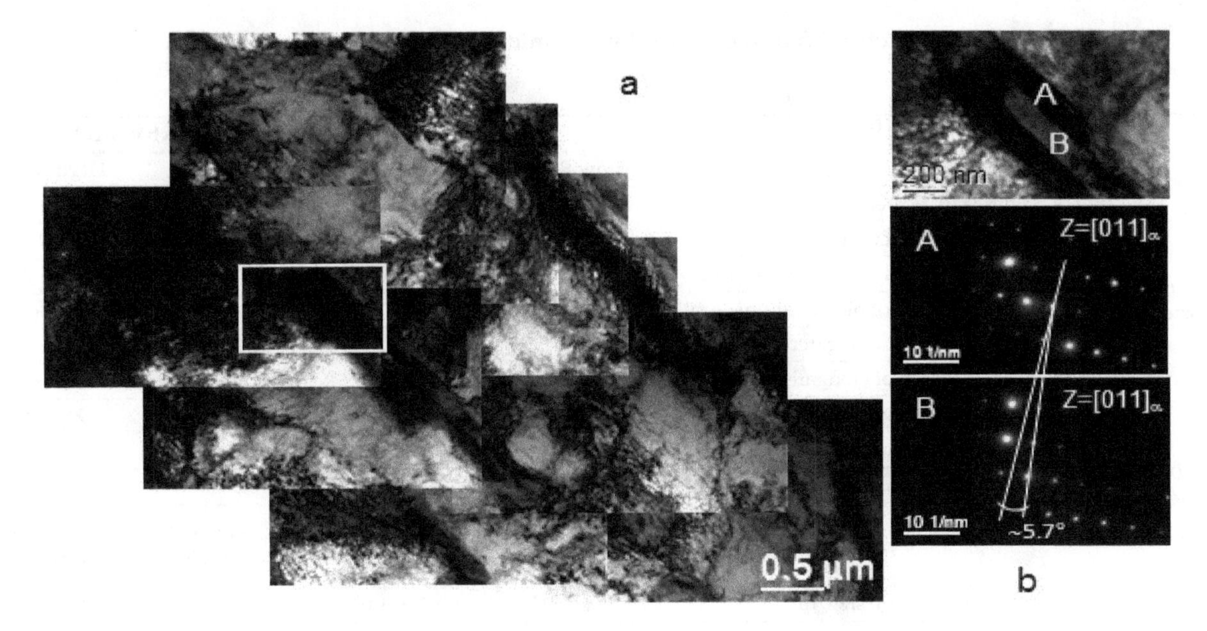

Figure 3. TEM (**a**) bright field image of microstructure and (**b**) dark field image of martensite in CrNbV-steel after low strain processing; (**b**) is from the white frame in (**a**); A and B diffraction patterns were taken from the corresponding points in the dark field image.

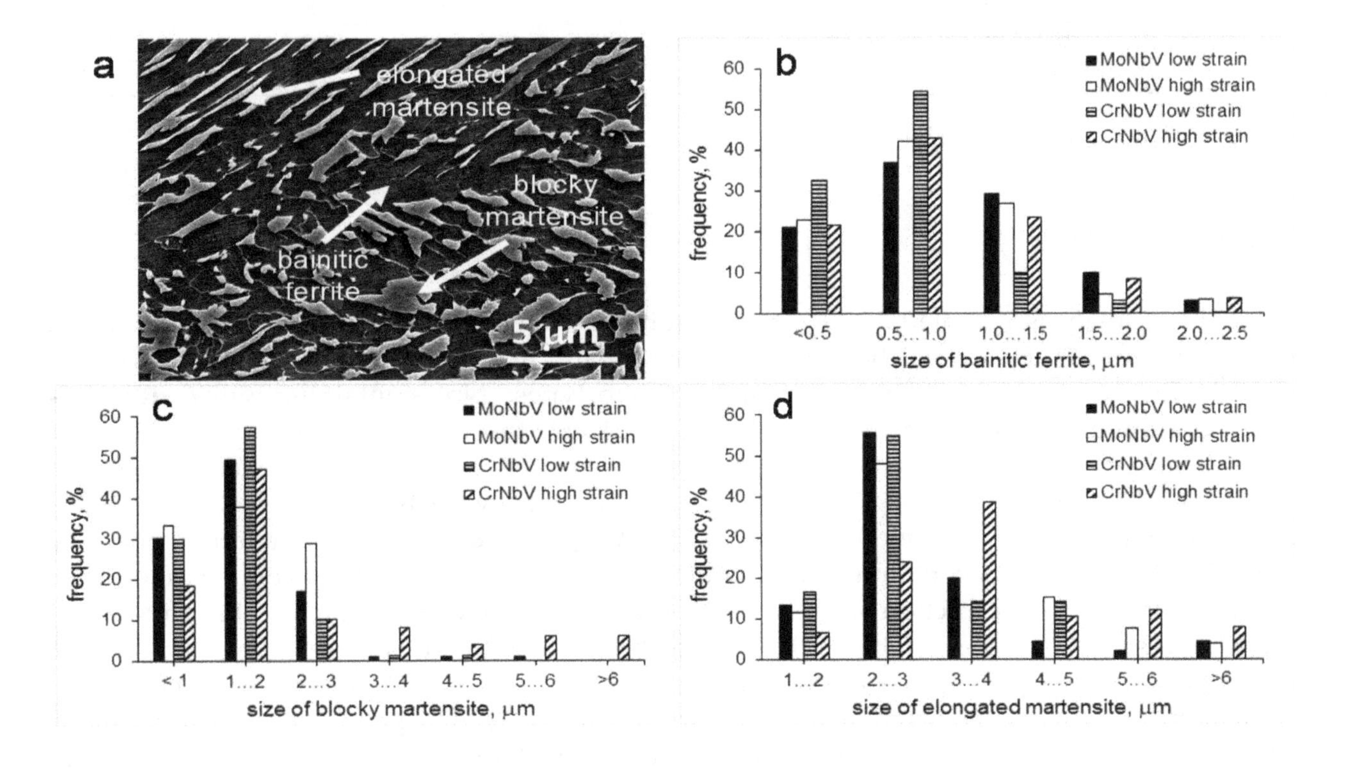

Figure 4. (**a**) A representative SEM image of microstructure in CrNbV-steel and size frequency distributions of (**b**) bainitic ferrite; (**c**) blocky martensite and (**d**) elongated martensite for four studied conditions.

Table 1. Microstructural parameters and mechanical properties of the studied steels.

Parameters		MoNbV Steel		CrNbV Steel	
		Low Strain	High Strain	Low Strain	High Strain
size of bainitic ferrite areas #, μm		0.95 ± 0.45	0.84 ± 0.42	0.72 ± 0.33	0.91 ± 0.45
martensite	fraction, %	20	11	20	17
	average size of blocky grains, μm	1.4 ± 0.7	1.4 ± 0.6	1.2 ± 0.7	2.3 ± 2.0
	maximum size of blocky grains, μm	5.0	2.8	4.2	10.4
	average length of elongated grains, μm	2.8 ± 1.2	3.0 ± 1.3	2.7 ± 0.9	3.7 ± 1.8
	maximum length of elongated grains, μm	8.3	8.0	4.8	10.4
>20 nm particles (SEM)	average size, nm	42 ± 15	21 ± 4	27 ± 12	26 ± 13
	number density, μm^{-2}	0.27	0.91	0.95	1.67
	area fraction	0.0003	0.0003	0.0006	0.0011
	chemistry	75% with MoNbV 25% with Mo	36% with MoNbV 64% with Mo	59% with NbV 41% with Nb	58% with NbV 42% with Nb
<20 nm particles (TEM)	average size, nm	2.4 ± 0.5	3.2 ± 1.0	2.7 ± 0.7	2.8 ± 1.0
	number density, μm^{-3}	15,667	9875	25,595	16,744
	volume fraction	0.0001	0.0002	0.0003	0.0003
	chemistry	Cementite, Fe_3C			
dislocation density, $\times 10^{15}$ m^{-2}		0.93 ± 0.15	0.43 ± 0.10	0.85 ± 0.11	0.41 ± 0.10
matrix unite cell size, nm		0.310	0.312	0.308	0.306
YS, MPa		850 ± 30	775 ± 35	765 ± 30	700 ± 20
UTS, MPa		1200 ± 45	1090 ± 40	1000 ± 25	975 ± 20
Elongation, %		14 ± 2	13.5 ± 2	16 ± 3	19.5 ± 2

The shortest distance between the martensitic grains. YS: yield stress; UTS: ultimate tensile strength.

3.2. Particle Precipitation

The SEM analysis revealed 20–70 nm precipitates in the bainitic ferrite in all four studied conditions (Figure 5). However, the particle chemistry and number density varied with condition. In the MoNbV-steel, the particles were mainly of two types: NbV-containing with/without Mo and Mo-containing without Nb or V. In the CrNbV-steel, the particles were also of two types: NbV-containing and Nb-containing (Figure 6). Detailed characterisation of the particle compositions was outside of this paper scope; however, on the basis of previously published data we believe that NbV-containing particles were MX type NbV(CN) [44–46], MoNbV-containing ones were MoNbVC [13,47], and Mo-containing ones could be complex FeMoC [48]. The average particle number density and area fraction were lower in the MoNbV-steel, compared to the CrNbV-steel, for both processing conditions (Table 1). In the MoNbV-steel, with an increase in deformation strain the average >20 nm particle size decreased (from 42 ± 15 to 21 ± 4 nm), number density increased (from 0.27 to 0.91 μm^{-2}), and the relative amount of Mo-containing particles to the total amount analysed also increased (from 25% to 64%). In the CrNbV-steel, the average >20 nm particle size and relative amount of Nb-containing particles did not show a significant variation with strain, although both the number density and area fraction increased (from 0.95 to 1.67 μm^{-2} and from 0.0006 to 0.0011, respectively) with strain.

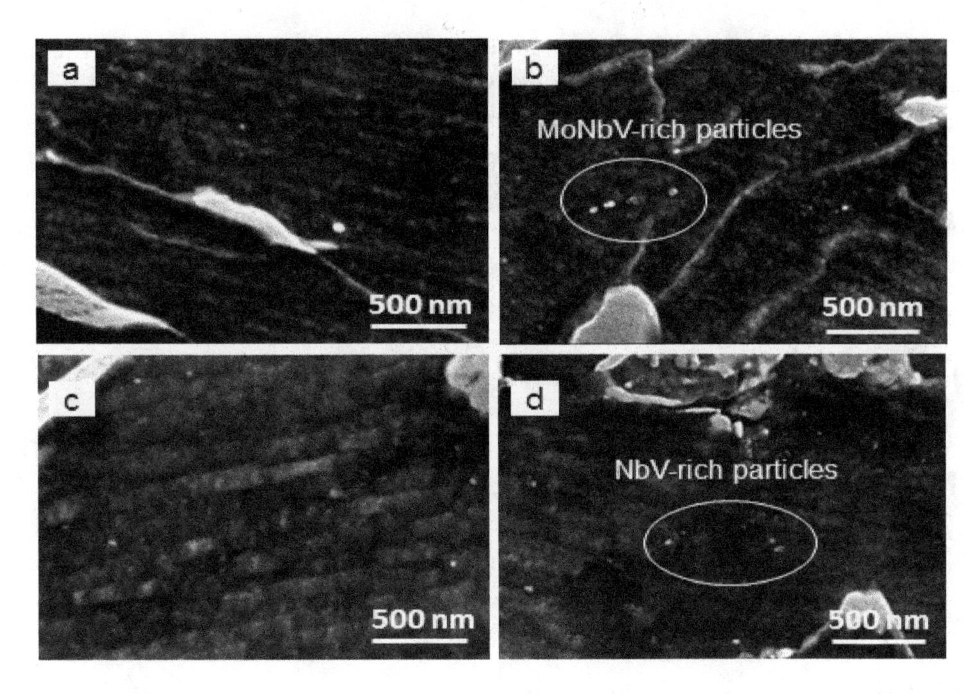

Figure 5. SEM images of precipitates in (**a**,**b**) MoNbV-steel and (**c**,**d**) CrNbV-steel after (**a**,**c**) low and (**b**,**d**) high strain processing.

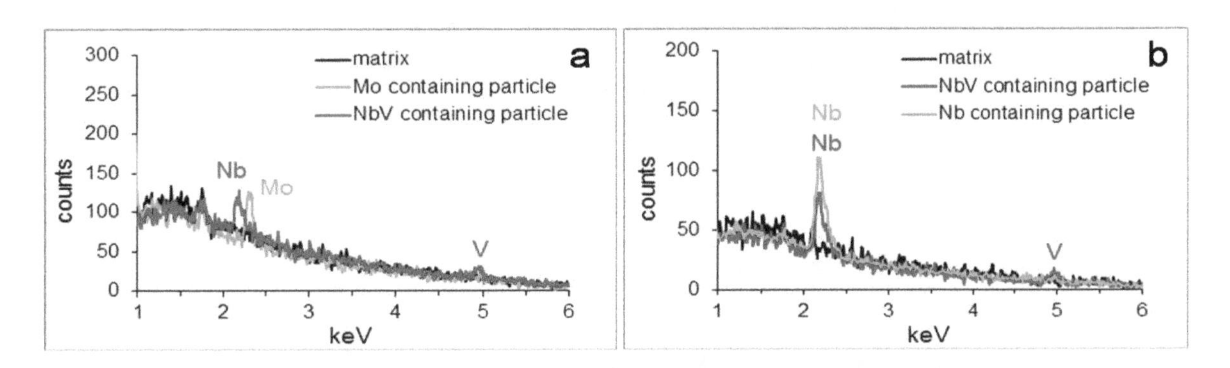

Figure 6. Energy dispersive X-ray spectroscopy (EDS) spectra of (**a**) Mo and NbV containing particles in the MoNbV-steel and (**b**) Nb and NbV containing particles in the CrNbV-steel.

The TEM investigation showed presence of 2–8 nm precipitates in the bainitic ferrite in all four conditions (Figure 7). As the particles were too small for EDS, their nature was analysed using the selected area diffraction technique. Numerous calculations (omitted here) suggested absence of Mo-, Cr-, Nb- or V-rich carbides or nitrides in the TEM studied particle size range in both steels. Thus, the particles were identified as Fe_3C exhibiting Bagaryatskii [49] orientation relationship to the bcc (bainitic ferrite) matrix: $[011]_{matrix} \parallel [001]_{Fe3C}$ and $[001]_{matrix} \parallel [321]_{Fe3C}$ (Figure 8). Measurements of d-spacing have shown $d_{0\bar{1}2}$ = 0.306 nm, $d_{1\bar{1}\bar{1}}$ = 0.317 nm and d_{200} = 0.358 nm, which was slightly larger than the theoretical values $d_{0\bar{1}2}$ = 0.281 nm, $d_{1\bar{1}\bar{1}}$ = 0.302 nm and d_{200} = 0.337 nm calculated using the Fe_3C lattice parameters a = 0.674 nm, b = 0.509 nm and c = 0.453 nm [50]. The unit cell size of bcc (bainitic ferrite) matrix, measured using the TEM diffraction patterns, was also expanded to 0.306–0.312 nm (Table 1) from the theoretical value of 0.286 nm. The Fe_3C expansion by 5–9% corresponds to this of matrix by 7–9%. It is important to note, that the matrix expansion was larger in the MoNbV-steel than that in the CrNbV-steel. The matrix expansion could result from an increased concentration of solid solute atoms [51]. The average Fe_3C size did not vary significantly with steel composition and processing (was within the measurement error). However, the average Fe_3C number density and area fraction were lower in the MoNbV-steel, compared to the CrNbV-steel, for both processing conditions (Table 1). With an increase in deformation strain the average 2–8 nm particle number density decreased from 15,667 to 9875 μm^{-3} in the MoNbV-steel and from 25,595 to 16,744 μm^{-3} in the CrNbV-steel. Within the 2–8 nm size range an opposite trend was observed for 2–3 nm and 3–8 nm particles: amount of 2–3 nm ones decreased with strain and this of 3–8 nm ones increased with strain (compared Figure 9a,b). An opposite trend with strain was also observed for the 2–8 nm particles (studied by TEM) compared to the >20 nm ones (studied by SEM): with an increase in strain the number density of >20 nm particles increased and this of 2–8 nm ones decreased.

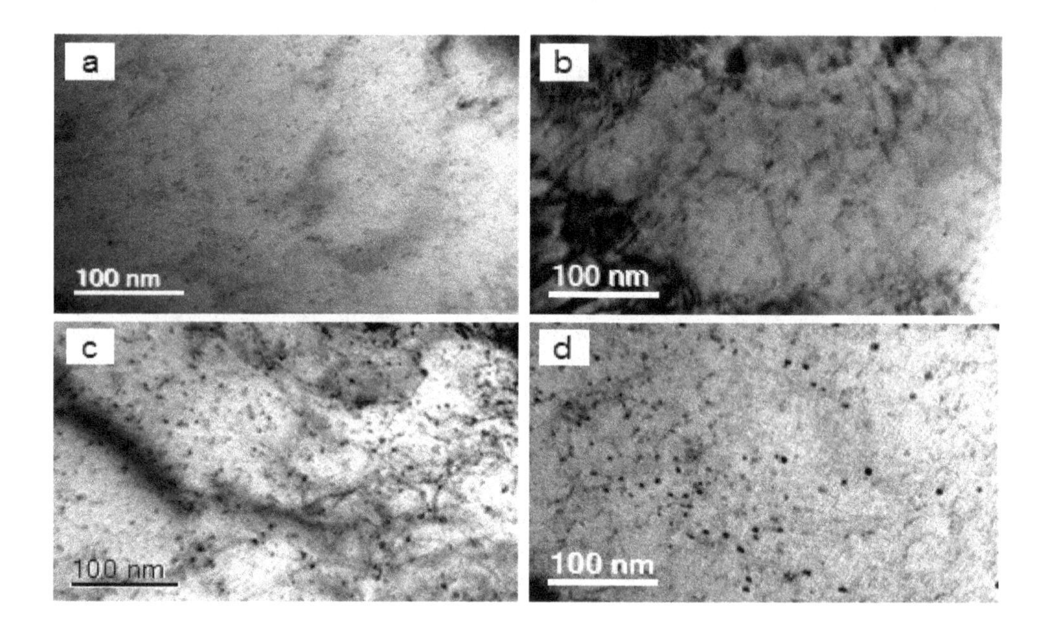

Figure 7. TEM bright field images of precipitates in (**a**,**b**) MoNbV-steel and (**c**,**d**) CrNbV-steel after (**a**,**c**) low and (**b**,**d**) high strain processing.

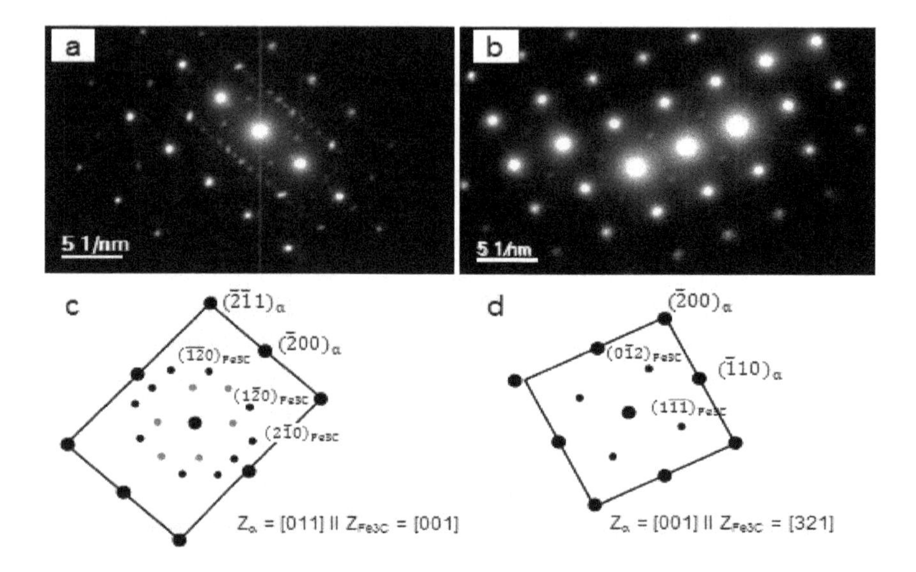

Figure 8. Selected area diffraction patterns of Fe_3C precipitates in (**a**) MoNbV-steel and (**b**) CrNbV-steel; (**c**) determination of the matrix-particle orientation relationship for image (**a**) and (**d**) this for image (**b**).

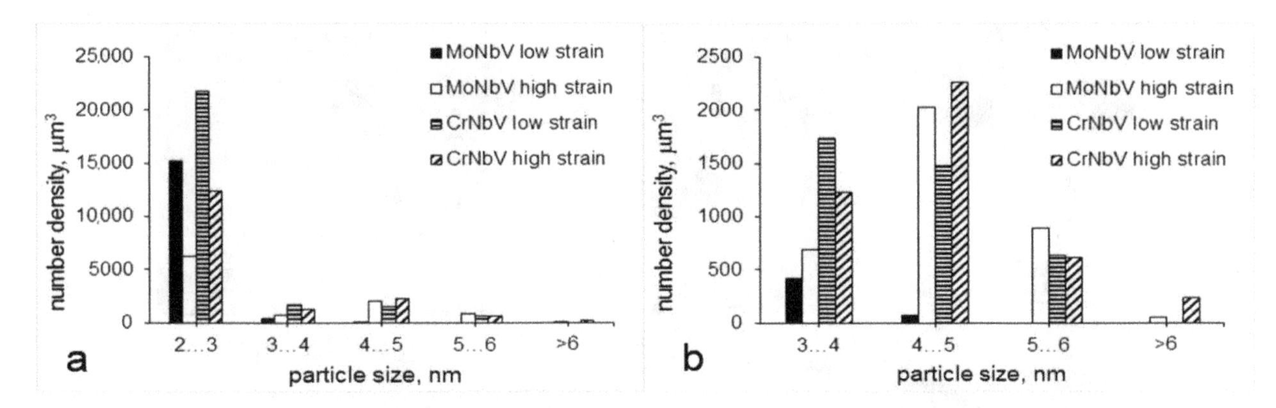

Figure 9. Number density distributions of precipitates studied by TEM for (**a**) >2 nm size range and (**b**) >3 nm size range.

3.3. Dislocation Structure

Typical dislocation structure in the middle of bainitic ferrite areas is shown in Figure 10 and some selected features are presented in Figure 11. In both steels, the average dislocation density in bainitic ferrite was at the level of $(0.9 \pm 0.15) \times 10^{15}$ m^{-2} after low strain processing and $(0.4 \pm 0.10) \times 10^{15}$ m^{-2} after high strain processing (Table 1). These values correspond to the reported in the literature for bainitic microstructures [52–54]. In the MoNbV-steel very high density dislocation walls surrounding a low density interior (arrangements resembling cells) where occasionally observed (Figure 11a), although they were not present in the CrNbV-steel. Bainitic ferrite areas closer to the martensite grains exhibited a higher local dislocation density than the overall average (Figure 11b). In both steels the dislocation arrays (Figure 11c,e), disintegrated walls (Figure 11d) and tangles (Figure 11f) were also observed, mainly after high strain processing.

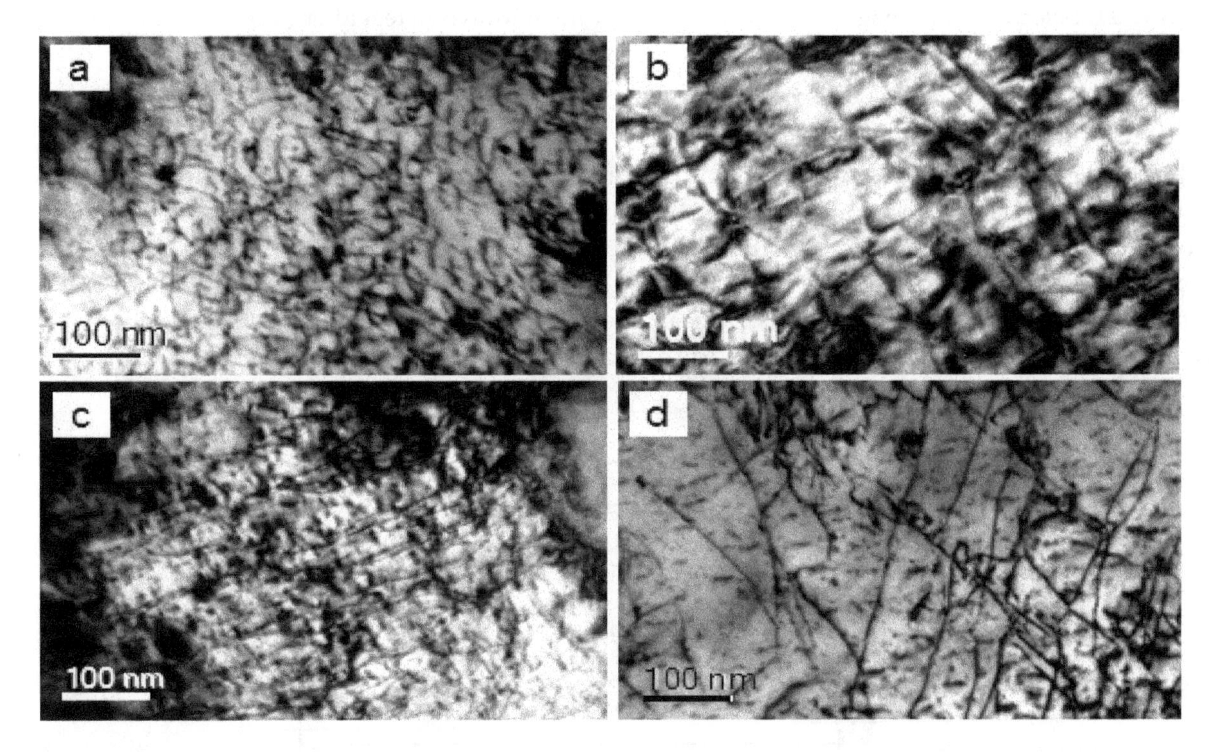

Figure 10. Representative TEM images of dislocation structure in (**a**,**b**) MoNbV-steel and (**c**,**d**) CrNbV-steel after (**a**,**c**) low and (**b**,**d**) high strain processing.

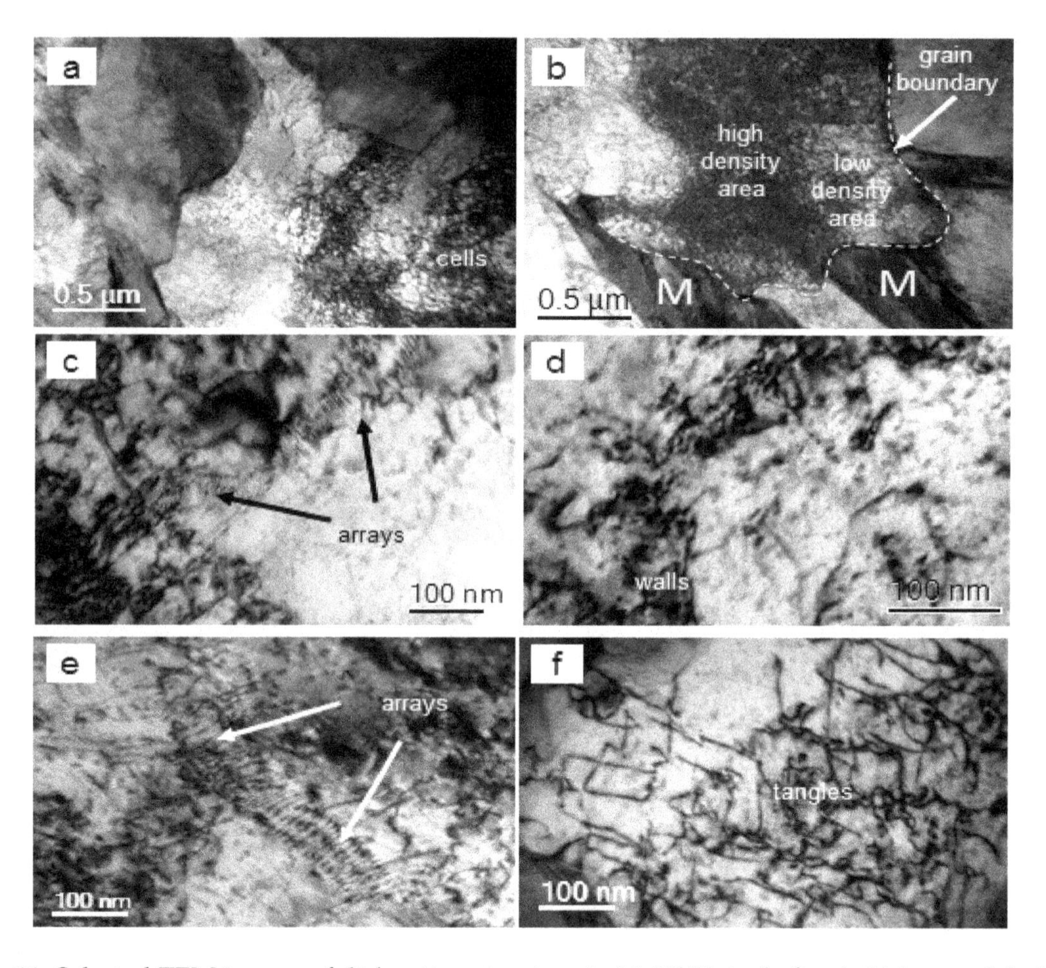

Figure 11. Selected TEM images of dislocation structure in MoNbV-steel after (**a**,**b**) low and (**c**,**d**) high strain processing; and in CrNbV-steel after (**e**) low and (**f**) high strain processing.

3.4. Mechanical Properties

The MoNbV-steel showed higher strength, and slightly lower elongation, than the CrNbV-steel (Table 1, Figure 12). The variations in yield stress (YS) and ultimate tensile strength (UTS) with steel composition were higher for the low strain schedule: 85 MPa in YS and 200 MPa in UTS for the low strain and 75 MPa in YS and 115 MPa in UTS for the high strain schedule. It is worth to note an opposite trend in the elongation variation with strain: in the MoNbV-steel elongation slightly decreased with strain, and in the CrNbV-steel it increased with strain.

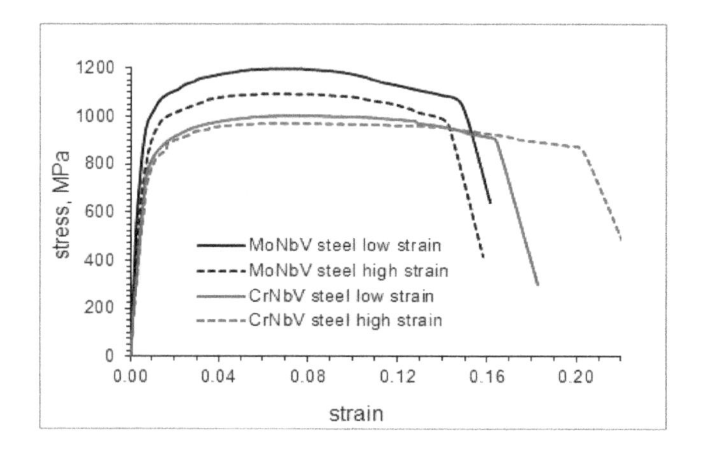

Figure 12. Engineering stress-strain curves for four studied conditions.

4. Discussion

According to various empirical equations [55–57]:

$$B_s = 830 - 270C - 90Mn - 37Ni - 70Cr - 83Mo,$$

$$B_s = 732 - 202C + 216Si - 85Mn - 37Ni - 47Cr - 39Mo,$$

$$B_s = 745 - 110C - 59Mn - 39Ni - 68Cr - 106Mo + 17MnNi + 6Cr^2 + 29Mo^2,$$

the bainite transformation start temperature, B_s, was similar in both steels: 605–628 °C in the MoNbV-steel and 612–631 °C in the CrNbV-steel. These values can decrease by 40–70 °C, if 0.06 wt. % of Nb additions and 30 $°C{\cdot}s^{-1}$ cooling rate are taken into account [58,59], reaching ~560 °C in the MoNbV-steel and ~565 °C in the CrNbV-steel. A possible effect of deformation on B_s is difficult to assess quantitatively. Although it is known that pre-strain may increase B_s [60], due to an increase in the number of bainite nucleation sites, and retard the bainite transformation rate following mechanical stabilisation of austenite [61,62]. Thus, it is obvious that for 500 °C finish cooling/holding temperature we observed the bainitic microstructure in both steels. However, the Mo and Cr additions, and strain variation did show some effects on: (i) dislocation structure in the bainitic ferrite and morphology of martensite; and (ii) particle precipitation. Consequently, the mechanical properties varied.

4.1. Strain Effect on Phase Transformation and Precipitation

In both steels, higher strains should have enhanced DRX (dynamic recrystallization) and strain induced precipitation. Although, the absolute values of grain size, particle number density and solid solute concentrations could have been expected to differ with Mo and Cr contents. Thus, with strain increase: (i) the average dislocation density in bainitic ferrite decreased in both steels; (ii) dislocation cell arrangements did not form in the MoNbV-steel and disintegrated walls were present instead; (iii) the fraction of martensite decreased in both steels, although by a different value: by 1.8 times in the MoNbV-steel and by 15% in the CrNbV-steel; and (iv) the average and maximum sizes of blocky and elongated crystals of martensite either remained constant or decreased in the MoNbV-steel, although they have increased in the CrNbV-steel. All these could be explained if after higher strain processing and more intense dynamic recrystallization of austenite (DRX) the prior austenite grain size (PAGS) was smaller in the MoNbV-steel, due to more effective grain boundary pinning by Mo solute atoms, and coarser in the CrNbV-steel, due to grain growth. Smaller PAGS would increase B_s temperature and help nucleation of the bainitic ferrite. With sufficient holding time, this would result in a low retained austenite fraction available for the martensitic transformation. A slightly lower dislocation density in bainitic ferrite after high strain schedule compared to the low strain one (Table 1), could be explained by the increased B_s temperature and longer time at high temperature available for re-arrangement of dislocations after bainitic ferrite formation.

With strain increase the >20 nm particles area fraction and number density increased and the <20 nm volume fraction and number density decreased in both steels. This indicates faster nucleation and growth of precipitates for the higher strain schedule. In addition, the amount of Mo-containing particles in the MoNbV-steel increased with strain. All these support the expected intensification of strain induced precipitation of NbV-containing particles in both steels and Mo-containing ones in the MoNbV-steel with strain increase.

Enhancement of strain induced precipitation should have resulted in decreased element concentrations in solid solution and possible prior austenite strength decrease. If this occurred, low strength austenite would be faster transforming to bainite (faster growth of the bainitic ferrite would take place) [54], resulting in a lower fraction of retained austenite available for the transformation to martensite during cooling to room temperature after holding. This could be another reason, in addition to PAGS size variation, leading to a decreased fraction of martensite after the higher strain processing.

4.2. Mo and Cr Effects on Phase Transformation

For the low strain schedule, we observed insignificant effect of 0.2 wt. % Mo or Cr additions on phase characteristics, in particular, the average size of bainitic ferrite areas, dislocation density in bainitic ferrite, size of blocky and elongated grains of martensite, and martensite fraction. Although, the maximum sizes of blocky and elongated martensite slightly increased with Mo content. This could result from the variation in recrystallization stop temperature, T_{nr}, and prior austenite grain size (PAGS) in the studied steels. Our measurements have shown T_{nr} to be higher in the MoNbV-steel, ~1000 °C, compared to the CrNbV-steel, ~975 °C, which is in-line with Mo being a stronger recrystallization retarding element than Cr [36,37]. Therefore, after 0.3 strain at 1175 °C and 0.35 strain at 1100 °C (modest strain levels with respect to the rate of DRX) the PAGS could be slightly larger in the MoNbV-steel as a result of partial DRX. A larger PAGS would result in a larger size of austenite retained after holding at 500 °C and, subsequently, a coarser martensite formed during the final cooling to the room temperature. Presence of diverse dislocation structure (cell walls) in the MoNbV-steel after low strain processing could also result from a larger PAGS. Larger PAGS was observed accelerating the bainite growth rate in low carbon steels [63].

For the high strain schedule, the effect of steel composition on microstructure was more pronounced. Mo addition led to a decreased size (average and maximum) and lower fraction of martensite. This could be explained if after a higher strain PAGS was smaller in the MoNbV-steel (opposite trend to the low strain schedule). High strain levels would increase the rate of DRX in both steels. However, in the MoNbV-steel the recrystallized fine grain size would be preserved by Mo solute atoms pinning the grain boundaries and preventing the grain growth. In contrast, the solute drag effect of Cr atoms was weaker and the grain growth took place in the CrNbV-steel. Smaller PAGS (larger grain boundary area) can facilitate nucleation of bainitic ferrite and increase the Bs temperature [64]. Therefore, in the MoNbV-steel smaller PAGS resulted in smaller size and lower fraction of the martensite.

4.3. Mo and Cr Effects on Precipitation

The effect of 0.2 wt. % Mo addition on precipitation was more pronounced than that of 0.2 wt. % Cr. In the MoNbV-steel, Mo was present in >20 nm particles, although in the CrNbV-steel Cr did not precipitate. The number density and area fraction of >20 nm Mo/MoNbV-containing particles were 3.5 and 2 times lower, for the low strain schedule, and 1.8 and 3.7 times lower, for the high strain schedule, in the MoNbV-steel compared to the corresponding parameters of Nb/NbV-containing particles in the CrNbV-steel. This indicates a stronger potential of Mo to increase solubility of C and possibly of Nb and V in austenite than that of Cr, thus delaying the precipitation. The effect of Mo addition on the C solubility was reported previously [65].

The number density and volume fraction of <20 nm Fe_3C particles were 1.6 and 3 times lower, for the low strain schedule, and 1.7 and 1.5 times lower, for the high strain schedule, in the MoNbV-steel. This suggests a stronger ability of Mo to retard Fe_3C precipitation than this of Cr. A combination of several factors might be responsible for this. The tendency for C atoms to form Mo-C dipoles in preference to Fe-C ones was reported previously [66]. This is linked to a higher binding energy between Mo-C atoms (0.45–0.5 eV [67]) compared to the binding energy between the Fe and C in cementite (0.40–0.42 eV [68,69]). A higher binding energy means that a higher activation energy is required for C atom to jump into another position. The carbon diffusivity in Fe also changes in the presence of different solutes and it was reported for both Cr and Mo that the activation enthalpy for carbon diffusion in iron increases [70–72]. However, the carbon diffusivity in ferrite was decreased more in the presence of Mo than of Cr [73]. Absence of Mo-, Cr-, Nb- or V-rich particle precipitation in the <20 nm size range in both steels increased concentrations of these elements, carbon and, maybe, nitrogen in solid solution. In addition, the solid solute concentrations could be higher in the MoNbV-steel, due to less developed precipitation. This would correspond to a larger unit cell expansion of the bainitic ferrite in the MoNbV-steel (0.310–0.312 nm) than that in the CrNbV-steel (0.306–0.308 nm). A larger effect of Mo than Cr on Fe lattice expansion is related to larger atom radius mismatch between Fe (126 pm) and Mo (139 pm) than between Fe and Cr (128 pm) [74].

4.4. Microstructure-Mechanical Properties Relationship: Role of Solute Atoms

In spite of similar phase balance, grain size and dislocation density, and less pronounced precipitation, the MoNbV-steel exhibited higher strength than the CrNbV-steel for both processing conditions (Table 1). This could result from higher solid solution strengthening from Mo, Cr, Nb, V, C and N atoms in the MoNbV-steel. To clarify this the contributions to yield stress from various microstructural parameters have been calculated as follows:

- from grain boundaries using the Hall–Petch equation:

$$\sigma_{gs} = \sigma_0 + k \cdot d^{-1/2},$$

  where σ_0 = 15 MPa and k = 21.4 MPa·mm$^{1/2}$ are accepted for pure iron [75], and d is the size of bainitic ferrite areas (the shortest distance between the martensitic grains);
- from precipitation of >20 nm particles using the Ashby-Orowan equation [76], which assumes the dislocation looping around relatively large particles:

$$\Delta\sigma_{ps1} = \frac{10.8\sqrt{f}}{D} \ln\left(\frac{D}{6.125 \times 10^{-4}}\right),$$

  where f is the particle volume fraction and D is the particle diameter in μm;
- from precipitation of <20 nm particles using the order strengthening relationship [77], which assumes the dislocation cutting of relatively small, coherent particles:

$$\Delta\sigma_{ps2} = 0.81 \cdot M \cdot \frac{\gamma}{2b} \cdot \left(\frac{3\pi f}{8}\right)^{0.5},$$

  where M = 3 is the matrix orientation factor, b = 0.312–0.306 nm is Burgers vector accepted according to the measured unite cell size of bainitic ferrite (Table 1), γ is the matrix-particle interface energy assumed for the Fe-Fe_3C interface to be γ = 0.5 J·m^{-2} [78], and f is the particle volume fraction;
- from dislocations using the long range work hardening theory [79]:

$$\Delta\sigma_{wh} = \frac{\alpha}{2\pi} G b \sqrt{\rho},$$

  where α = 0.5 is a constant, G = 85,000 MPa is the shear modulus, b = 0.312–0.306 nm is the Burgers vector and ρ is the measured dislocation density;
- the solid solution strengthening contribution from Mn and Si was estimated using the matrix concentrations of these elements and the following relationship [76]:

$$\Delta\sigma_{ss(Si,Mn)} = 83C_{Si} + 32C_{Mn},$$

  where C_{Si} = 0.3 wt. % and C_{Mn} = 1.5 wt. % are Si and Mn concentrations in the bainitic ferrite matrix, respectively.

A possible effect of martensite on the yield stress of studied steels was neglected in the calculation, due to the martensite requiring higher stresses and strains, than bainitic ferrite, to start yielding.

As can be seen from Table 2, for both steels the major contribution to the YS was coming from grain boundary strengthening, which is quite reasonable for the size of bainitic ferrite areas being below 1 μm. Contributions from dislocations were in the same range of values, which is in-line with similar dislocation densities measured in the studied steels. The solid solution strengthening from Si and Mn was the same for all four conditions, because the contents of these elements were similar in both steels and their precipitation was not observed. For three cases, namely the CrNbV-steel in both processing conditions and the MoNbV-steel after high strain processing, calculations overestimated the yield stress (a negative value for the difference between the measured and total calculated yield stress, Δ, was observed, Table 2). This can be related to two major reasons: (i) incomplete number of dislocation-obstacle interactions really occurring, compared to the theoretical maximum assumed in the applied equations; and (ii) the material inhomogeneity. Some volumes of the bainitic ferrite matrix could be softer, due to lower solute atom concentrations or/and lower number density of precipitates or/and lower dislocation density. These softer volumes would start yielding first. However, for the MoNbV-steel processed according to the lower strain schedule the measured yield stress was 89 MPa (~12%) higher than the calculated value. This is in contrast to the lowest precipitation strengthening contribution calculated for the MoNbV-steel subjected to the lower strain processing. The additional strengthening in this condition could originate from two sources: (i) solid solute atoms of Mo, C and, possibly, N; and (ii) atom clusters of Mo, Nb and V. Substantial strengthening from atom clusters was recently reported for microalloyed steels [80–83]. In spite of qualitatively similar effects of Mo and Cr on solubility of other elements, in the CrNbV-steel a possible strengthening from solute atoms and atom clusters did not exhibit itself. Obviously, a quantitatively weaker effect of Cr on solubility resulted in more pronounced precipitation and lower concentrations of microalloying elements available for solid solution and cluster strengthening.

Table 2. Calculated contributions to the yield stress in the studied steels.

Steel	Strain	Calculated Contributions						Experimental YS	Δ *
		Grain Boundaries #	Particles		Dislocation Density	Solid Solute (Si, Mn)	Total		
			>20 nm	>20 nm					
MoNbV	Low	586	19	21	62	73	761	850	+89
	High	618	31	30	42	73	794	775	−19
CrNbV	Low	675	37	37	58	73	880	765	−115
	High	595	52	37	40	73	797	700	−97

For the largest areas of bainitic ferrite; * Δ is the difference between the experimental and total calculated yield stress values.

For both steels, the measured yield stress decreased with an increase in strain during processing. In the CrNbV-steel this corresponds to the decrease in grain size strengthening contribution (in trend and value). However, in the MoNbV-steel the grain size and precipitation strengthening contributions increased with strain. Obviously, in the MoNbV-steel these contributions did not compensate for the decreased solid solution/cluster strengthening. This indicates a significant role of solid solute atoms and atom clusters as strengthening agents in bainitic microalloyed steels. With an increase in dislocation density, associated with bainitic microstructure, a potential number of dislocation-atom/cluster interaction sites may increase faster than the number of dislocation-precipitate interaction sites, due to a very high density of atoms/clusters. Thus, the role of solute concentrations increases in bainitic microstructures compared to the ferritic.

In the CrNbV-steel coarser martensite could be expected to decrease elongation for the higher strain processing. However, an opposite trend was observed: the elongation has increased in the CrNbV-steel for the higher strain schedule. This could have been related to more intensive dislocation generation in the CrNbV-steel during tensile testing, leading to more homogeneous slip and delated local micro-crack formation. The dislocation generation in the CrNbV-steel would be associated with the increased number density of >20 nm particles for this processing condition. Nano-precipitates were reported to stimulate dislocation generation [84,85].

Based on these a final conclusion can be made with respect to the design of steel composition: if a bainitic steel is required to exhibit superior strength, Mo should be used as a microalloying element, because it increases the solubility of other microalloying elements and increases the solid solution and cluster strengthening effects; if a bainitic steels is required to exhibit higher ductility, Cr can be used to facilitate bainite transformation but allow precipitation to take place.

5. Conclusions

Comparative study of Mo and Cr effects on microstructure and mechanical properties of newly developed NbV-microalloyed bainitic steels has shown the following:

1. Additions of either 0.2 wt. % Mo or 0.2 wt. % Cr (above 0.3 wt. % Cr) resulted in formation of bainite microstructure with mixed bainite morphologies and similar parameters of bainitic ferrite matrix and martensite as the second phase. However, for higher strain processing (when the DRX may be completed faster) the size and fraction of martensite were lower in the MoNbV-steel. This could be related to a smaller prior austenite grain size in the MoNbV-steel, which would correspond to a stronger solute drag effect of Mo during DRX. The influence of prior austenite grain size on bainite phase transformation in the studied steels requires further investigation.
2. The number density and area fraction of >20 nm NbV-containing particles were significantly lower in the MoNbV-steel. This supports a stronger effect of Mo than Cr on the increase of solubility of Nb and V in austenite, which results in their decreased precipitation. The number density and area fraction of <20 nm Fe_3C particles were also lower in the MoNbV-steel. This supports a stronger effect of Mo than Cr on the retardation of Fe_3C precipitation. Less pronounced precipitation in the MoNbV-steel, and related to this increased solid solute concentrations, corresponds to a wider expansion of the unit cell size of the bcc lattice of bainitic ferrite matrix.
3. In addition to grain size strengthening, the solid solution/atom cluster strengthening effect was the second dominating in the MoNbV-steel; although in the CrNbV-steel the precipitation strengthening from >20 nm particles, <20 nm particles and dislocations equally contributed to the overall steel strength.
4. Strain increase resulted in more pronounced strain induced precipitation of Mo, Nb and V in the MoNbV-steel and Nb and V in the CrNbV-steel. Solute depletion in microalloying elements could have contributed to decreased strength levels in both steels for the higher strain processing schedule.

Acknowledgments: This project was financially supported by the ARC Research Hub for Australian Steel Manufacturing and, in particular, by Bluescope Steel Ltd. (Melbourne, Australia). The microscopy was carried out using JEOL JSM-7001F FEGSEM (supported by grant No. LE0882613) and JEOL JEM-2011 TEM (supported by grant No. LE0237478) microscopes at the Electron Microscopy Centre at the University of Wollongong.

Author Contributions: Andrii Kostryzhev has conceived the idea, conducted the microstructure characterisation and analysis of the microstructure-properties relationship, and has written the paper. Navjeet Singh and Liang Chen carried out the sample processing in Gleeble. Navjeet Singh conducted the tensile testing. Chris Killmore and Elena Pereloma carried out the overall project management, participated in discussion of the results, and contributed to the paper writing.

References

1. Aaronson, H.I.; Reynolds, W.T., Jr.; Purdy, G.R. The incomplete transformation phenomenon in steel. *Metall. Mater. Trans. A* **2006**, *37*, 1731–1745. [CrossRef]
2. Humphreys, E.S.; Fletcher, H.A.; Hutchins, J.D.; Garratt-Reed, A.J.; Reynolds, W.T., Jr.; Aaronson, H.I.; Purdy, G.R.; Smith, G.D.W. Molybdenum accumulation at ferrite: Austenite interfaces during isothermal transformation of an Fe-0.24 pct C-0.93 pct Mo alloy. *Metall. Mater. Trans. A* **2004**, *35*, 1223–1235. [CrossRef]
3. Kong, J.; Xie, C. Effect of molybdenum on continuous cooling bainite transformation of low-carbon microalloyed steel. *Mater. Des.* **2006**, *27*, 1169–1173. [CrossRef]
4. Hu, H.; Xu, G.; Zhou, M.; Yuan, Q. Effect of Mo content on microstructure and property of low-carbon bainitic steels. *Metals* **2016**, *6*, 173. [CrossRef]

5. Sung, H.-K.; Lee, D.-H.; Shin, S.-Y.; Lee, S.; Yoo, J.-Y.; Hwang, B. Effect of finish cooling temperature on microstructure and mechanical properties of high-strength bainitic steels containing Cr, Mo and B. *Mater. Sci. Eng. A* **2015**, *624*, 14–22. [CrossRef]
6. Uemori, R.; Chijiiwa, R.; Tamehiro, H.; Morikawa, H. AP-FIM study on the effect of Mo addition on microstructure in Ti-Nb steel. *Appl. Surf. Sci.* **1994**, *76*, 255–260. [CrossRef]
7. Wan, R.; Sun, F.; Zhang, L.; Shan, A. Effect of Mo addition on strength of fire-resistant steel at elevated temperature. *J. Mater. Eng. Perform.* **2014**, *23*, 2780–2786. [CrossRef]
8. Dimitriu, R.C.; Bhadeshia, H.K.D.H. Hot strength of creep resistant ferritic steels and relationship to creep rupture data. *Mater. Sci. Technol.* **2007**, *23*, 1127–1131. [CrossRef]
9. He, R.; Jiang, L.; Dong, W. Development of high corrosion-resistant ferritic stainless steel and its application in the building cladding system. *Baosteel Tech. Res.* **2013**, *7*, 54–58.
10. Akben, M.G.; Bacroix, B.; Jonas, J.J. Effect of vanadium and molybdenum addition on high temperature recovery, recrystallization and precipitation behaviour of niobium-based microalloyed steels. *Acta Metall.* **1983**, *31*, 161–174. [CrossRef]
11. Andrade, H.L.; Akben, M.G.; Jonas, J.J. Effect of molybdenum, niobium, and vanadium on static recovery and recrystallization and on solute strengthening in microalloyed steels. *Metall. Trans. A* **1983**, *14*, 1967–1977. [CrossRef]
12. Schambron, T.; Chen, L.; Gooch, T.; Dehghan-Manshadi, A.; Pereloma, E.V. Effect of Mo concentration on dynamic recrystallization behavior of low carbon microalloyed steels. *Steel Res. Int.* **2013**, *84*, 1191–1195. [CrossRef]
13. Lu, J.; Omotoso, O.; Wiskel, J.B.; Ivey, D.G.; Henein, H. Strengthening mechanisms and their relative contributions to the yield strength of microalloyed steels. *Metall. Mater. Trans. A* **2012**, *43*, 3043–3061. [CrossRef]
14. Zhou, X.; Liu, C.; Yu, L.; Liu, Y.; Li, H. Phase transformation behavior and microstructural control of high-Cr martensitic/ferritic heat-resistant steels for power and nuclear plants: A review. *J. Mater. Sci. Technol.* **2015**, *31*, 235–242. [CrossRef]
15. Wang, Q.; Zhang, C.; Li, R.; Gao, J.; Wang, M.; Zhang, F. Characterization of the microstructures and mechanical properties of 25CrMo48V martensitic steel tempered at different times. *Mater. Sci. Eng. A* **2013**, *559*, 130–134. [CrossRef]
16. Chen, C.Y.; Chen, C.C.; Yang, J.R. Microstructure characterization of nanometer carbides heterogeneous precipitation in Ti–Nb and Ti–Nb–Mo steel. *Mater. Charact.* **2014**, *88*, 69–79. [CrossRef]
17. Mukherjee, S.; Timokhina, I.; Zhu, C.; Ringer, S.P.; Hodgson, P.D. Clustering and precipitation processes in a ferritic titanium-molybdenum microalloyed steel. *J. Alloy. Compd.* **2017**, *690*, 621–632. [CrossRef]
18. Wang, Z.; Zhang, H.; Guo, C.; Liu, W.; Yang, Z.; Sun, X.; Zhang, Z.; Jiang, F. Effect of molybdenum addition on the precipitation of carbides in the austenite matrix of titanium micro-alloyed steels. *J. Mater. Sci.* **2016**, *51*, 4996–5007. [CrossRef]
19. Larzabal, G.; Isasti, N.; Rodriguez-Ibabe, J.M.; Uranga, P. Evaluating strengthening and impact toughness mechanisms for ferritic and bainitic microstructures in Nb, Nb-Mo and Ti-Mo microalloyed steels. *Metals* **2017**, *7*, 65. [CrossRef]
20. Lee, W.-B.; Hong, S.-G.; Park, C.-G.; Park, S.-H. Carbide precipitation and high-temperature strength of hot-rolled high-strength, low-alloy steels containing Nb and Mo. *Metall. Mater. Trans. A* **2002**, *33*, 1689–1698. [CrossRef]
21. Jang, J.H.; Heo, Y.-U.; Lee, C.-H.; Bhadeshia, H.K.D.H.; Suh, D.-W. Interphase precipitation in Ti–Nb and Ti–Nb–Mo bearing steel. *Mater. Sci. Technol.* **2013**, *29*, 309–313. [CrossRef]
22. Funakawa, Y.; Shiozaki, T.; Tomita, K.; Yamamoto, T.; Maeda, E. Development of high strength hot-rolled sheet steel consisting of ferrite and nanometer-sized carbides. *ISIJ Int.* **2004**, *44*, 1945–1951. [CrossRef]
23. Park, D.-B.; Huh, M.-Y.; Shim, J.-H.; Suh, J.-Y.; Lee, K.-H.; Jung, W.-S. Strengthening mechanism of hot rolled Ti and Nb microalloyed HSLA steels containing Mo and W with various coiling temperature. *Mater. Sci. Eng. A* **2013**, *560*, 528–534. [CrossRef]
24. Bracke, L.; Xu, W. Effect of the Cr content and coiling temperature on the properties of hot rolled high strength lower bainitic steel. *ISIJ Int.* **2015**, *55*, 2206–2211. [CrossRef]
25. Lee, H.-J.; Lee, H.-W. Effect of Cr content on microstructure and mechanical properties of low carbon steel welds. *Int. J. Electrochem. Sci.* **2015**, *10*, 8028–8040.

26. Zhou, M.; Xu, G.; Tian, J.; Hu, H.; Yuan, Q. Bainitic transformation and properties of low carbon carbide-free bainitic steels with Cr addition. *Metals* **2017**, *7*, 263. [CrossRef]
27. Mishra, S.K.; Das, S.; Ranganathan, S. Precipitation in high strength low alloy (HSLA) steel: A TEM study. *Mater. Sci. Eng. A* **2002**, *323*, 285–292.
28. Janovec, J.; Svoboda, M.; Vyrostkova, A.; Kroupa, A. Time-temperature-precipitation diagrams of carbide evolution in low alloy steels. *Mater. Sci. Eng. A* **2005**, *402*, 288–293. [CrossRef]
29. Timokhina, I.B.; Hodgson, P.D.; Ringer, S.P.; Zheng, R.K.; Pereloma, E.V. Precipitate characterisation of an advanced high-strength low-alloy (HSLA) steel using atom probe tomography. *Scr. Mater.* **2007**, *56*, 601–604. [CrossRef]
30. Gorokhova, N.A.; Sarrak, V.I.; Suvorova, S.O. Solubility of titanium and niobium carbides in high-chromium ferrite. *Met. Sci. Heat Treat.* **1986**, *28*, 276–279. [CrossRef]
31. Tsai, S.-P.; Su, T.-C.; Yang, J.-R.; Chen, C.-Y.; Wang, Y.-T.; Huang, C.-Y. Effect of Cr and Al additions on the development of interphase-precipitated carbides strengthened dual-phase Ti-bearing steels. *Mater. Des.* **2017**, *119*, 319–325. [CrossRef]
32. Koyama, S.; Ishii, T.; Narita, K. Effects of Mn, Si, Cr and Ni on the solution and precipitation of Niobium carbide in iron austenite. *J. Jpn. Inst. Met.* **1971**, *35*, 1089–1094. [CrossRef]
33. Ray, A. Niobium microalloying in rail steel. *Mater. Sci. Technol.* **2017**, *33*, 1584–1600. [CrossRef]
34. Pereloma, E.V.; Bata, V.; Scott, R.I.; Smith, R.M. Effect of Cr on strain ageing behaviour of low carbon steel. *Mater. Sci. Forum* **2007**, *539–543*, 4214–4219. [CrossRef]
35. Lu, Q.; Xu, W.; van der Zwaag, S. Designing new corrosion resistant ferritic heat resistant steel based on optimal solid solution strengthening and minimisation of undesirable microstructural components. *Comput. Mater. Sci.* **2014**, *84*, 198–205. [CrossRef]
36. Yamamoto, S.; Sakiyama, T.; Ouchi, C. Effect of alloying elements on recrystallization kinetics after hot deformation in austenitic stainless steels. *Trans. ISIJ* **1987**, *27*, 447–452. [CrossRef]
37. De Abreu Martins, C.; Poliak, E.; Godefroid, L.B.; Fonstein, N. Determining the conditions for dynamic recrystallization in hot deformation of C–Mn–V steels and the effects of Cr and Mo additions. *ISIJ Int.* **2014**, *54*, 227–234. [CrossRef]
38. Hutchinson, B.; Siwecki, T.; Komenda, J.; Hagström, J.; Lagneborg, R.; Hedin, J.-E.; Gladh, M. New vanadium-microalloyed bainitic 700 MPa strip steel product. *Ironmak. Steelmak.* **2014**, *41*, 1–6. [CrossRef]
39. Kong, X.; Lan, L.; Hu, Z.; Li, B.; Sui, T. Optimization of mechanical properties of high strength bainitic steel using thermo-mechanical control and accelerated cooling process. *J. Mater. Process. Technol.* **2015**, *217*, 202–210. [CrossRef]
40. Abbasi, E.; Rainforth, W.M. Effect of Nb-Mo additions on precipitation behaviour in V microalloyed TRIP-assisted steels. *Mater. Sci. Technol.* **2016**, *32*, 1721–1729. [CrossRef]
41. Sourmail, T. Precipitation in creep resistant austenitic stainless steels. *Mater. Sci. Technol.* **2001**, *17*, 1–14. [CrossRef]
42. Williams, D.; Carter, C.B. *Transmission Electron Microscopy II—Diffraction*; Plenum Press: New York, NY, USA, 1996; pp. 321–323.
43. Pereloma, E.V.; Al-Harbi, F.; Gazder, A. The crystallography of carbide-free bainites in thermo-mechanically processed low Si transformation-induced plasticity steels. *J. Alloy. Compd.* **2014**, *615*, 96–110. [CrossRef]
44. Pandit, A.; Murugaiyan, A.; Saha Podder, A.; Haldar, A.; Bhattacharjee, D.; Chandra, S.; Ray, R.K. Strain induced precipitation of complex carbonitrides in Nb–V and Ti–V microalloyed steels. *Scr. Mater.* **2005**, *53*, 1309–1314. [CrossRef]
45. Cabibbo, M.; Fabrizi, A.; Merlin, M.; Garagnani, G.L. Effect of thermo-mechanical treatments on the microstructure of micro-alloyed low-carbon steels. *J. Mater. Sci.* **2008**, *43*, 6857–6865. [CrossRef]
46. Kostryzhev, A.G.; Strangwood, M.; Davis, C.L. Mechanical property development during UOE forming of large diameter pipeline steels. *Mater. Manuf. Process.* **2010**, *25*, 41–47. [CrossRef]
47. Miyata, K.; Omura, T.; Kushida, T.; Komizo, Y. Coarsening kinetics of multicomponent MC-type carbides in high-strength low-alloy steels. *Metall. Mater. Trans. A* **2003**, *34*, 1565–1573. [CrossRef]
48. Jack, D.H.; Jack, K.H. Carbides and nitrides in steel. *Mater. Sci. Eng.* **1973**, *11*, 1–27. [CrossRef]
49. Bagaryatskii, Y.A. Probable mechanism of martensite decomposition. *Dokl. Acad. Nauk SSSR* **1950**, *73*, 1161–1164. (In Russian)
50. Andrews, K.W. Tabulation of interplanar spacings of cementite Fe_3C. *Acta Crystallogr.* **1963**, *16*, 68. [CrossRef]

51. Agrawal, B.K. *Introduction to Engineering Materials*; Tata McGraw-Hill: New Delhi, India, 1988.
52. Irvine, J.; Baker, T.N. The influence of rolling variables on the strengthening mechanisms operating in Niobium steels. *Mater. Sci. Eng.* **1984**, *64*, 123–134. [CrossRef]
53. Khalid, E.A.; Edmonds, D.V. Mixed structures in continuously cooled low-carbon automotive steels. *J. Phys. IV* **1993**, *3*, 147–152. [CrossRef]
54. Garcia-Mateo, C.; Caballero, F.G.; Capdevila, C.; Garcia de Andres, C. Estimation of dislocation density in bainitic microstructures using high-resolution dilatometry. *Scr. Mater.* **2009**, *61*, 855–858.
55. Stevens, W.; Haynes, A.G. The temperature of formation of martensite and bainite in low-alloy steel. *J. Iron Steel Inst.* **1956**, *183*, 349–359.
56. Kunitake, T.; Okada, Y. The estimation of bainite transformation temperatures in steels by the empirical formulas. *J. Iron Steel Inst. Jpn.* **1998**, *84*, 137–141. [CrossRef]
57. Lee, Y.-K. Empirical formula of isothermal bainite start temperature of steel. *J. Mater. Sci. Lett.* **2002**, *21*, 1253–1255. [CrossRef]
58. Lee, Y.-K.; Hong, J.-M.; Choi, C.-S.; Lee, J.-K. Continuous cooling transformation temperatures and microstructures of niobium bearing microalloyed steels. *Mater. Sci. Forum* **2005**, *475–479*, 65–68.
59. Carpenter, K.R.; Killmore, C.R. The effect of Nb on the continuous cooling transformation curves of ultra-thin strip CASTRIP© Steels. *Metals* **2015**, *5*, 1857–1877. [CrossRef]
60. Lambers, H.G.; Tschumak, S.; Maier, H.J.; Canading, D. Role of austenization and pre-deformation on the kinetics of the isothermal bainite transformation. *Metall. Mater. Trans. A* **2009**, *40*, 1355–1366. [CrossRef]
61. Larn, R.H.; Yang, J.R. The effect of compressive deformation of austenite on the bainitic ferrite transformation in Fe-Mn-Si-C steels. *Mater. Sci. Eng. A* **2000**, *278*, 278–291. [CrossRef]
62. Bhadeshia, H.K.D.H. *Bainite in Steels*; IOM Communications: London, UK, 2001; p. 221.
63. Matsuzaki, A.; Bhadeshia, H.K.D.H. Effect of austenite grain size and bainite morphology on overall kinetics of bainite transformation in steels. *Mater. Sci. Technol.* **1999**, *15*, 518–522. [CrossRef]
64. Kang, S.; Yoon, S.; Lee, S.-J. Prediction of bainite start temperature in alloy steels with different grain sizes. *ISIJ Int.* **2014**, *54*, 997–999. [CrossRef]
65. Wada, H.; Pehlke, R.D. Nitrogen solubility and nitride formation in austenitic Fe-Ti alloys. *Matall. Trans. B* **1985**, *16*, 815–822. [CrossRef]
66. Tagashira, K.; Mutsuji, T.; Endo, T. Effect of Mo-C dipole on the development of {111} recrystllisation texture in Mo added low carbon steels. *Tetsu-to-Hagané* **2000**, *86*, 466–471. [CrossRef]
67. Bata, V. Minimising Ageing due to Carbon in Low Carbon Sheet Steel with the Aid of Mn, Cr, Mo or B Addition. Ph.D. Thesis, Monash University, Melbourne, Australia, June 2006.
68. Johnson, R.A.; Diens, G.I.; Damask, A.C. Calculations of the energy and migration characteristics of carbon and nitrogen in α-iron and vanadium. *Acta Metall.* **1964**, *12*, 1215–1224. [CrossRef]
69. Johnson, R.A. Clustering of carbon atoms in α-iron. *Acta Metall.* **1967**, *15*, 513–517. [CrossRef]
70. Liu, P.; Xing, W.; Cheng, X.; Li, D.; Li, Y.; Chen, X.-Q. Effects of dilute substitutional solutes on interstitial carbon in α-Fe: Interactions and associated carbon diffusion from first-principles calculations. *Phys. Rev. B* **2014**, *90*, 024103. [CrossRef]
71. Golovin, I.S.; Blanter, M.S.; Magalas, L.B. Interactions of dissolved atoms and carbon diffusion in Fe-Cr and FeAl alloys. *Defect Diffus. Forum* **2001**, *194–199*, 73–78. [CrossRef]
72. Pereloma, E.V.; Bata, V.; Scott, R.I.; Smith, R.M. Effect of Cr and Mo on strain ageing behaviour of low carbon steel. *Mater. Sci. Eng. A* **2010**, *527*, 2538–2546. [CrossRef]
73. Sandomirskij, M.M.; Grigorkin, V.I.; Zemskij, S.V. Alloying element effect on carbon diffusion in ferrite of pearlitic steel at tempering. *Izv. Vysshikh Uchebnykh Zavedenij. Chernaya Metall.* **1985**, *5*, 116–119.
74. Fleischer, R.L. Solution hardening. *Acta Metall.* **1961**, *9*, 996–1000. [CrossRef]
75. Dingly, D.J.; McLean, D. Components of the flow stress of iron. *Acta Metall.* **1967**, *15*, 885–901. [CrossRef]
76. Gladman, T. *The Physical Metallurgy of Microalloyed Steels*; The Institute of Materials, Cambridge University Press: Cambridge, UK, 1997.
77. Seidman, D.N.; Marquis, E.A.; Dunand, D.C. Precipitation strengthening at ambient and elevated temperatures of heat-treatable Al(Sc) alloys. *Acta Mater.* **2002**, *50*, 4021–4035. [CrossRef]
78. Zhang, X.; Hickel, T.; Rogal, J.; Fähler, S.; Drautz, R.; Neugebauer, J. Structural transformations among austenite, ferrite and cementite in Fe–C alloys: A unified theory based on ab initio simulations. *Acta Mater.* **2015**, *99*, 281–289. [CrossRef]

79. Argon, A.S. Mechanical properties of single-phase crystalline media: Deformation at low temperature. In *Physical Metallurgy*; Cahn, R.W., Haasen, P., Eds.; North-Holland Publishing: Amsterdam, The Netherlands, 1996.
80. Timokhina, I.B.; Enomoto, M.; Miller, M.K.; Pereloma, E.V. Microstructure-property relationship in the thermomechanically processed C-Mn-Si-Nb-Al-(Mo) transformation-induced plasticity steels before and after prestraining and bake hardening treatment. *Metall. Mater. Trans. A* **2012**, *43*, 2473–2483. [CrossRef]
81. Xie, K.Y.; Zheng, T.; Cairney, J.M.; Kaul, H.; Williams, J.G.; Barbaro, F.J.; Killmore, C.R.; Ringer, S.P. Strengthening from Nb-rich clusters in a Nb-microalloyed steel. *Scr. Mater.* **2012**, *66*, 710–713. [CrossRef]
82. Kostryzhev, A.G.; Al Shahrani, A.; Zhu, C.; Cairney, J.M.; Ringer, S.P.; Killmore, C.R.; Pereloma, E.V. Effect of niobium clustering and precipitation on strength of a NbTi-microalloyed ferritic steel. *Mater. Sci. Eng. A* **2014**, *607*, 226–235. [CrossRef]
83. Zhang, Y.-J.; Miyamoto, G.; Shinbo, K.; Furuhara, T.; Ohmura, T.; Suzuki, T.; Tsuzaki, K. Effects of transformation temperature on VC interphase precipitation and resultant hardness in low-carbon steels. *Acta Mater.* **2015**, *84*, 375–384. [CrossRef]
84. Hansen, N.; Barlow, C.Y. Microstructure evolution in whisker- and particle-containing materials. In *Fundamentals of Metal-Matrix Composites*; Surech, S., Mortensen, A., Needleman, A., Eds.; Elsevier: Boston, MA, USA, 1993; pp. 109–118.
85. Humphreys, F.J.; Hatherly, M. *Recrystallisation and Related Annealing Phenomena*; Elsevier: Oxford, UK, 2004.

Permissions

 Every chapter published in this book has been scrutinized by our experts. Their significance has been extensively debated. The topics covered herein carry significant findings which will fuel the growth of the discipline. They may even be implemented as practical applications or may be referred to as a beginning point for another development.

The contributors of this book come from diverse backgrounds, making this book a truly international effort. This book will bring forth new frontiers with its revolutionizing research information and detailed analysis of the nascent developments around the world.

We would like to thank all the contributing authors for lending their expertise to make the book truly unique. They have played a crucial role in the development of this book. Without their invaluable contributions this book wouldn't have been possible. They have made vital efforts to compile up to date information on the varied aspects of this subject to make this book a valuable addition to the collection of many professionals and students.

This book was conceptualized with the vision of imparting up-to-date information and advanced data in this field. To ensure the same, a matchless editorial board was set up. Every individual on the board went through rigorous rounds of assessment to prove their worth. After which they invested a large part of their time researching and compiling the most relevant data for our readers.

The editorial board has been involved in producing this book since its inception. They have spent rigorous hours researching and exploring the diverse topics which have resulted in the successful publishing of this book. They have passed on their knowledge of decades through this book. To expedite this challenging task, the publisher supported the team at every step. A small team of assistant editors was also appointed to further simplify the editing procedure and attain best results for the readers.

Apart from the editorial board, the designing team has also invested a significant amount of their time in understanding the subject and creating the most relevant covers. They scrutinized every image to scout for the most suitable representation of the subject and create an appropriate cover for the book.

The publishing team has been an ardent support to the editorial, designing and production team. Their endless efforts to recruit the best for this project, has resulted in the accomplishment of this book. They are a veteran in the field of academics and their pool of knowledge is as vast as their experience in printing. Their expertise and guidance has proved useful at every step. Their uncompromising quality standards have made this book an exceptional effort. Their encouragement from time to time has been an inspiration for everyone.

The publisher and the editorial board hope that this book will prove to be a valuable piece of knowledge for researchers, students, practitioners and scholars across the globe.

List of Contributors

Mingxue Sun and Yang Xu
School of Mechanical and Automotive Engineering, Qingdao University of Technology, Qingdao 266520, China

Wenbo Du
National Key Laboratory for Remanufacturing, Academy of Army Armored Forces, Beijing 100072, China

Mohamed Soliman
Institute of Metallurgy, Clausthal University of Technology, Robert-Koch-Straße 42, 38678 Clausthal-Zellerfeld, Germany

Chunyu He, Jianguang Wang, Yulai Chen, Wei Yu and Di Tang
National Engineering Research Center for Advanced Rolling Technology, University of Science and Technology Beijing, Beijing 100083, China

Jean-Yves Maetz and Matthias Militzer
Centre for Metallurgical Process Engineering, The University of British Columbia, Vancouver, BC V6T 1Z4, Canada

Yu Wen Chen and Jer-Ren Yang
Department of Materials Science and Engineering, National Taiwan University, Taipei 10617, Taiwan

Nam Hoon Goo and Soo Jin Kim
Technical Research Center, Hyundai-Steel Company, Dangjin 167-32, Korea

Bian Jian
Niobium Tech Asia, Singapore 068898, Singapore

Julio C. Villalobos
Instituto Tecnológico de Morelia, Avenida Tecnológico No. 1500, Col. Lomas de Santiaguito, Morelia 58120, México

Adrian Del-Pozo and Sergio Serna
CIICAp, Universidad Autónoma del Estado de Morelos, Av. Universidad 1001, Col. Chamilpa, Cuernavaca 62609, Mexico

Bernardo Campillo
Instituto de Ciencias Físicas-UNAM, Av. Universidad 1001, Col. Chamilpa, Cuernavaca 62609, Mexico
Facultad de Química-UNAM, Circuito de la Investigación Científica S/N, Mexico City 04510, Mexico

Jan Mayen
CONACYT, CIATEQ, Unidad San Luis Potosí, Eje 126 No. 225, Zona Industrial, San Luis Potosí 78395, Mexico

Hadi Torkamani, Shahram Raygan and Jafar Rassizadehghani
School of Metallurgy and Materials Engineering, College of Engineering, University of Tehran, 111554563 Tehran, Iran

Carlos Garcia Mateo, Javier Vivas and David San-Martin
Materalia Research Group, National Center for Metallurgical Research (CENIM), Consejo Superior de Investigaciones Científicas (CSIC), E-28040 Madrid, Spain

Yahya Palizdar
Research Department of Nano-Technology and Advanced Materials, Materials and Energy Research Center, 3177983634 Karaj, Iran

Guobiao Lin
School of Materials Science and Engineering, University of Science and Technology, Beijing 100083, China

Pengfei Wang
School of Materials Science and Engineering, University of Science and Technology, Beijing 100083, China
Department of Structural Steels, Central Iron and Steel Research Institute, Beijing 100081, China

Zhaodong Li, Caifu Yang and Qilong Yong
Department of Structural Steels, Central Iron and Steel Research Institute, Beijing 100081, China

Shitong Zhou
Department of Structural Steels, Central Iron and Steel Research Institute, Beijing 100081, China
Department of Materials Science and Engineering, Kunming University of Science and Technology, Kunming 650093, China

Hardy Mohrbacher
Department of Materials Engineering (MTM), KU Leuven, 3001 Leuven, Belgium
NiobelCon bvba, 2970 Schilde, Belgium

Gorka Larzabal, Nerea Isasti, Jose M. Rodriguez-Ibabe and Pello Uranga
CEIT, Materials and Manufacturing Division, 20018 San Sebastian, Basque Country, Spain
Universidad de Navarra, Tecnun, Mechanical and Materials Engineering Department, 20018 San Sebastian, Basque Country, Spain

Chuanfeng Wu, Minghui Cai, Peiru Yang and Junhua Su
School of Materials Science and Engineering, Northeastern University, Shenyang 110819, China

Xiaopeng Guo
School of Mechanical Engineering and Automation, Northeastern University, Shenyang 110819, China

Xianguang Zhang
School of Metallurgical and Ecological Engineering, University of Science and Technology Beijing (USTB), Beijing 100083, China

Kiyotaka Matsuura and Munekazu Ohno
Division of Materials Science and Engineering, Faculty of Engineering, Hokkaido University, Kita 13 Nishi 8, Kita-ku, Sapporo 060-8628, Hokkaido, Japan

Lena Eisenhut, Jonas Fell and Christian Motz
Department for Materials Science and Engineering, Saarland University, 66123 Saarbrücken, Germany

Andrii Kostryzhev and Navjeet Singh
Australian Steel Research Hub, University of Wollongong, Wollongong, NSW 2500, Australia

Liang Chen
School of Mechanical, Materials, Mechatronic and Biomedical Engineering, University of Wollongong, Wollongong, NSW 2500, Australia

Chris Killmore
BlueScope Steel Limited, Five Islands Road, Port Kembla, NSW 2505, Australia

Elena Pereloma
School of Mechanical, Materials, Mechatronic and Biomedical Engineering, University of Wollongong, Wollongong, NSW 2500, Australia
UOW Electron Microscopy Centre, University of Wollongong, Wollongong, NSW 2519, Australia

Index

Printed in the USA
CPSIA information can be obtained
at www.ICGtesting.com
JSHW050846251023
50683JS00018B/73

9 781647 254384